THE SECRET SOCIAL LIVES OF REPTILES

THE Secret Social Lives OF
REPTILES

J. Sean Doody, Vladimir Dinets & Gordon M. Burghardt

JOHNS HOPKINS UNIVERSITY PRESS | *Baltimore*

Johns Hopkins University Press
2715 North Charles Street
Baltimore, Maryland 21218-4363
www.press.jhu.edu

Library of Congress Cataloging-in-Publication Data

Names: Doody, J. Sean, 1965– author. | Dinets, Vladimir, author. |
    Burghardt, Gordon M., 1941– author.
Title: The secret social lives of reptiles / J. Sean Doody, Vladimir Dinets,
    Gordon M. Burghardt.
Description: Baltimore, MD : Johns Hopkins University Press, 2021. |
    Includes bibliographical references and index.
Identifiers: LCCN 2020028521 | ISBN 9781421440675 (hardcover) |
    ISBN 9781421440682 (ebook)
Subjects: LCSH: Reptiles—Behavior.
Classification: LCC QL669.5 .D66 2021 | DDC 597.9—dc23
LC record available at https://lccn.loc.gov/2020028521

A catalog record for this book is available from the British Library.

Frontispiece: Communal egg-laying in the giant Amazon river turtle (*Podocnemis expansa*) along the Purus River, Abufari Biological Reserve, Amazonas, Brazil. Photograph by Camila Ferrara.

Special discounts are available for bulk purchases of this book. For more information, please contact Special Sales at specialsales@jh.edu.

For Stan Rand
A pioneer whose foundational discoveries in herpetological social
behavior and inspiring mentorship fed research streams through many
diverse conceptual and geographical habitats

For Mike Bull
Whose ability to link broad concepts of natural history within
the context of long-term research on reptiles left a legacy that
we should embrace

What does it mean to be social? Members of a wolf pack stealthily hunting a wary elk, a flock of sandhill cranes migrating gracefully in formation in the crisp autumn sky, a mother wildebeest anxiously coaxing its calf to enter and cross a river, or even termites, in the many thousands, hastily working together to repair their damaged mound, all are activities we call social. Clearly, the sexual acts of humans, or of any other animal, are categorized as social. We keenly relate to all of these behaviors because we do them too. Remarkably, we express social behavior and have social lives in ways not unlike our close and distant relatives in the tree of life.

For a number of reasons, however, we have difficulty interpreting and relating to the activities of certain animals, such as reptiles—in particular, non-avian reptiles (crocodylians, turtles, snakes, lizards, amphisbaenians, and the tuatara). Starting from one end of a continuum, I am fairly confident I could persuade most people, albeit with some effort, that a group of Cagle's map turtles sun-basking on a log is an example of social behavior. Perhaps, that could be extended to a group of Nile crocodiles lounging on a river bank, waiting for prey. But, on the other end, I have serious doubts that I could be as persuasive when talking about a group of snakes (by the way, what is a group of snakes called?) or amphisbaenians, a close lizard relative that spends most of its life underground. With regard to the latter, my hunch is that most folks have never seen a photo of an amphisbaenian, let alone a live one . . . indeed, most are unaware that this group of reptiles even exists. This naïveté or ignorance, a brand of intellectual poverty, is part and parcel of an ever-growing problem in our lack of understanding and appreciation of Earth's tremendous biodiversity and its indispensable qualities (Greene, 2005, 2013; Wilson, 2009, 2014; McKeon et al., 2020).

Why do we distance ourselves from nature? Especially, why do we disconnect our lives from experiencing the full breadth of its biodiversity? Answers to these questions might give us better insights to my next question, the very basis of this book: why do we have difficulty perceiving reptiles as having complex social lives and other behavioral attributes, such as personality, reciprocity, and family? We regularly use these concepts and terms for mammals, birds, and even insects. Indeed, recent progress has been made in this field, but it is sparse. Warm-blooded chauvinism still abounds (Burghardt, 2020).

J. Sean Doody, Vladimir Dinets, and Gordon M. Burghardt, the authors of this important and captivating new book, enthusiastically, bravely, and masterfully tackle these questions, and many others, concerning the social lives of reptiles. In grappling with this imbalance of knowledge, the authors take us along on a wonderful intellectual journey.

Right off, the authors get to the core question: what is social behavior? They use the definition by Allaby (2009). Social behavior, broadly defined, is "any interaction between two or more members of the same species." I like this simpler and direct approach because it relaxes the tight and restrictive grip of past definitions. It drops unnecessary jargon, and it makes clear that reptiles are social. Importantly, throughout the book, the authors break the social–nonsocial dichotomy (for example, social mammals and birds versus nonsocial reptiles) that has utterly paralyzed past research and progress. We now have a clean slate and are given permission to operate under a continuum of possibilities (Schuett et al., 2016).

Accordingly, with this robust platform, we become immersed in an array of ideas, questions, examples, and future research goals as they pertain to the social lives of reptiles. Proximate and ultimate questions concerning the ramifications of social behavior are carefully presented, and numerous examples are provided. Among them, and most intriguing to me, is the notion of reptile societies and social networks. I immediately envision a few obvious examples, marine iguanas, rattlesnakes, and the giant tortoises of the Galápagos (Krause & Ruxton, 2002). What about species that do not form permanent aggregations—can they form a society, however subtle? If so, do these societies have culture?

Using an evolutionary comparative framework, the authors present examples of social behavior of all groups of reptiles. Importantly, the early chapters provide the reader with an abundance of introductory information on the biology of reptiles pertinent to better understand social behavior. Evolutionary history, sensory anatomy and physiology, mating systems, and other topics all are discussed in a manner easy to comprehend.

Appreciating the diversity of social behavior of all organisms across the tree of life has far-reaching ramifications for humans. As a result of the COVID-19 pandemic, we have become painfully aware of the negative impacts of limiting social and physical contact. Isolation, especially long-term, can have enormous deleterious effects on individuals and, as we are seeing, our communities. Whether we like it or not, we have been abruptly awakened to a new era of social practices. We are redefining and readjusting our own social orbit and mores. To be certain, many studies will be conducted on the psychological effects of this global virus on our own species and how it influenced the social lives of nonhuman animals.

With a broad and more inclusive understanding of social behavior, especially the subtle forms, we can approach conservation biology and captive maintenance (e.g., zoos, laboratories, and pets) of reptiles with new insights and tools. Essentially, a new ethics is born with new responsibilities.

The authors close with a few wise words: "Our changing world opens up many opportunities to study the five aims of ethology (Burghardt, 1997) in terms of sociality—mechanisms, development, function, evolution, and private experience. . . . While we have documented, often only partially, much of the information acquired in recent years on non-avian reptile social lives, we have just scratched the surface of many mysteries and their illuminating, as well as delightful, solutions."

Read this enjoyable book. It will alter your perspectives not only on reptiles but on all other creatures including *Homo sapiens*. To that end, Doody, Dinets, and Burghardt have admirably achieved their goal.

Gordon W. Schuett

References

Allaby, M. (2009). *Ecology: Plants, animals, and the environment.* New York: Infobase Publishing.

Burghardt, G. M. (1997). Amending Tinbergen: A fifth aim for ethology. In R. W. Mitchell, N. S. Thompson & H. L. Miles (Eds.), *Anthropomorphism, anecdotes, and animals* (pp. 254–276). Albany, NY: SUNY Press.

Burghardt, G. M. (2020). Insights found in century-old writings on animal behaviour and some cautions for today. *Animal Behaviour, 164,* 241–249. https://doi.org/10.1016/j.anbehav.2020.02.010.

Doody, J. S., Burghardt, G. M., & Dinets, V. (2013). Breaking the social–non-social dichotomy: A role for reptiles in vertebrate social behavior research? *Ethology, 119,* 95–103.

Greene, H. W. (2005). Organisms in nature as a central focus for biology. *Trends in Ecology & Evolution, 20,* 23–27.

Greene, H. W. (2013). *Tracks and shadows: Field biology as art.* Berkeley & Los Angeles: University of California Press.

Krause, J., & Ruxton, G. D. (2002). *Living in groups.* Oxford, UK: Oxford University Press.

McKeon, S., Weber, L., Adams, A. J., & Fleischner, T. L. (2020). Human dimensions: Natural history as the innate foundation of ecology. *Bulletin of the Ecological Society of America, 101,* e01656.

Schuett, G. W., Clark, R. W., Repp, R. A., Amarello, M., Smith, C. F., & Greene, H. W. (2016). Social behavior of rattlesnakes: A shifting paradigm. In G. W. Schuett, M. J. Feldner, C. F. Smith & R. S. Reiserer (Eds.), *Rattlesnakes of Arizona* (Vol. 2, pp. 161–244). Rodeo, New Mexico: ECO Publishing.

Wilson, E. O. (2009). *Biophilia.* Cambridge, MA: Harvard University Press.

Wilson, E. O. (2014). *The Meaning of Human Existence.* New York: W. W. Norton & Company.

In common parlance, reptiles are a class of vertebrates along with fishes, amphibians, birds, and mammals. Today we know that birds, along with non-avian dinosaurs, are one of the "reptilian" clades. However, our focus in this book is on non-avian reptiles, including the turtles, crocodiles, lizards, and snakes—animals that everyone recognizes but that have received relatively little scientific attention. This lack of attention is especially true of their behavior, social behavior in particular. Many reptiles are dismissed by those studying social behavior in humans, birds, carnivores, nonhuman primates, horses, and other animals as solitary, dull, slow moving, and having tiny brains and simple behavior. All these claims are disputable with actual data. It is our goal in this book to (a) show that reptiles engage in a remarkable diversity of often complex social behavior, (b) explain the reasons for their neglect, (c) explore, with examples, the burgeoning scientific observations and research of which even seasoned animal behavior scholars are unaware, and (d) present the case that the great diversity in the evolutionary "experiments" found among reptiles offers some of the best opportunities evolutionary biology, ethology, behavioral ecology, and comparative psychology have available to answer profound questions on the factors underlying how and why terrestrial vertebrate sociality in birds and mammals has evolved the way it has.

The three authors of this book came together through their diverse interests in the behavior of different reptilian groups and the shared conviction that social behavior in reptiles has been greatly underappreciated. The first symposium on reptile social behavior, organized by David Crews and Neil Greenberg and published in the *American Zoologist* in 1977, was held in New Orleans in 1976. In spite of the fine and provocative studies reported there, the interest in reptile social behavior did not take off, although perhaps in a delayed fashion it had an effect. Exactly 40 years later, the authors of this book organized a second symposium on social behavior of reptiles at the Joint Meetings in Ichthyology and Herpetology, also held in New Orleans. Building on the work described in those two meetings, and on the hundreds of studies published before and since, we have written what we think is the first detailed integrated treatment of social behavior in all the traditional reptile groups.

Our goal is to present the social behavior of reptiles in a comparative framework embedded in the biology and evolution of the various diverse groups. Given the fact that so much is still unknown, we do not attempt to present any new phylogenetic or ecological analyses using the many modern tools available that have been applied to systematic, morphological, physiological, or behavioral traits (such as those involved in foraging, displays, social organization, or predator defense). Too much awaits discovery. We do hope that our efforts will facilitate or inspire future studies.

We have organized the book in 11 chapters. The first briefly reviews some historical and conceptual issues in the study of animal social behavior and sociality, documenting how reptiles have been neglected and why they have an important part to play in any comprehensive understanding of social behavior, especially in terrestrial vertebrates. Chapter 2 provides an overview of current views on reptile evolution and relationships and discusses aspects of their biology and behavior, including physiology, sensory and motor systems, reproduction, and other topics, as a prelude to focusing on their social behavior. Chapter 3 presents current ideas on social organization and mating relationships among reptiles. What is a society and do any reptiles have one? What are the mating relationships found in reptile groups? The answers in these often long-living vertebrates may surprise you. Chapter 4 reviews the many ways that reptiles communicate with conspecifics and also with predators and prey. Chapter 5 goes into the variety of ways in which reptiles court and mate. After mating, offspring appear. While many reptiles nest or give birth singly, this is often not the case. Chapter 6 gives a detailed look at communal breeding, which is surprisingly diverse and is where the most elaborate construction feats of reptiles take place.

The actual appearance of offspring has led to much work on parental care (Chapter 7) and social synchrony of hatching and emergence (Chapter 8). Behavioral development is perhaps the least studied of the major ethological aims in reptiles, but some fascinating work is presented in Chapter 9. Social aspects of behavior outside reproductive settings are reviewed in Chapter 10 with coverage of feeding, thermoregulation, predator avoidance, and habitat choices. The final chapter, Chapter 11, presents emerging data on the cognitive and emotional lives of reptiles, including topics on plasticity, social learning, and play. This chapter ends, and so also the book, by identifying some of the conceptual, empirical, methodological, and experimental frontiers that lie ahead and that will allow for more sustained comparative treatment with avian and mammalian radiations. The book has been kept relatively compact in terms of text but the references are extensive to facilitate exploring the now vast and rapidly growing literature as well as important earlier work that helped get us to where we are today in understanding non-avian reptile sociality. We have each worked on various aspects of reptile social behavior. While we cover many diverse phenomena and species and attempt for balance, we often draw upon the systems with which we have most experience and knowledge, rather than exhaustively covering all of the remarkable research going on around the world, though we do discuss a significant amount.

Reptiles, alien in so many ways to our mammalian psychology and behavior, have (if we put aside our anthropocentric and mammalocentric biases) more similarity with mammals, as well as birds, than we have allowed ourselves to appreciate. Even the most eminent biologists, including, as we shall see, Charles Darwin and Carl Linnaeus, had blinders on when describing reptiles. Unfortunately, such attitudes continue to the present day in both public and scientific discourse. Prejudice, fear, and discrimination need to stop and be countered by enlightened views, even when dangerous or invasive species are present. Our species' rapid destruction of our biological home continues unabated, and hundreds of reptile species, including some of the most charismatic, fascinating, and socially complex ones, may be exterminated before we even begin to unravel the true extent of their secret societies.

All three of the authors have mentors, colleagues, and students to which we owe much, as well as friends and families who aided, supported, and endured our passion and the preparation of this book. Our work has been supported over the years by many sources. We want to thank Vince Burke, formerly of Johns Hopkins University Press, who encouraged our plan to write a book and then had to endure our many delays with forbearance and patience. We also thank the editorial staff at the Press and the anonymous manuscript reviewer who provided much useful information, as well as Stephanie Drumheller, Harry Greene, and Gordon Schuett for reviewing the initial draft of the book. We thank Carrie Love for meticulous and invaluable copyediting and Michael Taber for preparing the indexes.

GMB wants to thank his wife, Sandra Twardosz, for her support and the students who helped in reference checking the manuscript, including Mary-Elisabeth Gibbons, Matthew Jenkins, Ariel Lathan, and Harry Pepper. He also is indebted to the many colleagues and students who contributed to his work with reptiles over more than 50 years, as well as funding sources, especially the National Institute of Mental Health and the National Science Foundation. JSD would like to thank his parents and Lizzy Coleman for their encouragement and support. VD would like to thank his wife for support and endurance. We thank the University of South Florida for the purchase of two photographs.

THE SECRET SOCIAL LIVES OF REPTILES

1 |

Social Behavior Research

Its History and a Role for Reptiles

1.1. What Is Social Behavior?

Social behavior, broadly defined as any interaction between two or more members of the same species (Allaby, 2009), was, we suspect, present in the first animals, at least 600 million years ago (Budd, 2008; Maloof et al., 2010). Since then, a remarkable diversity of animals has evolved, giving rise to an equally remarkable diversity of social interactions and repertoires. Classic examples are foraging animals aggregating to dilute predation risk, colorful signaling or boastful posturing aimed at potential mates or territorial rivals, and parents nurturing their vulnerable offspring until they can fend for themselves.

Animals can use social behavior to solve problems that are impossible or difficult for individuals to resolve alone (Krause et al., 2010). For example, a sea eagle that is gliding gracefully along the treetops of a river in tropical Australia suddenly veers into the canopy, plunging into a colony of 500,000 terrified flying foxes (bats). Although many of the bats scatter in chaotic fashion, the eagle easily plucks a bat from the roost trees—a sure evening meal. Yet, the chance of any one flying fox being captured by the eagle is minuscule due to the social phenomenon called predator dilution.

Now imagine a flying fox roosting by itself. It would stand little chance against this predator. Problems that animals face in survival or reproduction, in this case the management of predation risk, can be solved evolutionarily through social means—in this instance, colonial roosting.

In a dramatic example of pair bonding, deep sea anglerfish live in sparse populations in complete darkness up to 2.5 km below the ocean surface, making finding mates particularly challenging. The broad evolutionary solution to this problem lies in making sure that any social encounter with the opposite sex becomes a successful mating, but how do the fish accomplish this? When it encounters a female, the male, which is 10 times smaller than the female, bites into the female's skin, attaching himself to her permanently (Monk, 2000, and references therein). His jaws, eyes, brain, and all other internal organs—except the testes—are gradually absorbed into the female over time, and his circulatory system is apparently incorporated into hers. The male now becomes a degenerate parasite, and the female carries him (and up to five other males) around until she is ready to extract sperm from him. The deep sea anglerfish's evolutionary solution to the problem of mate-finding in total darkness was an intimate mating encounter that invariably ends the male's tenure as a free-living organism.

Another type of social behavior is found in penguins, who achieve precise thermoregulation in groups by forming huddles and "slow dancing" synchronously. The emperor penguin is the only animal capable of breeding in winter on the Antarctic continent, owing to physiological and (social) behavioral adaptations to extreme conditions. As conditions deteriorate to −25°C, with a wind speed of 15 m/sec., almost all the penguins, which are already clumped rather tightly into a colony, form compact huddles of 8–10 birds/m² (Prévost, 1961; Ancel et al., 2015). Individual penguins step in a synchronized way within these huddles, performing coordinated "dance-step" movements that create wave-like patterns, to stay warm in temperatures up to −50°C and in wind speeds up to 200 km/hr (Zitterbart et al., 2011; Gerum et al., 2013). Paradoxically, birds in the huddles face the risk of overheating (with ambient temperatures reaching 37°C, which is significantly above the birds' critical thermal maximum of 20°C), and so they break the huddles after 40–60 minutes to dissipate heat (Ancel et al.,

2015). Clearly, social behavior underpins survival under harsh conditions, and yet social solutions create new evolutionary problems.

Perhaps the signature examples of complex and adaptive sociality involve cooperation in solving problems such as obtaining, sharing, or defending resources, which may include food, prey, nest sites, and rearing offspring. The highest, that is, most complex, developments here are typically attributed to eusocial insects such as ants, bees, and termites, which have caste systems, among other features (E. O. Wilson, 1975). Although there is much variation among non-avian reptiles, and many live rather solitary lives, reptiles can and do solve a multitude of problems using social behavior (e.g., Boersma, 1982). Indeed, they display an impressive array of social behaviors that perhaps span a wider chunk of the solitary-social continuum than any other vertebrate group. Because evolution tends to modify existing blueprints rather than reinvent them, and because reptiles essentially gave rise to vertebrates that evolved even more socially and cognitively complex organization and behavior, reptiles are essential for a comprehensive understanding of how social behavior evolves in vertebrates. Gaining such knowledge for the social repertoire of reptiles, however, is a formidable challenge given that many live secret lives we are only beginning to discover.

1.2. Breaking the Social–Nonsocial Dichotomy

Strictly speaking, most animal species interact to some degree with conspecifics, and most bisexual species at least engage in courtship and mating. Non-avian reptiles (simply "reptiles" from here on out) are often secretive animals and thus their social repertoire has been little studied, historically, as compared to birds and mammals. Nonetheless, many authors continue to label reptilian species as "asocial" or "nonsocial" (see examples in Doody, Burghardt & Dinets, 2013; Pérez-Cembranos & Pérez-Mellado, 2015). In some cases, this label provided a fallacious paradigm that could be challenged by the authors' own findings (Wilkinson et al., 2010; Pérez-Cembranos & Pérez-Mellado, 2015). A more useful approach would be to recognize a continuum along which each species sits that ranges from weakly to strongly social (Doody, Burghardt & Dinets, 2013). At the "strongly social" end of the continuum sits obligatory eusociality: caste systems involving division of labor in reproduction, defense, and foraging, overlapping generations maintaining prolonged contact with each other, and cooperative care of young. At the "weakly social" end sit animals that rarely interact with conspecifics, other than during brief, simple mating encounters, and of course solitary asexual animals. Many years ago, Leyhausen (1965) wrote a compelling case for the often complex communal lives of "solitary" mammals in a classic paper too often neglected. Thus, even deciding where a species sits on a weakly-to-strongly-social continuum is a difficult task, and not to be confounded with the solitary-to-group-living continuum, in spite of frequent overlap. Levels and types of sociality are more like a tree than a linear system. Different levels of complexity can take multiple forms.

1.3. Current Frameworks for Classifying Animal Sociality

A generally accepted framework that adequately classifies the magnitude of social behavior among all animal taxa has not been developed (but see Kappeler, 2019). Nevertheless, attempts to subdivide sociality, or the degree to which individuals tend to associate in social groups, from families and herds to cooperative societies, have been articulated (e.g., E. O. Wilson, 1975; Wittenberger, 1981). Michener (1969) developed a classification for social insects that included "solitary" (no parental care), "subsocial" (adults care for their own young), "communal" (adults use the same nest but do not care for the brood cooperatively), "quasisocial" (cooperative brood care in the same nest), "semisocial" (quasisocial plus a worker caste), and "eusocial" (semisocial plus an overlap in adult generations). While these terms make distinctions among important biological processes that allow evolutionary insights, they do not address variation among animals with different social repertoires. For example, even among species that do not exhibit parental care after birth or hatching, there is considerable variation in the presence of other social behaviors such as dominance hierarchies, territoriality, male-to-male combat, complex courtship, group

vigilance, signaling, posturing, eavesdropping, communal nesting, cooperative hunting, pair bonding, sexual selection, and social monogamy (Doody,
Burghardt & Dinets, 2013). Even among insects, there are highly social species, such as migratory locusts, that exhibit no parental care after laying
eggs. Thus, coining terms for the degree of sociality based around one
major type of social behavior—e.g., parental care—has little broad utility.
Quite removed from insect societies are the attempts to classify and
analyze nonhuman primate societies, attempts with a long history (e.g.,
Chance & Jolly, 1970; Crook, 1970). Some proposed types are a single
female and her offspring (e.g., orangutans), a one-male-several-females
group (harem or polygamous as in some baboons), a monogamous family
group (e.g., gibbons), a multimale-multifemale group (many macaque
monkeys), a polyandrous family group, and a fission-fusion society (chimpanzees). These are further embedded in hierarchical or territorial systems that may be seasonal or permanent. Several of these systems are also
found in reptiles.

Replacing a one-dimensional scale with a multidimensional character
space would be an improvement (Kappeler, 2019), but it still does not solve
all problems. Developing a framework that can adequately facilitate the
classification of all animals would require a broad perspective and intensive review, and it would also be subject to much debate and disagreement
as to which animals fit where on the continuum. Are polygamous species
more social than monogamous ones? Are those with dominance hierarchies more social than those with territoriality? What about species that
can transition from territorial to hierarchical systems as a function of ecology or density? Are large herding species more social than their group-
hunting predators? Simple metrics are hard to apply across disparate taxa
(Doody, Burghardt & Dinets, 2013). We will not address such conundrums
but do note that all these complications can be found among reptiles. Kappeler (2019) offers an encouraging putative framework for studying (and
thus quantifying and comparing) social behavior among animals. Kappeler
recognizes four core components of a social system: social organization,
social structure, mating system, and care system. Disappointingly, the
word "reptile" was used just once in that paper. Nevertheless, the paper

will likely stir a debate leading to improved frameworks, concepts, and definitions that will facilitate comparative study of social behavior among vertebrates and animals (see also D. R. Rubenstein & Abbot, 2017).

1.4. A Brief History of the Science of Social Behavior

Social behavior has intrigued humans for thousands of years, as seen in paintings on cave walls dating back at least 30,000 years (Valladas et al., 2001). Aristotle might be credited with the first thoughtful treatment of social behavior in *Historia Animalium*, or *History of Animals*, circa 350 BCE:

Now some simply like plants accomplish their own reproduction according to the seasons; others take trouble as well to complete the nourishing of their young, but once that is accomplished they separate from them and have no further association; but those that have more understanding and possess some memory continue the association, and have a more social relationship with their offspring. (Balme, 1991, p. 67)

Among the fishes some form shoals with each other and are friends, while those that do not shoal are at war. (Balme, 1991, p. 233)

While descriptions of social behavior became more abundant over the next two millennia, the first theory of the origin or social behavior, and thus a means of explaining behavior in general, must be credited to Charles Darwin:

Under changed conditions of life, it is at least possible that slight modifications of instinct might be profitable to a species; and if it can be shown that instincts do vary ever so little, then I can see no difficulty in natural selection preserving and continually accumulating variations of instinct to any extent that may be profitable. It is thus, as I believe, that all the most complex and wonderful instincts have originated. (Darwin 1859, p. 209)

Darwin (1859, 1871) asserted that behavior, including social behavior, evolves by natural selection and sexual selection, although he never completely gave up on inheritance of acquired traits (now resurfacing under the guise of epigenetics and gene activation and silencing). He also foreshadowed kin selection by proposing that social behavior can evolve through natural selection on "families" rather than just individuals. Despite the apparent road forward paved by Darwin for future generations of researchers, there was resistance to the idea that behaviors evolve, even among those convinced that evolution should replace "creation" explanations for organismal diversity.

Comparative ethologists such as Lorenz and Tinbergen, building on foundational studies by Whitman, Heinroth, and Julian Huxley, accumulated many examples of phylogenetic factors in behavior, especially social behavior (Tinbergen, 1951; Burghardt, 1973, 1985; Burkhardt, 2005). Some of this early ethological work involving social behavior, largely courtship, parental care, and fighting, was summarized in books for general readers (e.g., Lorenz, 1950, 1966; Tinbergen, 1953a, 1953b; Etkin, 1964). Two towering achievements were the comprehensive analyses of social displays in ducks and related waterfowl (Lorenz, 1970; original in 1941) and the detailed comparison of the cliff nesting black-legged kittiwake (*Rissa tridactyla*) with other gull species, showing how ecology drove behavioral adaptations (Cullen, 1957). By the 1960s theoretical models, phylogenetic computer programs, and conceptual advances helped to situate behavioral evolution as a key aspect of the evolutionary synthesis (Hamilton, 1964a, 1964b; Lack, 1966, 1968a; J. A. Wiens, 1966; G. C. Williams, 1966).

To elaborate a bit, the early to mid-1900s saw the development of (overlapping) frameworks for studying and understanding behavior. These were comparative psychology and behaviorism (the study of animal behavior as a branch of psychology, often focused on learning rather than natural behavior), ethology (the naturalistic study of animal behavior from an evolutionary perspective), and later sociobiology/behavioral ecology (the study of the ecological and current adaptations of animals, largely in field settings); see reviews in Burghardt (1973), Dewsbury (1984a,b), E. O. Wilson (1975), and Burkhardt (2005).

Perhaps the most significant development that propelled in-depth studies of social behavior research was sociobiology, integrating many of the above principles into a coherent approach and promoted, almost single-handedly, by E. O. Wilson (1975). Sociobiology formally extended neo-Darwinism into the study of social behavior and animal societies. Although it was primarily concerned with the adaptive significance of social behavior in nonhuman animals (Wittenberger, 1981), its treatment of social biology in humans was particularly controversial, drawing criticism from social scientists as well as political groups (M. S. Gregory et al., 1978). Nevertheless, the major achievements of the sociobiological approach endured, and although today referred to most often as behavioral ecology, it remains the cornerstone for much current social behavior research.

Over the years controversies erupted, such as on the role of nature versus nurture; the analysis and importance of instinct; competing visions of group, individual, and gene evolution; niche construction; animal culture; and the evolutionary significance of kin selection and altruism (e.g., Wynne-Edwards, 1962; Hamilton, 1964a, 1964b; J. L. Brown, 1966; J. A. Wiens, 1966; G. C. Williams, 1966; Laland & Janik, 2006; D. S. Wilson & E. O. Wilson, 2007). These topics lurk in the background of the many areas covered in this book but are not central to our mission. The shift to theoretical analysis and hypothesis testing in social behavior did have a downside: the neglect, even denigrating and pushing aside, of descriptive accounts of what animals actually do in natural settings. Even the journal *Ethology*, cofounded by Konrad Lorenz and formerly coedited by GMB, no longer publishes naturalistic descriptive ethology (pers. comm. to GMB by current *Ethology* editor, Wolfgang Goymann, December 2019). This trend has led to debates and controversy on the scientific value of natural history data collected without a hypothesis-testing agenda (H. W. Greene, 2005). This book is, to some extent, an amalgam of various threads and emphases in the modern study of animal behavior.

1.5. Social Behavior in Reptiles

Birds, mammals, and, to a lesser extent, teleost fishes are widely recognized as containing the most social vertebrate lineages due to the widespread incidence of large, stable aggregations, the complexity of the interactions within these social groupings, and the prevalence of prolonged parental care of neonates in some groups (Tinbergen, 1953a, 1953b; E. O. Wilson, 1975; P. M. Bennett & Owens, 2002; Krause & Ruxton, 2002). Similarly, amphibians are renowned for having complex acoustic (frogs) and chemical (salamanders) communication systems and social systems that involve breeding aggregations, parental care, and pair bonds (Heatwole & Sullivan, 1995). In contrast, reptiles are often stereotyped as solitary and aggressive, lacking parental care or stable social aggregations, with little diversity to their social behavior beyond territoriality and dominance hierarchies (E. O. Wilson, 1975; MacLean, 1990). This perspective has dominated the literature, despite the view by some that reptiles have an important role in social behavior research (Brattstrom, 1974; Burghardt, 1977b; C. C. Carpenter & Ferguson, 1977; Chapple, 2003; Fox et al., 2003; Doody, Freedberg & Keogh, 2009; Doody, Burghardt & Dinets, 2013; Schuett et al., 2016; Whiting & While, 2017). Even prestigious and accomplished herpetologists have underestimated the social behavior of reptiles. For example, Branch (1989) states that "parental care is completely absent in most reptiles, and only in crocodiles does the male contribute anything more than gametes." This is unambiguously false (see Chapter 5).

The documented prevalence of different kinds of social behavior and social interactions does differ among vertebrate lineages, and this may be a source of the view that virtually all reptiles are asocial. For instance, the proportion of species that provide parental care to their eggs or young is higher in mammals and birds and lower in fishes, amphibians, and reptiles (Fig. 1.1).

Despite this pattern and the fact that each vertebrate group includes species that have varying levels of sociality and parental care, researchers have often dichotomized social behavior in vertebrates by labeling species as either "social" or "nonsocial" (Sharp et al., 2005; Bellemain et al., 2006; Höner et al., 2007), with reptiles often labeled as "nonsocial" (Wilkinson et al., 2010; Wilkinson & Huber, 2012). We contend that this social dichotomy is simplistic, scientifically misleading, neglects the diversity of social systems that is evident within genera, families, and vertebrate classes, and thus impedes our understanding of social behavior in

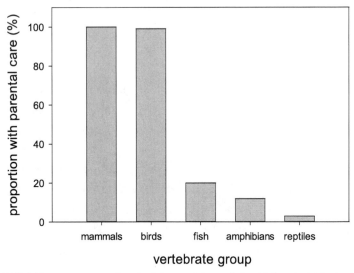

Fig. 1.1. Relative occurrence of postnatal parental care (after birth or hatching) across "traditional" vertebrate groups. From Doody, Burghardt & Dinets (2013).

reptiles and its evolution across vertebrates. For example, spatial and temporal variation in the social organization of several vertebrate species (Leyhausen, 1965; Lott, 1991) presents challenges when assigning species to particular categories. Conversely, regarding reptiles as "nonsocial" ignores their diverse and often complex social behaviors. This perspective ultimately results in reptiles being overlooked by researchers examining vertebrate social behavior, limits the scope of the reptile social behavior studies, and creates the impression that reptile species provide little opportunity for studying the mechanisms underlying the evolution of complex social behavior. In reality, the types and extent of sociality differ within all of the classes of vertebrates, including reptiles. Moreover, the effects of this dismissal of reptiles as having complex social behavior, emotions, cognition, and phenomena, such as play, social learning, and deception, have been ensconced in the literature by leading neuroscientists based on brain structures (e.g., MacLean, 1985, 1990). However, recent comparative neuroscience is documenting the ancient evolutionary origins of neural structures underlying complex social behavior (e.g.,

Bass & Chagnaud, 2012), meaning that reptiles have the necessary "hardware" for significant behavioral complexity.

1.6. Why Has Sociality in Reptiles Been Largely Neglected?

Reptiles are less studied than other vertebrates; research effort is not evenly distributed among the vertebrate groups, with reptiles receiving far less attention compared to birds, mammals, and fishes (Gaston & May, 1992; Bonnet et al., 2002; Pawar, 2003). For example, a survey of 1,000 scientific articles revealed that, although species richness is greater in reptiles than in mammals, the behavioral literature contains four times as many articles on mammals (Bonnet et al., 2002). There is also a particularly strong bias toward birds (44% of all journal articles on vertebrates), despite birds representing only 20% of the world's vertebrate diversity (Bonnet et al., 2002).

But this is only part of the story. The neglect of reptile social behavior may be a consequence of the broader bias against reptiles. The appearance, life history, ecology, and behavior of reptiles might contribute to their neglect in research (Kellert, 1993; Pawar, 2003; Doody, Burghardt & Dinets, 2013). Humans exhibit less affinity to "scaly" reptiles compared to "cute and cuddly" mammals and birds, and many people fear and actively avoid reptiles, especially snakes and crocodylians. To many people (including some researchers), snakes, crocodiles, and lizards are abhorrent creatures (in contrast, turtles are usually well liked). Charles Darwin, for one, was not impressed with the appearance or disposition of reptiles:

[The Marine Iguana] is a hideous-looking creature, of a dirty black colour, stupid and sluggish in its movements. . . . I threw one several times as far as I could, into a deep pool left by the retiring tide; but it invariably returned in a direct line to the spot where I stood. . . . Perhaps this singular piece of apparent stupidity may be accounted for by the circumstance, that this reptile had no enemy whatever on shore. (Darwin, 1845, p. 390)

Like their brothers the sea-kind, [Galápagos Land Iguanas] are ugly animals . . . from their low facial angle they have a singularly stupid appearance. (Darwin, 1845, p. 392)

In reference to observing a pitviper in Argentina, he declared: "I do not think I ever saw anything more ugly, excepting, perhaps, some of the vampire bats" (Darwin, 1845, p. 103). To be fair to Darwin, he at least insightfully considered the reason for this repugnance: "I imagine this repulsive aspect originates from the features being placed in positions, with respect to each other, somewhat proportional to those of the human face; and thus we obtain a scale of hideousness" (Darwin, 1845, p. 103).

Darwin is acknowledging that we view facial features of animals through biased eyes by interpreting them as human facial expressions. Thus, the strong brows of pitvipers and crocodiles make us anthropomorphically (and inaccurately) interpret their gaze as "mean," "angry," or "of evil intent." This is not helped by the fact that some crocodylians and many venomous snakes are indeed a lethal threat to humans. Bears and large cats can be equally dangerous, but stuffed bears and kittens delight most children. Reptiles also lack the facial expressions used by many mammals and do not use vocal signals as often as birds do. Squamate reptiles, in particular, tend to communicate using chemosensory cues that are not detectable by the human senses (Burghardt, 1970a; Pianka & Vitt, 2003). This may lead researchers to focus on vertebrate groups (e.g., birds, mammals, and amphibians) whose communication systems (e.g., visual and auditory) are more salient to human sensory perceptions (see a discussion of this bias in Chapter 4 and in Rivas & Burghardt, 2001, 2002).

Another major contributing factor is conspicuousness. The secretive nature of many reptiles relative to other vertebrates makes them not only difficult to study but also results in a lack of the initial natural history observations that draw the quantitative researcher to the species in the first place. This is compounded by findings that marking and capturing animals in field studies can often alter their behavior (Rodda et al., 1988). For example, the reproductive behavior of birds, mammals, and even amphibians is often conspicuous compared to that of reptiles, and this has consequences for the relative contribution of vertebrate groups to theory in areas such as reproductive ecology and behavioral ecology. The disparity between our knowledge of the breeding behavior of birds and that of reptiles is evidenced by the availability of complete field guides to the identification of birds' eggs and nests from at least three continents versus no such guides for reptiles. In fact, nest sites have not been found for a great many reptiles (reviewed in Doody, Freedberg & Keogh, 2009). In an illuminating example, the proportion of Australian lizards known to nest communally rises more than 10-fold (from 0.06 to 0.86) when species for which nests have not been discovered are excluded (Doody, Freedberg & Keogh, 2009), suggesting that communal breeding (and possibly other social behaviors) has been vastly underestimated in reptiles due to the inconspicuousness of their nests. Moreover, the perceived absence of such social behaviors in reptiles may generate a feedback loop that acts to perpetuate the bias toward studying social behavior in birds, mammals, and fishes, rather than in reptiles. In this way, confirmation bias may afflict the student of social behavior.

There are several additional obstacles that may inhibit the completion and publication of studies on reptile social behavior. For instance, it is often more difficult to obtain funding for reptile research compared to studies on birds and mammals (Czech et al., 1998). In addition, "taxonomic chauvinism" may occur among referees for scientific journals and presumably also the referees for funding bodies, which leads to reptile researchers using more space in the introduction of their articles "justifying" how their study makes a contribution to the field (Bonnet et al., 2002). As a consequence, most studies on reptile social behavior are published in taxon-specific journals (e.g., herpetology journals) or natural history outlets rather than broader journals in the fields of behavioral ecology and evolutionary biology. Perhaps relatedly, texts on animal behavior, behavioral ecology, the evolution of behavior, comparative cognition, and the like largely ignore the scientific literature on reptiles. Thus, we believe that many behavioral researchers are unaware of the complexity and diversity of social systems in reptiles (Lott, 1991) and of the opportunities that reptiles provide for examining evolutionary hypotheses in vertebrate social behavior (While, Uller & Wapstra, 2009a, 2009b; A. R. Davis et al., 2011; Doody, Burghardt & Dinets, 2013). The opposite also appears to be true: until recently, many herpetologists were not trained in ethology and field techniques for behavioral studies, and they often did not even consider such research, focusing instead on taxonomy, faunal composition, ecology, physiology, and other subjects. Less than 10% of papers on crocodylians published during 1900–2010 were devoted primarily to behavior, as opposed to more than 60% of papers on songbirds published during the same period. This percentage is even lower for some other groups of reptiles, approaching 0% for worm lizards (VD, unpubl. data).

1.7. Advances in Reptile Social Behavior Research

In 1976, a handful of researchers held the first-ever symposium on the social behavior of reptiles at the 64th annual meeting of the American Society of Zoologists (now the Society for Integrative and Comparative Biology) in New Orleans, Louisiana (N. Greenberg & Crews, 1977). The purpose of the symposium, according to participant Burghardt (1977b), was to address the question "What types of social behavior do extant reptiles engage in?" As such, the symposium would "help dispel the common view, even held by competent biologists and ethologists, that reptiles have an extremely limited and uninteresting lifestyle" (Burghardt, 1977b, p. 177). Highlights from the symposium publication (in the *American Zoologist*, now *Integrative and Comparative Biology*) included group vigilance and social facilitation in neonate iguanas; social aggregations in neonate snakes; dominance in neonate turtles; signaling and social behavior of crocodylians; acoustic and visual displays in geckos; neuroanatomy, signaling, species recognition, and behavior in anoles; display behavior in tortoises; communication and display behavior in snakes; and hormones, reproductive behavior, and speciation in reptiles. Although these were (collectively) an eclectic treatment, they presented novel findings based on excellent scientific endeavors that undoubtedly foreshadowed this review of social behavior in reptiles.

Exactly four decades later, the three authors of this book organized another symposium on social behavior in reptiles, in the same city, at the joint meetings of the American Society for Ichthyologists and Herpetologists. Surprisingly, there still remained no detailed review of social behavior in reptiles (let alone across all vertebrates). The new symposium highlighted the surge of social behavior research on reptiles in recent years. This work has revealed a remarkable suite of social behaviors, including embryo-embryo and parent–embryo communication, environmentally cued hatching, underwater vocalizations, multiparental crèches, biparental care, cooperative construction of homes, formation and maintenance of stable social groups, group vigilance, kin recognition, social and genetic monogamy, inbreeding avoidance mechanisms, cooperative feeding and hunting, alternative mating strategies, mate guarding, courtship dances, social learning, and play behavior. Collectively, along with advances in previously known social behaviors (e.g., signaling, territoriality, courtship, mating, communal nesting, and postnatal care), these examples demonstrate that reptile social behavior is now a burgeoning subdiscipline that should reposition the repertoire of reptiles into a more prominent role for understanding the evolution of social behavior in vertebrates. In this book, we highlight these areas.

Our goal is to illuminate a garden of fertile research areas that will attract cultivators, including behavioral ecologists, ethologists, comparative psychologists, evolutionary biologists, herpetologists, and especially students interested in social behavior and its origins. While we have kept this review compact and do not discuss in detail many fascinating phenomena, we provide numerous citations for further exploration in a list of references almost half as long as the text itself. We now invite you to enter the world of the cool reptiles.

2 |

Reptile Evolution and Biology

The term "reptiles" has been in use since the eighteenth century. Originally it included all tetrapods (vertebrates that had limbs and so were not fishes) except birds and mammals, so frogs and salamanders were also considered reptiles. In the early nineteenth century, tetrapods were divided into four classes: amphibians, reptiles, birds, and mammals (Latreille, 1825), and this classification remained almost universally accepted until very recently. The latter three classes were thought of as being more "advanced" than amphibians because they possessed numerous anatomical and physiological innovations, particularly the amniotic egg (see below). Collectively these three groups are known as Amniota. When large Mesozoic tetrapods such as *Iguanodon* and other dinosaurs were discovered in large numbers during the nineteenth century, they were also classified as reptiles.

By the end of the twentieth century, it gradually became clear that reptiles (class Reptilia) were not a monophyletic group. Birds are just one of numerous dinosaur lineages (Heilmann, 1927; Prum, 2008); the line leading to mammals branched off very early and probably had little in common with subsequent reptilian evolution (Watson, 1957). The term "reptiles" is so widespread and familiar, however, that it is still widely used and will probably remain in use for decades to come. In this book we discuss reptiles, or more accurately non-avian reptiles, in the traditional sense (including dinosaurs), but we will also mention birds, mammals, and extinct relatives of mammals to provide the reader with a more universal picture of social behavior in amniotes.

2.1. Evolution of Reptiles

The first amniotes evolved from ancient tetrapods (which were also ancestral to lissamphibians—modern frogs, salamanders, and their relatives—and are informally called "amphibians" together with lissamphibians) at least 310 million years ago, during the Carboniferous (Laurin & Reisz, 1995). The main differences between amniotes and their ancestors are novel adaptations to living on dry land. In addition to desiccation-resistant amniotic eggs (see below), amniotes have watertight skin often with scales of varying structure; we know that the earliest known amniotes had scales because scale prints can be seen in their fossilized tracks (Falcon-Lang et al., 2007).

At the end of the Carboniferous, the climate became drier, perhaps giving amniotes an advantage over amphibians, and amniotes became dominant in the Permian, 299-252 million years ago (Sahney et al., 2010). By that time, numerous amniote lineages already existed, and, as described in the following chapters, there is some evidence for their social behavior. Two groups were particularly important: synapsids and diapsids (Fig. 2.1). Synapsids, first known from the Early Carboniferous (Mann et al., 2019), flourished during the Permian and evolved spectacular diversity of sizes and feeding strategies (Hotton et al., 1986). There is now a growing tendency to exclude synapsids from "reptiles" (Angielczyk, 2009), but we will mention them briefly whenever relevant behavioral information is available for them. The other major group, the diapsids, first known from the Late Carboniferous (De Braga & Reisz, 1995), was at that time represented by small, lizard-like predatory animals; it split into two main branches, Archosauromorpha and Lepidosauromorpha (E. H. Colbert et al., 2001).

The Permian ended with the greatest mass extinction in the history of our planet, and numerous amniotes, including most synapsids, went extinct

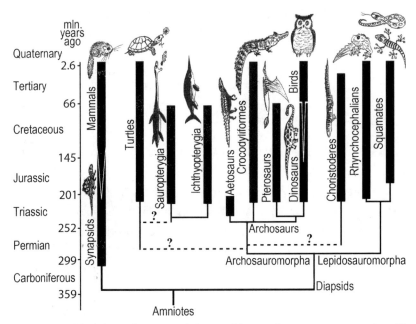

Fig. 2.1. A simplified chart of amniote phylogeny. Many extinct groups not mentioned in the text are omitted. Based on Nesbitt (2011) and M. S. Y. Lee (2013).

(Sahney & Benton, 2008). In the Triassic, synapsids were generally rare, but one lineage began accumulating important physiological innovations and eventually produced the mammals (Kermack & Kermack, 1984). Archosauromorphs largely replaced the synapsids and branched into numerous spectacular lineages, such as crocodyliformes (crocodiles and their relatives), aetosaurs (heavily armored herbivores), pterosaurs ("flying reptiles"), and dinosaurs; all of these, except for some early groups, are collectively known as archosaurs. Pterosaurs were the only flying vertebrates of the time; many other groups became marine, including marine crocodyliformes, choristoderes (small aquatic reptiles, mostly somewhat crocodile-like), ichthyosaurs, and sauropsids such as plesiosaurs (E. H. Colbert et al., 2001). The origins of many marine lineages and the first turtles are still controversial (Zardoya & Meyer, 2004; Bever et al., 2015; Reeder et al., 2015; Sues, 2019). Recent whole-genome molecular studies place turtles into Archosauria (Z. Wang et al., 2013), but studies combining molecular and morphological studies find them to be a sister group to all other major diapsid lineages (Schoch & Sues, 2015, 2018). Lepidosauromorphs included the ancestors of modern lizards, snakes, and the tuatara.

During the Jurassic (201–146 million years ago) and the Cretaceous (146–66 million years ago), archosaurs achieved spectacular diversity, arguably representing the pinnacle of tetrapod evolution. Some (perhaps many) of them were effectively endothermic, and certain lineages evolved hair-like structures or feathers. One feathery group of dinosaurs, known today as birds, evolved powered flight and began to compete with pterosaurs (Prum, 2008). Lepidosauromorphs were also successful during that time, particularly one group called Squamata. Squamates that evolved during that time included lizards, snakes, and giant marine reptiles called mosasaurs that were apparently related to monitor lizards (Gauthier, 1994). Snakes appear in the fossil record in the Middle Jurassic, but until the end of the Cretaceous, their fossils are remarkably scarce; many, if not most, snakes of that time apparently still had hind legs, such as the recently documented *Najash* (Garberoglio et al., 2019). Turtles began to look very much like those living today (Schoch & Sues, 2015; Sues, 2019). Meanwhile, non-mammalian synapsids gradually became extinct, and mammals remained small and generally rare (E. H. Colbert et al., 2001). *Arboroharamiya*, a Jurassic mammal, already had hair (Zheng et al., 2013), while *Spinolestes*, a mammal from the Early Cretaceous, had hair, hedgehog-like spines, and external ears (T. Martin et al., 2015).

At the end of the Cretaceous, another mass extinction extinguished much of reptile diversity including all non-avian dinosaurs. Only mammals, birds, turtles, crocodiles, and squamates survived it relatively well, with more than one family still representing each group (Sahney et al., 2010). In addition, the ancient lepidosaurian lineage Rhynchocephalia survived in New Zealand and (for a short while) in South America (Apesteguía et al., 2014), but apparently not elsewhere (Jones et al., 2009); one genus of choristoderes survived until about 20 million years ago (S. E. Evans & Klembara, 2005). There are claims (hotly disputed) that a few non-avian dinosaurs survived for at least a million years after the mass extinction (Fassett et al., 2002).

From that time until less than 100,000 years ago, mammals, birds, turtles, and squamates kept diversifying; crocodylians were also common and widespread wherever the climate allowed (Sahney et al., 2010; Mannion et al., 2015). As humans began colonizing the planet, they caused extinctions in one geographical region after another. At first, this "new mass extinction" affected mostly large animals and those living on oceanic islands, but then entire ecosystems began to collapse, and the diversity of all major taxa is now in decline (Barnosky et al., 2011). Among the recently extinct reptiles are the last marine (Molnar, 1982) and fully terrestrial (Mead et al., 2002) crocodiles, the largest lizard (the megalania, *Varanus priscus*) (Flannery, 2002), and almost all large tortoises (van Dijk et al., 2014). Countless others are endangered; these include almost all large turtles, the last surviving rhynchocephalian (the tuatara, *Sphenodon punctatus*), and the last two species of gharials (one of the three surviving crocodylian families) (IUCN, 2020).

Below is a very brief overview of living diapsids (reptiles and birds). Their sister group, the mammals, is outside the scope of this book; it is sufficient to mention that living mammals are represented by three groups that are very different in morphology and, particularly, reproductive biology. The egg-laying mammals (Monotremata) live only in Australia, New Guinea, and adjacent islands and include four species of echidnas (Tachyglossidae) as well as the platypus (*Ornithorhynchus anatinus*). These share a number of "reptilian" traits besides being oviparous, including a cloaca, low body temperatures, and partial heterothermy (reviewed in Burghardt, 2005). Moving forward, more comparisons with the reptiles in terms of their social behavior will be important. Bayard Brattstrom, a herpetologist who focused on reptile evolution and behavior, did an analysis of social behavior in short-beaked echidna (*Tachyglossus aculeatus*) and concluded that it was not nearly as complex as that of the many lizards he had studied (Brattstrom, 1973). Marsupials (Marsupialia) include about 340 species inhabiting Australia, New Guinea, and surrounding islands (west to Sulawesi), as well as the Americas, where only one species, the Virginia opossum (*Didelphis virginiana*), occurs north of Mexico. The remaining 94% of extant species are placental mammals (Placentalia); they have worldwide distribution on land and at sea (see Nowak, 2018, for more details).

2.2. Diversity of Extant Reptiles

Reptiles are the largest group of tetrapods: according to the regularly updated Reptile Database (Uetz et al., 2020), as of 2020 there are over 11,100 described species. The number of recognized bird species is similar but is growing more slowly, and much of this growth is due to splitting of "old" species that is not sufficiently justified (Dinets, 2014). There are over 8,000 described species of amphibians (AmphibiaWeb, 2019) and less than 6,000 described species of mammals (IUCN, 2020). The number of known species of reptiles grows by 200–300 per year due to both splitting "old" species and discovering new ones, and this growth shows no sign of slowing down—in fact, it is accelerating (Uetz, 2010).

According to Uetz (2015), Eurasia and adjacent islands (including New Guinea) have over 3,000 species of reptiles. Of these, only ~250 occur in Europe (including Turkey) and 455 in the Middle East. There are more than 2,000 species in South America and over 1,700 in North and Central America (433 of them found north of Mexico). The West Indies have >640 species despite a small land area. Mainland Africa has almost 1,700 species, plus ~500 on Madagascar. Australia has over 1,000 species, with lizards being particularly diverse. Oceania (without species-rich New Guinea and adjacent islands) has only ~350. There are no reptiles in the Antarctic, and they do not inhabit the northernmost parts of Eurasia and North America, but two species—the European common viper (*Vipera berus*) and the common lizard (*Zootoca vivipara*)—occur above the Arctic Circle in Scandinavia, while the Magellanic tree iguana (*Liolaemus magellanicus*) almost reaches the southern tip of South America (IUCN, 2020). Reptiles occur on many oceanic islands and have been recently introduced to many islands where they did not previously occur, such as Hawaii (McKeown, 1996). The redtail toadhead agama (*Phrynocephalus erythrurus*) occurs above 5,000 m in Tibet (E. Zhao & Adler, 1993), while the hot springs keelback snake (*Thermophis baileyi*) reaches 4,450 m, also in Tibet (IUCN,

2020). *Liolaemus* tree iguanas occur above 4,500 m in the Andes (Olave et al., 2011). The painted saw-scaled viper (*Echis coloratus*) and various agamid lizards occur on the shores of the Dead Sea at 430 m below sea level (VD, pers. obs.).

There are over 70 species of marine reptiles. Most of them—including the only marine lizard, the marine iguana (*Amblyrhynchus cristatus*)—inhabit coastal waters. A few out of over 60 species of sea snakes and six out of seven species of sea turtles are fully pelagic, while the estuarine crocodile (*Crocodylus porosus*) can cross vast stretches of open ocean (Rasmussen et al., 2011). Most sea snakes are viviparous, while all other extant marine reptiles lay eggs on land. Marine reptiles are generally confined to warm waters, but leatherback sea turtles (*Dermochelys coriacea*) regularly forage in cold waters and have been recorded in the Bering Strait area, in the Barents Sea, and at the southernmost tip of New Zealand. Loggerhead sea turtles (*Caretta caretta*) have also been recorded in cold waters of the North Atlantic and the North Pacific (Anan'eva et al., 1998).

Living reptiles are thus represented by four lineages with remarkably different levels of diversity. According to the Reptile Database (Uetz et al., 2020), over 96% are squamates, with over 6,680 species of lizards (including ~350 species of worm lizards, formerly considered a separate lineage from lizards) and over 3,600 species of snakes. There are over 350 species of turtles, ~30 species of crocodylians, and just one species of rhynchocephalian, the tuatara.

2.3. Extant Archosaurs

Archosaurs are represented by two surviving lineages: crocodylians and birds. Crocodiles have often been used as models for reconstructing unknown aspects of the biology of non-avian dinosaurs (see, for example, Brazaitis & Watanabe, 2011). But this approach has limited utility: as discussed below, modern crocodylians are a highly derived group with natural history and, particularly, physiology greatly modified from the ancestral condition. Non-avian dinosaurs were extremely diverse in all aspects of their biology, including reproduction and energetics, and also highly modified in almost all cases. So, extrapolating anything from extant crocodylians to extinct dinosaurs is more likely to be misleading than informative, just as extrapolating back from birds to their extinct dinosaur ancestors is.

Extant crocodylians belong to three families (Fig. 2.2). Their taxonomy is still changing a bit, as relationships between extant and extinct taxa are better understood (Brochu, 2003). The following account is based on Grigg and Kirshner (2015).

Crocodiles (Crocodylidae), the largest family, include at least 17 species distributed in freshwater and coastal habitats of tropical Asia, New Guinea, northern Australia, Africa, Madagascar, and the Americas from Mexico and Florida to Peru and Venezuela. A few species are capable of long-distance migration over the open ocean. The most "marine" of them, the estuarine crocodile (*Crocodylus porosus*), has the widest distribution, from Vanuatu, Fiji, and Fraser Island off eastern Australia to India and Sri Lanka

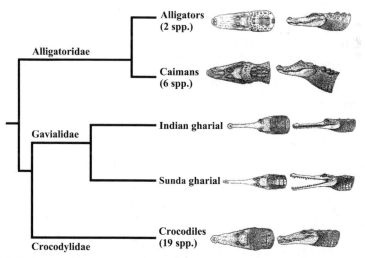

Fig. 2.2. Phylogeny of extant crocodylians. Modified from Dinets (2013b); species outlines from Wermuth & Fuchs (1978).

(historically also southern China and the Seychelles); a migrating individual was caught off Japan.

Alligators and caimans (Alligatoridae) are less tolerant of seawater and have more fragmented distributions. Three genera of caimans live in South America; the spectacled caiman (*Caiman crocodilus*) also lives in Central America and southernmost Mexico. Alligators are more cold-tolerant than other crocodylians and have a subtropical distribution: the American alligator (*Alligator mississippiensis*) inhabits the southeastern United States, while the Chinese alligator (*A. sinensis*) lives in eastern China.

Gharials (Gavialidae) are particularly narrow-snouted, highly aquatic crocodylians, now represented by just two species; a third one, small and apparently more marine, occurred in Oceania and went extinct soon after human arrival (Molnar, 1982). The Indian gharial (*Gavialis gangeticus*) is now limited to a handful of populations in rivers of northern India and adjacent parts of Nepal, while the Sunda gharial (*Tomistoma schlegelii*) survives in remote areas of Malaysia and western Indonesia.

Birds, the only surviving dinosaurs, are relatively uniform in anatomy, physiology, and breeding biology despite the large number of species. Most living lineages have appeared in an explosive evolutionary spike following the mass extinction at the end of the Cretaceous (Prum et al., 2015). They all belong to three groups known to predate that massive branching event: (1) Palaeognathae, which include ostriches and similar large flightless birds as well as kiwis (*Apteryx*) and quail-like Neotropical birds called tinamous (Tinamidae); (2) Galloanserae, which includes chickens, quail, pheasants, peacocks, turkeys, grouse, megapodes, ducks, geese, and swans, among others; and (3) Neoaves ("new birds"), which includes the rest of living birds, about 95% of extant species (Harshman, 2007).

2.4. Turtles and Tortoises

Although the origin of turtles has been one of the greatest mysteries of reptile paleontology, the phylogeny of living groups (Fig. 2.3) is relatively well agreed upon. The following account is based on van Dijk et al. (2014), unless noted otherwise.

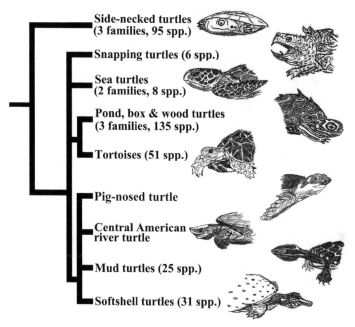

Side-necked turtles
(3 families, 95 spp.)

Snapping turtles (6 spp.)

Sea turtles
(2 families, 8 spp.)

Pond, box & wood turtles
(3 families, 135 spp.)

Tortoises (51 spp.)

Pig-nosed turtle

Central American
river turtle

Mud turtles (25 spp.)

Softshell turtles (31 spp.)

Fig. 2.3. Simplified phylogeny of extant turtles. Based on Sterli (2010) and Thomson & Shaffer (2010).

Living turtles (including tortoises) belong to two very distinctive groups that differ in many aspects of anatomy. They can be easily distinguished by the way they retract their heads into their shells. The so-called side-necked turtles (Pleurodira) bend their necks horizontally, so that the head is positioned sideways when inside the shell. All other turtles (Cryptodira) bend their necks vertically in an S shape, so that their head is still facing forward when pulled in; their scientific name means "hidden-neck turtles."

Pleurodires are a freshwater group (some extinct species were marine) that are common and widespread in Australia, New Guinea, sub-Saharan Africa (with one species occurring also in Yemen), and South America. There are three extant families and over 90 described species, including the famously bizarre mata mata (*Chelus fimbriata*).

Cryptodires are much more diverse. In addition to numerous fresh-water taxa, they include all extant marine and terrestrial species, and even one that is partially arboreal. Some families that are now small and local used to be distributed much more widely. Many species belonging to this group, particularly those living in Southeast Asia and Central America, are now declining catastrophically due to overhunting. A few species are now captive-bred in the millions for human consumption and pet trade.

Snapping turtles (Chelydridae) are a particularly ancient lineage, and they look the part. Recent molecular studies have shown that the group includes six to eight species occurring from southern Canada to Ecuador (Turtle Taxonomy Working Group, 2017).

Sea turtles are today represented by seven to eight species; all of them belong to the family Cheloniidae except the largest one, the leatherback sea turtle (Dermochelyidae). That species is the largest and most cosmopolitan fully marine reptile of today, reaching over 2.5 m in length and 650 kg in weight (McClain et al., 2015) and occurring all over the Atlantic, Indian, and Pacific Oceans.

The turtles most familiar to residents of temperate countries belong to two large families: Emydidae, which includes most North American species and the European pond turtle (*Emys orbicularis*), and Geoemydidae, which includes most Eurasian turtles. Both families also include species adapted to terrestrial life, such as box turtles (*Terrapene*, Emydidae) of North America and Neotropical wood turtles (*Rhinoclemmys*, Geoemydidae) of South America. The Emydidae family also includes one brackish-water species, the diamondback terrapin (*Malaclemys terrapin*) of eastern North America. Two more families belonging to the same lineage are fully terrestrial. The Platysternidae includes one species, the big-headed turtle (*Platysternon megacephalum*) of Southeast Asia, which is the only turtle known to occasionally climb trees. Testudinidae include all living tortoises, plus many large tortoise species recently driven to extinction by humans (van Dijk et al., 2014).

Another turtle lineage includes four families highly adapted to freshwater life. The pig-nosed turtle (*Carettochelys insculpta*) lives in northern Australia and southern New Guinea; it has flippers like a sea turtle but a soft skin like a softshell turtle. The Central American river turtle (*Dermatemys mawii*) is a large, herbivorous turtle of Belize, Guatemala, and tropical Mexico that is now almost extinct in the wild. Mud and musk turtles (Kinosternidae) are widespread in the Americas; there are about 25 species, mostly small and secretive. Finally, the softshell turtles (Trionychidae) are an ancient group of unique turtles that live in freshwater and brackish habitats of Asia, New Guinea, and Africa, with one genus (*Apalone*) occurring in North America. Many Asian softshells are now vanishingly rare; the global population of the Yangtze giant softshell turtle (*Rafetus swinhoei*), the largest freshwater turtle in the world, is now down to just four individuals, three in Vietnam and one in China (Long Nguyen pers. comm.).

2.5. Living Squamates

The name Squamata means "scaly ones" and refers to the scales that cover the bodies of lizards and snakes. These scales are formed in the outer layer of skin, which is periodically shed to allow growth. Snakes usually shed the outer layer of skin all in one piece, while in lizards and the tuatara it usually tears apart before falling off. "Scales" of turtles and crocodylians are very different: they are permanent structures that form deeper in the skin and are properly called scutes.

The phylogeny of squamates (Fig. 2.4) is still controversial; numerous fossorial (burrowing through the soil and living mostly or entirely underground) groups present particular challenges since it is often unclear if their similarities result from relatedness or from adapting to a similar lifestyle. The following account is compiled from J. J. Wiens et al. (2012), Pyron et al. (2013), Hsiang et al. (2015), Martill et al. (2015), and Streicher and Wiens (2017).

A small, little-known family called blind lizards (Dibamidae) is one of the reptile groups most thoroughly adapted to underground life. Their phylogenetic position has been controversial since their discovery; the most recent analysis (Streicher & Wiens, 2017) found blind lizards to be the most basal lineage among extant squamates, while some other studies (e.g., Reeder et al., 2015) find them to be related to Gekkota, a large, well-defined and likely ancient group that includes true geckos (Gekkonidae,

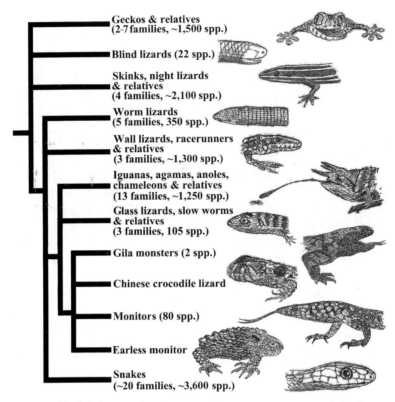

Geckos & relatives
(2-7 families, ~1,500 spp.)

Blind lizards (22 spp.)

Skinks, night lizards
& relatives
(4 families, ~2,100 spp.)

Worm lizards
(5 families, 350 spp.)

Wall lizards, racerunners
& relatives
(3 families, ~1,300 spp.)

Iguanas, agamas, anoles,
chameleons & relatives
(13 families, ~1,250 spp.)

Glass lizards, slow worms
& relatives
(3 families, 105 spp.)

Gila monsters (2 spp.)

Chinese crocodile lizard

Monitors (80 spp.)

Earless monitor

Snakes
(~20 families, ~3,600 spp.)

Fig. 2.4. Simplified phylogeny of extant squamates. Based on J. J. Wiens et al. (2012).

recently often split into up to four families) and their relatives. Gekkota are particularly diverse in the Old World tropics. Some geckos are very good at colonizing human-modified landscapes; a few species now occur almost worldwide as human commensals. Some, such as the leopard gecko (*Eublepharis macularius*), have been essentially domesticated and are among the most popular pets.

Another old, very large branch of the squamate tree includes skinks (Scincidae) and their relatives such as plated (Gerrhosauridae), spinytail (Cordylidae), and night (Xantusiidae) lizards. This cosmopolitan group is remarkable for including numerous lineages showing varying stages of limb and digit loss (which apparently evolved dozens of times in lizards), and also for its ability to colonize remote islands.

For people living in Europe and northern Asia, by far the most familiar lizards are wall lizards (Lacertidae). They are also widespread in savannas and deserts of Africa, in arid parts of Asia, and in Southeast Asia. Surprisingly, their closest living relatives appear to be the worm lizards (Amphisbaenidae), yet another ancient group particularly well adapted to subterranean life; they are more broadly distributed than blind lizards but also poorly known. Their relatedness to wall lizards was first claimed based on molecular data and was only recently confirmed by fossil evidence (Müller et al., 2011). Wall and worm lizards are also related to teiids (Teiidae) of the New World, a diverse family in which smaller species such as racerunners (*Aspidoscelis*) resemble wall lizards, while larger species, commonly called tegus, have converged in appearance and behavior with monitor lizards (Varanidae, see below).

A major lineage called Iguania includes agamids (Agamidae), chameleons (Chamaeleonidae), and iguanids (Iguanidae, but recently split into up to 10 families). Iguanids (in the broad sense) are by far the most diverse group of lizards in the New World; they include such familiar lizards as the green iguana (*Iguana iguana*), anoles (Polychrotidae), fence lizards (*Sceloporus*), and horned lizards (*Phrynosoma*). In the Old World they are replaced by agamid lizards, which often lead very similar lifestyles and sometimes look amazingly similar to certain iguanids. One of the greatest mysteries of biogeography is the presence of a few species of iguanids in Madagascar and Fiji; it is still unclear if they somehow got there from the Americas or represent the last survivors of a near-total replacement by agamids (Burghardt & Rand, 1982). Chameleons inhabit Africa and particularly Madagascar, with only a few species in Eurasia. This group includes the smallest currently known lizard, the tiny leaf chameleon (*Brookesia micra*) of Madagascar, less than 3 cm in total length (Glaw et al., 2012).

Anguimorph lizards include the glass lizards (*Anguidae*), the family of mostly legless lizards that includes the familiar slow worm (*Anguis fragilis*) of Europe and the largest legless lizard, sheltopusik (*Pseudopus apodus*) of the Mediterranean region, as well as the American glass lizards (*Ophisaurus*) and two related New World families. More robust lizards belonging to anguimorph clade include two species of beaded lizards (*Heloderma*) of

North and Central America, the crocodile lizard (*Shinisaurus crocodilurus*) of China and Vietnam, and the monitor lizards (*Varanus*); the latter include the largest living lizard—the Komodo dragon (*V. komodoensis*) (Murphy et al., 2002)—and the even larger, closely related megalania, which became extinct just a few thousand years ago, probably due to human-induced habitat changes or hunting in its native Australia (Molnar, 2004). Australia and Sunda Islands have the highest varanid diversity; a few species occur in mainland Asia and Africa. The same clade includes the small, partially fossorial earless monitor (*Lanthanotus borneensis*), an enigmatic species of remote Bornean rainforests.

The phylogeny of snakes (Fig. 2.5) is also far from clear; the following account is based on the most recent studies (Head, 2015; Figueroa et al.,

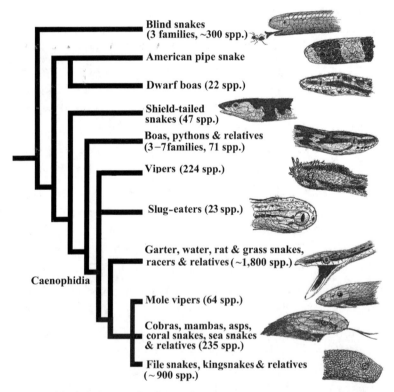

Fig. 2.5. Simplified phylogeny of extant snakes. Based on M. S. Y. Lee et al. (2007), Figueroa et al. (2016), and Streicher & Wiens (2016).

2016; Head et al., 2016; Streicher & Wiens, 2016; Zaher et al., 2019). It is generally agreed that snakes descend from some lizards, but their exact position among lizards is controversial: morphological studies suggest that they are most closely related to the earless monitor, while all recent molecular studies place them as a sister group to a branch that includes Iguania and Anguimorpha. One hotly contested study (Fry et al., 2006) claims that the common ancestor of snakes and these lizards was venomous and suggests that all of them together should be called Toxicofera ("venom-carriers").

The oldest living lineages of snakes are mostly represented by small, fossorial or semifossorial species such as blind snakes (Typhlopidae) and threadsnakes (Leptotyphlopidae); the latter family includes the world's smallest snake, the 10 cm long Barbados threadsnake (*Leptotyphlops carlae*) (Hedges, 2008). These snakes (sometimes collectively called Scolecophidia, although this grouping might be polyphyletic) do not have hinged jaws, so they can feed only on small prey; many specialize on ants and termites. The more recent lineages (collectively called Alethinophidia) are more diverse in morphology and ecology; every large group includes fossorial, terrestrial, arboreal, and aquatic forms. Some features of snake eye and brain morphology seem to indicate that their eyes were once almost lost and then re-evolved (Walls, 1940). Snakes have highly modified anatomy with a single lung, detachable lower jaw, and complete loss of legs and supporting bones, although in a few older groups the hind leg bones are partially preserved and end in spurs in males (more about these spurs in Chapter 5, section 5.7).

Boas (Boidae) and pythons (Pythonidae) are very similar. This similarity has long been thought to be a result of convergence (both groups being adapted to killing prey by constriction), but recently they are increasingly thought to be actually related. Boas are much more diverse and pantropical in distribution, while pythons naturally occur only in the Old World and reach their highest diversity in Australia and New Guinea. The reticulated python (*Python reticulatus*) may be the longest extant reptile, possibly reaching 7 m in length (Fredriksson, 2005), while the green anaconda (*Eunectes murinus*) is the heaviest living snake (Rivas, 2020). Much larger snakes belonging to extinct family Madtsoiidae survived in Australia until

50,000 years ago but became extinct soon after (and probably due to) human arrival (Flannery, 2002).

Vipers (Viperidae) include the true vipers of Eurasia and Africa and pit-vipers that live on all continents except Antarctica and Australia. These snakes have a very efficient venom delivery mechanism, with front teeth resembling folding syringe needles, and are mostly ambush hunters.

Most remaining snakes, perhaps 70%, have traditionally been lumped into one huge family, Colubridae, but we now know this group is not monophyletic. The majority of colubrids are nonvenomous, but there are many venomous taxa, including some of the deadliest snakes known. The traditional "colubrids" are by far the most diverse snakes on all continents except Australia, where their diversity and distribution are limited. More recent studies split colubrids into two groups (still very large): Colubroidea proper, 7–12 families, including many highly aquatic lineages, and Elapoidea, a diverse group of 6–14 families that includes kingsnakes (*Lamprophis*), highly aquatic file snakes (*Acrochordus*) of tropical Asia and Australia, and all venomous snakes other than vipers. These include mole vipers (family Atractaspididae) and elapid snakes (Elapidae). Mole vipers (also known as stiletto snakes) are a bizarre group of mostly African snakes, usually fossorial and sometimes highly venomous, with venomous teeth so long that they can be used even when the mouth is closed. Elapids are the largest group of venomous snakes; they include cobras, mambas, and kraits and are very diverse in the tropics of the Old World, but they are represented only by three genera of coral snakes in the Americas. Some Australian species have evolved close resemblance to vipers, which are absent there. Elapids also include sea snakes that are common in the Indian Ocean and much of the tropical western Pacific, but they are represented by just one species—the highly pelagic yellow-bellied sea snake (*Hydrophis platurus*)—in the central and eastern Pacific and are completely absent from the Atlantic.

2.6. General Anatomy and Physiology of Reptiles

Modern reptiles belong to two very different tetrapod lineages that split almost 300 million years ago: turtles and crocodylians (as well as birds) are archosauromorphs, while snakes, lizards, and the tuatara are lepidosauro-morphs (Schoch & Sues, 2015). For that reason, they have relatively few unifying anatomical and physiological features. Their bodies can be en-closed in skin, scales, shells, bony plates, and combinations thereof. They can be colorful or nondescript to human eyes. The role of color, color changes, and the underlying physiology has been the focus of much work (Cooper & Greenberg 1992).

One common feature, however, is being ectothermic, meaning that, un-like birds and mammals, reptiles cannot metabolically maintain a con-stant body temperature. It has often been claimed that ectothermy is "primitive" while endothermy (being capable of maintaining constant body temperature) is "advanced." This is correct in the sense of ectothermy being an ancestral condition of tetrapods, but the reality is much more complex (for example, crocodylians appear to be secondarily ectothermic, see below). Moreover, many larger reptiles are not completely ectothermic: they can maintain a body temperature a few degrees higher than the envi-ronment, at least temporarily. Such abilities have been found in sea turtles (see below), larger snakes (see Chapter 7), and the Argentine black and white tegu (*Salvator merianae*), a large lizard (Tattersall et al., 2016). As de-scribed below, females of some snakes can raise their body temperature above ambient when brooding eggs, a process termed thermogenesis. And all reptiles (except maybe some fossorial or fully aquatic taxa) routinely maintain near-constant body temperature for prolonged periods of time by moving between warmer (usually sunlit) and cooler (usually shady) patches of their habitat.

Maintaining constant body temperature by increased metabolism re-quires expending more energy; birds and mammals have much higher metabolic rates and need about 10 times more food than reptiles of the same body weight (Huey, 1982). This difference is particularly pronounced for smaller animals because they lose heat faster. Not surprisingly, the smallest reptiles are much smaller than the smallest birds or mammals: some geckos and chameleons weigh 0.15 g as adults (Hedges & Thomas, 2011; Glaw et al., 2012), while the smallest birds and mammals all weigh more than 1.5 g. Crocodylians are a notable exception. Although some extinct species became adults at around 30 cm total length, all extant crocodylian species mature at lengths of more than 100 cm, with adult males always being longer than 120 cm (Dinets, 2013a). This is also the minimum length at which male crocodylians become capable of producing infrasound signals (see Chapter 4, section 4.2), a finding that led Dinets (2013a) to suggest that by mating only with males capable of producing infrasound, female crocodylians have imposed an evolutionary limit on their minimal size.

Lower food consumption gives reptiles an advantage in areas where food is scarce; they outnumber birds and mammals and exceed them in biomass in many tropical environments, but particularly in tropical deserts (Wallace, 2011). It also means that they are better suited for certain lifestyles, particularly ambush hunting, which is a frequently used way of obtaining food for many squamates, some turtles, and all crocodylians.

Interestingly, there is compelling anatomical and physiological evidence that modern crocodylians had endothermic ancestors and possibly switched back to being ectothermic as an adaptation to ambush hunting and/or in order to survive prolonged dry seasons with limited food availability (Grigg & Kirshner, 2015). Large crocodiles and pythons have been known to fast for a year or more; some researchers have suggested that this ability allowed them to survive the mass extinction at the end of the Cretaceous (McCue, 2010). Since birds are endothermic, it has been suggested that this is the ancestral condition for all archosaurs, and there is some paleontological evidence that some archosauromorphs were already endothermic in the Triassic (M. F. Williams, 1997). Interestingly, the ancestors of mammals appear to also have evolved endothermy during the Triassic, despite the exceptionally warm climate of that time (Crompton et al., 2017; Benton, 2020); a warmer climate probably allowed endothermy to evolve more easily since its metabolic costs were lower. Anomodontia, a group of mostly herbivorous synapsids that was very diverse and abundant in the Permian and the Triassic, might also have been endothermic (Botha-Brink & Angielczyk, 2010). There is some evidence (Benton et al., 2019) that hair (in mammalian ancestors) and feathers (in archosaurs) also appeared in the Triassic, presumably for insulation.

The idea that all or most non-avian dinosaurs were endothermic was initially controversial but is now increasingly accepted (Persons & Currie, 2015). An alternative theory is that the larger dinosaurs were gigantothermic (Paladino et al., 1990), meaning that they were so large that they could maintain a constant or at least elevated body temperature during cool periods without high levels of metabolism, but there is some evidence against this, including the fact that even the largest living reptile, the estuarine crocodile, which attains a body mass of over 1,000 kg (Britton et al., 2012), is not gigantothermic (Seymour, 2013). The only extant reptiles known to be partially gigantothermic are the sea turtles, particularly the largest of them, the leatherback turtle, which can maintain body temperature a few degrees higher than that of the seawater while swimming (Standora et al., 1984; Sato, 2014). This ability has apparently allowed the leatherback turtle to be the only extant marine reptile regularly occurring outside tropical and subtropical seas, and it implies that many extinct marine reptiles may have been gigantothermic or endothermic (Surmik & Pelc, 2012). Isotope data suggest that ichthyosaurs, plesiosaurs, and mosasaurs were fully endothermic (Bernard et al., 2010; Harrell et al., 2016), although this has been disputed for mosasaurs, which may have been partially endothermic or gigantothermic (Motani, 2010), while marine crocodylian lineages had different degrees of endothermy, possibly explaining their differential survival during the global cooling at the end of the Jurassic (Séon et al., 2020).

Although modern reptiles cannot actively maintain a constant body temperature, they are remarkably good at using their environment to warm up or cool down. Basking is very common, particularly in cool weather; many reptiles living in particularly cold areas have evolved dark coloration to better absorb solar heat (Clusella Trullas et al., 2007). A good example is the large black subspecies of western fence lizard (*Sceloporus occidentalis*) that inhabits the high elevations of the Sierra Nevada in California (Leache et al., 2010). A study of spinytail lizards in South Africa found that black populations are more common in areas with frequent fog and cloud cover (Janse van Rensburg et al., 2009). In addition to adaptations for heating, reptiles have evolved numerous methods of avoiding overheating, such as urinating or salivating on their legs, as used by *Terrapene* box turtles (Alderton, 1988). Some species have evolved remarkable abilities to tolerate low or high temperatures: European common vipers and common gartersnakes from the northernmost populations can be seen basking on the snow or ice at 1°C (VD and Robert Mason, pers. obs.), rubber boas (*Charina bottae*) can be active at 4°C (Dorcas & Peterson, 1998), Okinawan pitvipers (*Ovophis okinavensis*) can actively hunt frogs at 10°C to 15°C (Mori et al., 2002), and tuataras have an optimal body temperature of 16°C to 21°C (Thompson & Daugherty, 1998), while some desert lizards can remain active at temperatures exceeding 40°C (Huey, 1982). Many reptiles survive freezing of their entire body in winter; the largest animal known to have such ability is the common box turtle (*Terrapene carolina*) of North America (Storey & Storey, 1992). Although reptiles move more slowly at low temperatures, there is evidence of some unknown physiological mechanisms allowing them to perform very fast movements when needed; Whitford et al. (2020) documented this ability in rattlesnakes (*Crotalus* spp.) performing defensive strikes.

Of course, being ectothermic puts certain restrictions on reptiles. Both diversity and biomass of reptiles are usually low at higher latitudes and altitudes, and the coldest parts of the planet have either very few reptiles or none at all. Herbivorous reptiles seem to be particularly constrained in geographical distribution, probably because digesting plant food is difficult at low temperatures (L. C. Zimmerman & Tracy, 1989). The same reason has been proposed as an explanation of the overall rarity of herbivory among extant reptiles (there are no herbivorous snakes or crocodylians and only an estimated 2% of lizards are herbivorous), although some taxa such as tortoises and liolaemid lizards appear to have overcome these constraints (Espinoza et al., 2004), and Mesozoic crocodylomorphs included at least three herbivorous lineages (Melstrom & Irmis, 2019). Herbivory is particularly rare among small reptiles, and juveniles of most (but not all) herbivorous lizards are omnivorous (Pough, 1973). Some reptiles have active lifestyles such as those of wide-ranging active hunters or nomadic herbivores; the desert monitor (*Varanus griseus*) is an example of the former (Tsellarius, 1994) while the green sea turtle (*Chelonia mydas*) represents the latter (Bjorndal, 1980). Snakes run the gamut from sit-and-wait predators, such as most pitvipers, to active hunters and foragers, such as many colubrids. Some species have to spend nine months per year hibernating and suffer very high mortality during that time, up to 40% in northernmost populations of common vipers (Mallow et al., 2003). It has been suggested (Burghardt, 1984, 2005) that the relative rarity of play behavior in reptiles is at least partially due to their more stringent energy budget. However, low metabolism does not prevent reptiles from participating in various energetic activities, such as displaying, ritualized fighting, and even play.

All extant reptiles breathe air, although a few aquatic species have been found to supplement air breathing with direct absorption of dissolved oxygen from the water. Many turtles do this through their skin, particularly the lining of the cloaca; some also use specialized extensions of the cloaca called cloacal bursae (Belkin, 1968; Stone et al., 1992; FitzGibbon & Franklin, 2010), while a softshell turtle can breathe underwater through the lining of its mouth (Gage & Gage, 1886). Sea snakes also absorb some oxygen and shed carbon dioxide through their skin (Graham, 1974).

Lung anatomy of extant reptiles is remarkably diverse. Most squamates use skeletal muscles to compress and inflate their lungs and have to stop breathing while moving fast; however, some taxa, such as varanid and teiid lizards, have evolved various anatomical solutions to this problem (Klein et al., 2003). Breathing is even more difficult for turtles since they cannot move their body wall; they have evolved structures that function much like the mammalian diaphragm. Some aquatic species, such as sea turtles, still have to stop breathing while moving on land, but others can breathe while running and even while sealed inside the shell (Landberg et al., 2003). In legless lizards one lung is usually reduced. Snakes have just one functional lung left; in some species, such as the puff adder (*Bitis arietans*), there is an extension of the lung that penetrates the neck and is used for sound production (Kent, 1987). Crocodylians have a diaphragm that they move by moving the liver. Their lungs are more complex and structurally advanced than mammalian lungs: they are built in such a way that the air always moves through them in the same direction, so that there are no cul-de-sacs and the air is completely exchanged in every breathing cycle, with oxygen being absorbed continuously while both inhaling and exhaling (Grigg & Kirshner, 2015). Birds share this advantage, but they have pushed it even further by evolving multiple air sacs that work together with lungs to improve breathing efficiency (Maina, 2006).

Squamates, the tuatara, and turtles have three-chambered hearts, while the hearts of birds and mammals are four-chambered. Having a three-chambered heart allows for partial mixing of oxygen-rich and oxygen-poor blood, which can to some extent be regulated in some or most species. The adaptive significance of this feature (if any) is still unclear, except in aquatic species where it likely serves to increase diving times (Hicks, 2002). In crocodylians, the situation is even more complex: the crocodylian heart, perhaps the most advanced among all animals, is four-chambered with an additional opening allowing for blood admixture during diving; that opening makes the heart functionally three-chambered during mixing and creates structural flexibility unique among animal hearts (Grigg & Kirshner, 2015).

Like birds and some mammals (the monotremes), reptiles have just one opening, called a cloaca, through which urine and feces are excreted; it also serves as the genital opening in females and in tuatara males (the only living male reptiles without at least one penis, see below). Reptiles generally have more acidic digestive juices than birds and mammals; their feces resemble those of birds. The urine of mammals and turtles contains mainly urea, while other reptiles and birds excrete mostly uric acid. Interestingly, no reptile or bird is known to use urine for scent-marking as many mammals do, but it is unknown if urine composition is the reason for that difference. Using feces (independently of associated pheromones) for transmitting social information has been demonstrated (see Chapter 4).

2.7. Motor Functions

For a long time, conventional wisdom was that "splayed-out" reptiles, most of which have limbs attached to the sides of the body, were a perfect illustration of "primitive condition" compared to birds and mammals, which walk on limbs positioned underneath the body. This idea is still present in many school textbooks—the people who conceived it probably never tried to catch a lizard by hand. In addition, reptiles (both living and extinct) were commonly considered to be slow, bumbling, and incapable of sustained physical effort (see, for example, Owen, 1866).

In reality, many reptiles are outstandingly agile, perform remarkable feats of endurance, and can achieve spectacular precision in their movements. Squamates cannot breathe and run simultaneously (an ancestral condition shared with amphibians), but archosaurs (and synapsids) have overcome this limitation. The sprawling position of most reptiles has its own benefits, allowing more flexibility of movement and faster speed gain (Reilly & Delancey, 1997). Many lizards can run bipedally: basilisks (*Basiliscus*) are famous for their ability to run bipedally so fast that they can run on the surface of water for more than 20 m (Hsieh & Lauder, 2004), while hatchling green iguanas (*Iguana iguana*) can run over open water when migrating from island nesting sites (Burghardt et al., 1977). Migrating sea turtles can travel for over 11,000 km (Nichols et al., 2000) and can swim at up to 10 km/h almost continuously (Eckert, 2002). Crocodiles are capable of performing transoceanic migrations, presumably without or almost without feeding, and show an impressive variety of gaits on land, including galloping at up to 18 km/h. Female crocodiles can handle eggs and hatchlings with their massive jaws without damaging them (Grigg & Kirshner, 2015). Seebacher et al. (2003) found that, although crocodiles are slower swimmers than fish, they are as efficient as aquatic mammals and more efficient than semiaquatic mammals. Extinct ichthyosaurs were probably as fast as dolphins and thunniform fishes, the fastest swimmers of today (Maddock et al., 1994).

Snakes, in particular, are a marvel of mechanical evolution because they can achieve amazing speeds despite having no limbs at all. Black mambas (*Dendroaspis polylepis*) can reach 11 km/h when moving over rough terrain (H. W. Greene, 1997) and move at an astonishing speed through tree crowns (VD, pers. obs.). Smaller species, often having the word "racer" in their common names, can be even faster: coachwhips (*Masticophis flagellum*) have been clocked at 15 km/h (Bellairs, 1970). Some sea snakes are relatively fast swimmers, exceeding 5 km/h (Shine, Cogger, et al., 2003).

Strike speeds of pitvipers can exceed 5.5 m/s (LaDuc, 2002). Snakes have evolved some completely novel ways of movement, including walking on belly scales and "sidewinding" over loose sand (H. W. Greene, 1997). These abilities required some anatomical changes, such as extreme streamlining of the body, increasing the number of vertebrae (to more than 400 in some cases) and skeletal muscles, and evolving unique muscles that connect the ribs with large belly scales (Kent, 1987; Lillywhite, 2014). Rattlesnakes have specialized tailshaker muscles capable of sustained, high-frequency movements, similar to cardiac muscle (Martin & Bagby, 1973; Savitzky & Moon, 2008). Western diamondback rattlesnakes (*Crotalus atrox*) can rattle 90 times per second continuously for hours at a time, demonstrating an aerobic capacity rare among vertebrates (Schuett et al., 2016).

Of course, not all reptiles are fast. Some are famous for their ability to move slowly or not move at all for extended periods of time. But these abilities are by no means inferior: they are adaptations for ambush hunting and were achieved by evolving numerous morphological novelties. For example, chameleons have unique skeletal features that allow them to navigate the three-dimensional mazes of shrubs, trees, and vine tangles (J. A. Peterson, 1984), while crocodylians have numerous adaptations for conserving energy and maintaining fixed position without movement (Grigg & Kirshner, 2015).

In addition to independently evolving powered flight two or more times (in pterosaurs, in dinosaur lineage leading to birds, and maybe also in some other dinosaurs; see M. Wang et al., 2019), reptiles evolved the ability to glide multiple times: there are "flying" snakes (*Chrysopelea*), "flying" geckos (*Ptychozoon, Hemidactylus*), gliding lizards (*Holaspis*), and the so-called flying dragons (*Draco*). The latter are small arboreal lizards with folding wings, supported by long rib extensions, capable of very fast and maneuverable gliding for up to 60 m with only 10 m of altitude loss (McGuire & Dudley, 2005). There were also numerous extinct gliding taxa, many of which were remarkably similar to flying dragons; of these taxa, tiny *Icarosaurus siefkeri* from the Late Triassic of New Jersey was possibly the best vertebrate glider ever (McGuire & Dudley, 2011). For comparison, mammals evolved powered flight only once (in bats).

Of particular interest are the dexterity and precision of forelimb movements. There is evidence that the ability to perform highly skilled and precise movements, particularly of forelimbs, evolved very early in tetrapods (Iwaniuk & Whishaw, 2000). In fact, much of the skeletal and muscular anatomy of limbs is very similar across tetrapod taxa; even frogs show some remarkable similarities with mammals (Abdala & Diogo, 2010). Among living reptiles, the simplest forelimb anatomy is found in the tuatara, while turtles and crocodylians have various modifications for aquatic lifestyles, and many lizards have adaptations for greater flexibility, especially in arboreal taxa (Haines, 1946). Increased grasping abilities in arboreal lizards (and probably some extinct taxa) were later used for improved prey handling (Fontanarrosa & Abdala, 2016). Arboreal monitor lizards are particularly good at this: for example, black tree monitors (*Varanus beccarii*) use their long forelimbs to reach into tree cavities and grab and extract prey (Mendyk & Horn, 2011). Arboreal mammals achieved similar results repeatedly and by different means (Whishaw et al., 1992); humans owe their unique dexterity to their arboreal ancestors. Using dexterous forelimbs for prey acquisition and handling is just one of many remarkable behavioral similarities between monitor lizards and mammals (Sweet & Pianka, 2003).

2.8. Sensory Organs

Many reptiles, and probably all birds, have tetrachromatic vision; mammals have lost some visual pigments and their vision is dichromatic (primates are the only known exception, with their secondarily evolved trichromatic vision) (Bowmaker, 1998). Some reptiles and birds can see well into the ultraviolet (UV) part of the light spectrum (K. Arnold & Neumeyer, 1987); UV reflectance of certain parts of the body is used as an honest signal of status by many birds and at least one reptile, Broadley's flat lizard (*Platysaurus broadleyi*) (Whiting et al., 2006). It has been suggested that snakes might use the ability to see UV light to visually follow pheromone trails (Sillman et al., 1997). The ability to see light polarization has been demonstrated for some species (Meyer-Rochow, 2014). Some nocturnal reptiles, particularly crocodylians and geckos, have excellent low-light vision (Grigg & Kirshner, 2015). This is evident in bright eyeshine at night in response to a flashlight; these species possess a layer of crystals called the tapetum lucidum, which is located behind the retina and reflects the light, causing it to pass through the retina a second time (Kent, 1987). On the other hand, fossorial species often have reduced vision or are only capable of telling light from darkness since their eyes are tiny and covered with skin (Foureaux et al., 2010). Snakes might have partially lost vision during the early stages of their evolution and then re-evolved it (Walls, 1940); their eyes work differently from those of other amniotes in that focusing is achieved by moving the lens back and forth rather than stretching it (Lillywhite, 2014), and their retinas have unique cone-like rod cells that might have evolved to replace the previously lost cone cells (Schott et al., 2016). Most birds have poor low-light vision, but nocturnal birds such as owls and nightjars have excellent night vision (G. R. Martin, 1982). Differences in eye size and morphology suggest that Mesozoic archosaurs were represented by diurnal, crepuscular, and nocturnal species, with pterosaurs being mostly diurnal and predatory dinosaurs mostly nocturnal (L. Schmitz & Motani, 2011). Pterosaurs are believed to have been mostly visual predators (Witmer et al., 2003), while many ichthyosaurs had huge eyes and were probably nocturnal or deepwater visual hunters (Motani et al., 1999). Most mammals have good low-light vision, but it is relatively poor in most diurnal primates such as humans (*Homo sapiens*) (VD, pers. obs.).

One early archosauromorph possibly had a fully functional third eye (Stocker et al., 2016). Many lizards and, particularly, the tuatara have a so-called parietal eye on the nape; its functions are still not fully understood, although it is known to be important for the animal's ability to regulate its temperature and light exposure (Hutchison & Kosh, 1974; Cree, 2014, has a compact review, focused on the tuatara but covering all reptiles). One extinct monitor lizard had two eyes of different origin on its nape (making it four-eyed), which is unique among known jawed vertebrates (K. T. Smith et al., 2018). The olive sea snake (*Aipysurus laevis*) and other *Aipysurus* species have photoreceptors in their tails, used to ensure that the tails do not remain exposed when the snake hides in a shelter (K. Zimmerman & Heatwole, 1990; Crowe-Riddell et al., 2019); it is unknown if this is a widespread trait in reptiles or a unique occurrence.

Hearing anatomy in reptiles is extremely diverse; many species are more or less similar to mammals and, particularly, birds in their hearing abilities (Wever,1974; Dooling & Popper, 2000), although there is no evidence of extremely acute hearing such as in owls (Strigidae) and the bat-eared fox (*Otocyon megalotis*) (Dyson et al., 1998; P. B. Grant & Samways, 2015). Most reptiles use tympanic (eardrum-based) hearing, which has evolved independently at least five times in tetrapods (in archosaurs, turtles, lepidosaurs, and mammals, as well as in frogs) and allows for directional hearing (Christensen-Dalsgaard & Carr, 2008). Such hearing was already present in early amniotes (Müller & Tsuji, 2007). Snakes and some lizards have lost external ears and all sensitivity to higher frequencies, although they can still perceive low-frequency sounds (Young et al., 2014). Worm lizards have a unique bone structure in their lower jaw that functions like an eardrum but is better adapted for registering sounds propagating through soil (Kent, 1987). In snakes, the sides of the jaw can move independently of one another, so a snake resting its jaws on a surface has sensitive stereo hearing that can detect the position of prey and is capable of detecting vibrations on the angstrom scale (Friedel et al., 2008). Snakes also have mechanoreceptors between scales that are sensitive to vibration of frequencies up to 800 Hz (Kent, 1987). Crocodylians use infrasound for communication, although how exactly they can hear it is still unclear (Todd, 2007). Unlike humans and many other nonaquatic animals, crocodylians are also capable of locating the sources of underwater sounds (Dinets, 2013d). Their brains have a special mechanism for detecting the differences in the time of arrival of the sound between left and right ears; that mechanism is shared with birds and so is likely ancestral in archosaurs, but it is absent in mammals (Kettler & Carr, 2019). Aquatic turtles hear much better in the water than in the air (Willis, 2016).

Chemical senses in reptiles are so complex that it is impossible to describe them simply as senses of smell and taste. Many reptiles have excellent chemoreceptive abilities; crocodiles and Komodo dragons can smell carrion from hundreds of meters away (Auffenberg, 1981; Burghardt et al.,

2002; Grigg & Kirshner, 2015). Some dinosaurs, particularly *Tyrannosaurus*, had expanded olfactory regions in the brain, indicating that their sense of smell was also acute (Brochu, 2000; Witmer & Ridgely, 2009). Squamates and the tuatara have a vomeronasal organ (VNO), also called the Jacobson's organ, located above the palate that is especially well developed in some lizards and all snakes, while being rather rudimentary in tuatara (Cree, 2014). It is also found in turtles. In many lizards and all snakes, it is used, in conjunction with the tongue, to analyze volatile and especially nonvolatile chemicals (Burghardt, 1970a, 1980; Halpern, 1992; Halpern & Martinez-Marcos, 2003). Numerous experimental studies, often using variants of tests with cotton swabs or marked objects in the animals' enclosures (e.g., Cooper & Burghardt, 1990a), show that lizards and snakes use chemical cues to identify prey, predators, mates, and conspecifics. Electrophysiological studies have confirmed that the flicking tongue transfers chemicals to the vomeronasal organ, where information is transferred to the accessory olfactory tract for analysis (Meredith & Burghardt, 1979). In fact, in snakes and many lizards, the accessory olfactory tract is much larger than the main olfactory tract associated with nasal olfaction. A deep fork in the tongue, found in all snakes and some lizards such as monitors and tegus, gives a directional sense of chemical cues (such as pheromone trails) by the vomeronasal organ, termed vomerolfaction (Cooper & Burghardt, 1990b). This system can be sufficiently sensitive to follow the scent trail of a particular female among the trails of numerous other females or the trail of an envenomated mouse among the trails of dozens of other mice, and can even determine the direction in which a struck rodent ran. Blind snakes can follow pheromone trails of ants, their main food, and distinguish between trails of different ant species (Gehlbach et al., 1971; Furry et al., 1991). Both nasal olfaction and vomerolfaction are widely used in communication (see Chapter 4); the latter is far more salient in snakes. Many mammals also have vomeronasal organs, but in some primates (including humans), birds, and crocodylians, embryonic vomeronasal organs disappear before birth or hatching (Kent, 1987). Reptiles have a vast array of glands and associated products that can be used in social communication, predator defense, and other contexts (Weldon et al., 2008), but their functional properties have not been well studied, with the primary exception of femoral glands of lizards (social recognition and marking) and cloacal scent glands in snakes (largely predator deterrents) (see Chapter 4).

Interestingly, most birds are believed to have a poor sense of smell, although there is a growing number of exceptions, and the importance of chemical perception in birds might be greatly underestimated (Steiger et al., 2008; Hiltpold & Shriver, 2018). Limited olfactory abilities might explain why visual and vocal displays are so widespread in birds. Those birds that do have an excellent sense of smell appear to specialize in particular odors, rather than have a universal sense of smell, as in many mammals. For example, some New World vultures (Cathartidae) can smell carrion from more than 1 km away (Houston, 1986), while spilling a few grams of fish oil on the ocean surface can attract storm-petrels (Hydrobatidae) from more than 1 km away (VD, pers. obs.). There are profound differences in diversity and repertoire of olfactory genes between squamates, turtles, crocodylians, birds, and mammals, and even between some closely related taxa (Vandewege et al., 2016).

Many reptiles have excellent senses of touch and muscular coordination: female crocodylians can transport eggs and hatchlings in their jaws without damaging them. Crocodylians also have sensory pits on some of their scutes; these pits, called integumentary sense organs (ISOs) are extremely sensitive to touch (more so than primate fingertips) and vibration, and they are somewhat sensitive to changes in pH and temperature (Grigg & Kirshner, 2015). The ISOs, which were also present in at least some non-avian dinosaurs (C. T. Barker et al., 2017) and extinct crocodylians, allow detection of prey in complete darkness (Soares, 2002), and they may allow detection of the tongue-action of drinking prey at the water's edge (Steer & Doody, 2009). Some snakes, particularly those living in turbid water, have tentacles, hair-like scale projections, and other organs that apparently allow them to sense vibrations from prey movements (Povel & Van Der Kooij, 1996); these organs are similar in function to vibrissae of mammals and hair-like feathers found on the heads of some birds, such as kiwis. Fossil evidence suggests that pterosaurs were particularly good at maintaining balance and received a lot of sensory information through their wing membranes (Witmer, Chatterjee, et al., 2003).

Boas, pythons, and pitvipers have independently evolved heat-sensing organs that allow them to "see" infrared radiation from mammals and birds and are sufficiently sensitive to locate a mouse at a distance of up to a few meters (Newman & Hartline, 1982). These organs (called lacrimal pits in pitvipers and labial pits in boas and pythons) can also detect objects that are colder than the environment (Kent, 1987), and they apparently work by detecting differences in temperature (Q. Chen et al., 2017); these organs resemble the most advanced human-made thermal imaging devices but have some structural advantages over those devices (Z. Zhang et al., 2015). The only other vertebrate known to have such organs is the common vampire bat (*Desmodus rotundus*), but its heat-sensing pits only allow it to locate large mammals at distances of less than 20 cm (Kürten & Schmidt, 1982).

Magnetoreception has been demonstrated for some species of sea turtles and crocodiles, and it is apparently used for navigation (Rodda, 1984; Grigg & Kirshner, 2015); it is also present in many (possibly almost all) birds and mammals but missing in humans (Wiltschko & Wiltschko, 2005). So far there is no evidence of electroreception in reptiles or birds, although it is present in some amphibians and fishes and has been found in a few mammals, including echidnas (Tachyglossidae), the platypus (*Ornithorhynchus anatinus*), and the Guiana dolphin (*Sotalia guianensis*) (Czech-Damal et al., 2012). It is possible that at least some reptiles possess yet unknown senses: navigational abilities of *Anolis* lizards cannot be explained by any known physical mechanism (Leal et al., in prep.).

2.9. Brains, Sensory Processing, and Behavioral Complexity and Cognition

As reptiles have highly evolved sensory organs, they need sufficiently complex brains to allow rapid processing of sensory information. Reptiles have smaller brains than many birds and most mammals of similar size, and for centuries this has been considered a sign of their limited intelligence. However, that difference in brain size has been historically overestimated for a number of reasons, such as failure to account for differences in overall body plan (Font et al., 2019; Font, Burghardt & Leal, 2021). For example, many reptiles are covered with heavy armor and have heavy bodies, whereas birds have evolved light bones and feathers, which facilitate flying. In addition, reptile brains have a very different structure from those of mammals, and, like in birds, the brain volume is utilized more efficiently, so direct comparisons of size are misleading (Jarvis et al., 2005; Font, Burghardt & Leal, 2021). In fact, as the following chapters will show, reptiles are capable of highly complex and flexible behavior. For example, crocodylians use small sticks as bait for egrets looking for nesting material (Dinets et al., 2014), rattlesnakes setting up ambushes for small mammals remove objects that might interfere with future strikes (Putman & Clark, 2015; Schuett et al., 2016), monitor lizards living near food stalls in Thailand stage sudden mock attacks on tourists to make them drop snacks, preferring women and children as targets (Wipatayotin, 2014), and at least one lizard species can follow bird flocks to fruiting fig trees (Whiting & Greeff, 1999). Many remarkable feats of learning and general cognition in reptiles have been documented (Burghardt, 1977a, 2013; Manrod et al., 2008; Szabo et al., 2020). Long-term memory has not been studied much, but it has recently been found that some tortoises can remember training for 10 years (Gutnick et al., 2020). The ability to quickly and efficiently process huge amounts of sensory information is particularly essential for flying animals, and pterosaurs were apparently good at it despite having smaller brains than birds (Witmer et al., 2003). The general anatomy of the reptilian brain and nervous system is summarized in Wyneken (2007) and described in comparison with other vertebrates in Kent (1987).

The patterns of sensory representation in the brain are remarkably similar in reptiles and mammals, suggesting that they were already present in the earliest amniotes (Gaither & Stein, 1979), even though the neuronal architecture of archosaurs, mammals, and squamates is different (Dugas-Ford et al., 2012). Indeed, fossil evidence suggests that early amniotes had complex hearing that required highly advanced sensory processing abilities

(Müller & Tsuji, 2007). Snakes have unique visual neuroanatomy (Wyneken, 2007), supporting the theory (Walls, 1940) that they nearly lost and then re-evolved vision.

Ectothermy probably had multiple effects on behavioral evolution of reptiles, but these are poorly understood (Burghardt, 1988). It has been claimed (Avery, 1976) that behavioral complexity and plasticity are reduced in reptiles inhabiting high latitudes. There has been no rigorous testing of this claim; it is possible that most known cases of complex behavior are known from warm climates simply because most species of reptiles are tropical and relatively few inhabit temperate zones. Alligators, subtropical in distribution, have behavior at least as complex (if not more so) as crocodiles, which are tropical. Some of the most striking examples of complex social behavior discussed in the following chapters come from species inhabiting subtropical and temperate areas: Australian skinks, North American rattlesnakes, and *Sceloporus* lizards.

2.10. Reptilian Reproduction

The following overview of reptilian reproduction is based on E. H. Colbert et al. (2001), Pough et al. (2015), and Grigg and Kirshner (2015), unless noted otherwise.

Unlike most amphibians, all living reptiles, as well as birds and mammals, have internal insemination: the male injects his sperm inside the female's body instead of sprinkling it over already laid eggs. Crocodylian and turtle males have one penis, while snakes and lizards have two, so-called hemipenes, of which only one is used at a time. The benefits of having two penises are poorly understood, although it is known that male brown anoles (*Anolis sagrei*) can increase sperm transfer by using them in alternating order for multiple matings (Tokarz & Slowinski, 1990), while common gartersnakes (*Thamnophis sirtalis*) preferentially use the larger right hemipenis, which allows them to create a larger postcopulatory plug, thus preventing the female from mating with other males for longer time (Shine, O'Connor & Mason, 2000). A tuatara male does not have a penis and performs insemination while pressing its cloaca to the female's, but this is a secondary condition as a tuatara embryo does have a penis (Sanger et al., 2015). Interestingly, while older groups of modern birds such as ostriches and ducks do have penises, the younger group, called Neoaves (to which most living birds belong), does not; instead, it uses cloacal contact (called a "cloacal kiss") (Briskie & Montgomerie, 1997). Since both crocodylians and older bird groups have penises, it is likely that extinct archosaurs such as pterosaurs and non-avian dinosaurs also had them, but there is no fossil evidence. The situation is complicated by the fact that the penises of mammals, birds, crocodylians, turtles, and squamates are all very different in histology and structure, and could have evolved independently (Kelly, 2002).

Shapes and structures of penises are amazingly diverse (for example, some turtles have gigantic penises with multiple tips and lobes, see Naish, 2012), but finding information on corresponding female organs is difficult. Female crocodylians and turtles, as well as some birds and all mammals, have a clitoris, but it is still unknown if it has any adaptive function. Squamate females have paired hemiclitorises (Böhme & Ziegler, 2009).

Some species of squamates have no males and reproduce asexually (by parthenogenesis); these include some whiptails (*Aspidoscelis*) and many other lizards in the family Teiidae, many species of Caucasian wall lizards (*Darevskia*), a few geckos, the Brahminy blind snake (*Indotyphlops braminus*), and possibly the Arafura file snake (*Acrochordus arafurae*) (Dubach et al., 1997). Females of parthenogenetic geckos and whiptails engage in mating-like behavior called pseudocopulation that stimulates them to produce eggs (Cuéllar & Kluge, 1972; Crews et al., 1986). Asexual species of whiptails and Caucasian wall lizards result from hybridization of sexually reproducing species; such hybridization events have been shown to be rare in Caucasian wall lizards due to the presence of efficient behavioral isolation mechanisms in parent species (Galoyan et al., 2019, 2020). Females of parthenogenetic species can later mate with males of sexually reproducing ones, producing triploid and tetraploid hybrid species (Reeder et al., 2002; Danielyan et al., 2008). Parthenogenetic geckos and the Brahminy blind snake are known as extremely successful invasive species, possibly due to the ability of a single animal to start a new population (Wynn et al.,

1987), but it should be noted that many invasive gecko species with world-wide distributions are not parthenogenetic (Uetz, 2015). Parthenogenetic wall lizards can be more successful than the ancestral sexual species and can replace them, particularly at higher elevations (Tarkhnishvili et al., 2010). Parthenogenetic lizard species are of hybrid origin, while snakes probably evolved the ability to breed without males because many of them have very low population densities and small home ranges, so they have difficulty finding mates (Rivas & Burghardt, 2005).

Recently a growing number of sexually reproducing species have been found to be capable of laying viable eggs or giving birth to living young without mating; these include snakes and monitor lizards (W. Booth et al., 2012; Reynolds et al., 2012; W. Booth & Schuett, 2016; L. Allen et al., 2018). The adaptive importance of this ability is uncertain: it is thought to either be accidental and inconsequential (van der Kooi & Schwander, 2015) or to represent an alternative reproductive strategy (Germano & Smith, 2010). Rare cases of parthenogenesis have been observed in birds, but they result in nonviable or abnormal offspring (Sarvella, 1974). Parthenogenesis in mammals is impossible due to a number of developmental and genetic constraints (Engelstaedter, 2008).

Another feature assisting reptiles in colonizing new lands is the ability of females of many species to store sperm for extended periods of time. The record (more than 45 years) belongs to Cuvier's dwarf caiman, *Paleosuchus palpebrosus* (Davenport, 1995); it is so exceptional that this record might actually represent the only known case of parthenogenesis in crocodylians. This ability might explain the transoceanic colonization of the New World by crocodiles (A. Schmitz et al., 2003). The longest recorded sperm storages in squamates are five and six years in the eastern (*Crotalus adamanteus*) and western (*C. atrox*) diamondback rattlesnakes, respectively (W. Booth & Schuett, 2011; Schuett et al., unpubl. data).

Multiple paternity is extremely widespread in reptiles. According to a review by Uller and Olsson (2008), it is present in all species investigated so far, and it occurs in about a half of clutches in most snakes and lizards but is less common in species with prolonged pair bonding, as well as in turtles and crocodylians. The authors suggest that multiple paternity is more common in reptiles than in birds and mammals due to "the combined effect of mate-encounter frequency and conflict over mating rates between males and females driven by large male benefits and relatively small female costs," the latter being the consequence of less developed or absent parental care in reptiles.

Most reptiles, all birds, and one small group of extant mammals (the monotremes) lay eggs. Unlike amphibian eggs, amniotic eggs are extremely complex and are the main biological feature uniting reptiles and birds with mammals. Each egg has a shell and membranes that protect the developing embryo, contain yolk (the embryo's source of food) and waste, and provide gas exchange. The innermost of the membranes is called the amniotic membrane; it is also present in viviparous reptiles and mammals and secretes the amniotic fluid (Kent, 1987). The outer membranes have to allow sufficient oxygen intake while preventing the egg from drying out; this is a difficult task, and many eggs can survive only within narrow ranges of temperature and humidity, making them dependent on parental care. The developing embryo's blood vessels spread out of an opening in the middle of its belly and form an extensive network outside its body to access various parts of the egg. By the time of hatching, the food stores are usually fully consumed, and all that remains of that outside network is a little scar on the hatchling's belly. In most mammals an umbilical scar (commonly called the belly button) also forms, as the blood vessels grow out of the embryo and into the placenta.

Eggs of most squamates, the tuatara, most turtles, and egg-laying mammals have leathery shells, while eggs of crocodylians, a minority of turtles, geckos of family Gekkonidae, and all birds have hard shells. Fossil evidence suggests that eggs of some (but not all) non-avian dinosaurs also had hard shells, while eggs of pterosaurs did not (Ji et al., 2004; Norell et al., 2020). Eggs of some reptiles need to remain still from ~24 hours after laying until the embryo is somewhat developed (about two weeks in crocodylians), but can be moved after that (Deeming, 1991; Grigg & Kirshner, 2015). Bird eggs need to be periodically rotated, although eggs of megapodes (Megapodiidae) are a notable exception (Shaffer et al., 2014). Bird eggs also require external warmth to initiate embryonic development and must be kept warm from that moment until hatching, making them dependent on near-constant parental care. Eggs of non-avian dinosaurs apparently represented various transitional stages between reptilian and avian modes of development: for example, coelurosaurian *Troodon formosus* laid two eggs simultaneously at daily or longer intervals and incubated them using a combination of soil and direct body contact (Varricchio et al., 1997).

Some small geckos can lay viable eggs every four to six weeks, but in the tuatara the process is exceptionally slow: a female needs one to three years to form eggs with yolk, and up to seven months to form the shell, so it takes up to five years from copulation to hatching. This means reproduction occurs at two- to five-year intervals, the slowest in any reptile; the sexual maturity is also reached slowly, at up to 15 years of age (Cree, 2002, 2014).

Embryonic development in reptiles (particularly crocodylians) is very similar to that of birds. One notable difference is that, although in all birds (and mammals) the sex of the offspring is determined by chromosomes, in many reptiles (including all crocodylians, most turtles, many lizards, and the tuatara) it is determined by incubation temperature (W. R. Branch, 1989; Valenzuela & Lance, 2004). In the Australian bearded dragon (*Pogona vitticeps*), sex determination is usually chromosomal, but very high temperatures can trigger sex reversal in adults with a subsequent switch to temperature-dependent sex determination (Holleley et al., 2015). Transitions between chromosomal and temperature-based sex determination are common in reptile evolution and have happened at least 17 times just in geckos (Gamble et al., 2015).

Hatchlings of most species have a tiny sharp blade on their snout, known as egg tooth; it is used to cut open the shell and is shed or resorbed soon after hatching. Bird, echidna, and platypus hatchlings also have it, and so did hatchlings of at least some non-avian dinosaurs (García, 2007). In some squamates it is a real tooth, but in most cases it is a simple keratinous structure. Some amphibian hatchlings also have it (Altig & McDiarmid, 2007), so possessing an egg tooth at hatching is likely the ancestral condition for amniotes.

Many squamates and virtually all mammals are viviparous; several lineages of extinct marine reptiles, such as ichthyosaurs and plesiosaurs, were also viviparous; but no species of turtle, crocodylian, or bird has evolved viviparity, possibly due to evolutionary constraints imposed by egg anatomy (see Chapter 7). In squamates, viviparity has evolved >100 times and is associated with cold climates (Shine & Bull, 1979; Blackburn, 1985, 2015) or shady habitats, such as closed-canopy forests in the case of chameleons (see D. F. Hughes & Blackburn, 2019), and, surprisingly, with the evolution of sociality (Halliwell, Uller, Holland & While, 2017). In at list one species, the yellow-bellied three-toed skink (*Saiphos equalis*), live newborns and eggs can be present in the same clutch (Laird et al., 2019). Squamates with fully evolved viviparity have also evolved placentae multiple times. It is often claimed that, unlike mammalian placentae, reptilian placentae provide the embryo only with oxygen, while the source of food is the yolk sac. However, at least three lineages of skinks have evolved fully functional placentae (Blackburn & Flemming, 2012), and there is evidence (H. Clark et al., 1955) of direct nutrient transfer from the mother to the embryos in the common gartersnake, one of the few reptile species in which this issue has been studied in detail. In mammals, the monotremes and marsupials do not have placentae (which is why marsupials are born so tiny), but other species do. Only one group of marsupials, the bandicoots (Peramelemorphia), have a tiny placenta-like structure connecting the embryo to the uterine wall (Padykula & Taylor, 1982). Interestingly, genes encoding some key proteins involved in placenta formation have been independently acquired by mammals and *Mabuya* skinks from retroviruses (Cornelis et al., 2017).

Whether they come from eggs or live births, young reptiles enter the world with their sensory and motor abilities highly developed, usually look similar to adults, and can, in most cases, take care of themselves without parental care—however, as we will see in Chapter 7, parental care can be highly beneficial. Early independence has multiple effects on the evolution of behavior of reptiles (as compared to parental-care-dependent birds and mammals), discussed in Chapter 9.

3 |

Mating Systems, Social Structure, and Social Organization

E. O. Wilson (1975) noted that while the average complexity of reptile social behavior is probably below that of mammals and birds, reptilian social life is considerably diverse, with a "few flashes of sophistication" and "some adaptations advanced even by mammalian standards." Some 45 research years later, reptiles have risen from this social obscurity, exhibiting many complex behaviors other than parental care, including those involved with mating systems and group living. Moreover, precursors to complex sociality such as monogamy and group interactions are necessary for understanding the evolution of, and constraints to, social complexity (Gardner et al., 2016; Halliwell, Uller, Holland & While, 2017; Whiting & While, 2017). For example, lizards have been used as a model for understanding the transition to complex sociality, due to delayed dispersal, parental care, monogamy, and family living (Whiting & While, 2017). Reptiles are thus a crucial pillar in the foundation underpinning social behavior evolution. Here we highlight the major patterns in mating systems, social structure, and social organization in reptiles that emerge, despite excessive knowledge gaps.

3.1. Mating Systems

Reptiles are traditionally thought to exhibit resource-defense polygyny, with intense male-to-male competition for females or resources that would support females (Stamps, 1977; Martins, 1994; Aldridge & Brown, 1995; C. M. Bull, 2000; McCoy et al., 2003; Shine, 2003). It was only after genetic technology unequivocally documented multiple paternity in snakes, and then other reptiles, beginning about 1990 (Schwartz et al., 1989) that a major reassessment of mating systems in reptiles, especially squamates (Rivas & Burghardt, 2005), took place. Today we know that multiple paternity is widespread, if not ubiquitous, across species (reviewed in Uller & Olsson, 2008; Wusterbarth et al., 2010; Wapstra & Olsson, 2014). Reptile mothers are often polyandrous. This can occur through two mechanisms. In sperm storage, sperm from prior mating seasons can fertilize eggs; this is known to occur in all major groups of reptiles, obscuring the magnitude of multiple mating within a mating season (Schuett, 1992; Davenport, 1995; Olsson & Madsen, 1998; Pearse & Avise, 2001; Pearse et al., 2002; Sever & Hamlet, 2002; Gist et al., 2008; Uller et al., 2013). Nevertheless, female reptiles may mate multiple times within a season as well as mate once and utilize stored sperm from previous seasons (Olsson & Madsen, 1998; Uller & Olsson, 2008). However, there is ample evidence that polyandry (and, thus, polygynandry) is common in reptiles (Rivas & Burghardt, 2005).

Among reptile groups, the highest frequency of multiple paternity was found in squamates (55%, $N=35$ spp.), followed by turtles (42%, $N=22$ spp.) and crocodylians (32%, $N=1$ spp.; but see below) (Uller & Olsson, 2008; J. A. Moore, Daugherty, Godfrey & Nelson, 2009; see also Wusterbarth et al., 2010, and Wapstra & Olsson, 2014, for reviews since 2008). Here we add to that summary. The tuatara (*Sphenodon punctatus*) is seasonally monogamous and polygynous across years, and it displays a very low (8%) level of multiple paternity (J. A. Moore, Daugherty, Godfrey & Nelson, 2009). Single paternity was also dominant in Mexican spinytail iguana (*Ctenosaura pectinata*), with a 9% level of multiple paternity (Faria et al., 2010). In crocodylians, multiple paternity ranged from 24% to 95% in nine different species (L. Davis et al., 2001; McVay et al., 2008; Lance et al.,

2009; Amavet et al., 2008, 2012; Hu & Wu, 2010; Muniz et al., 2011; Oliveira et al., 2014; Lafferriere et al., 2016; Ojeda et al., 2016; H. Wang et al., 2017).

What are the benefits of multiple paternity for mothers? Because parental care is rare in reptiles, any benefits of multiple paternity are likely indirect genetic effects (Uller & Olsson, 2008). Uller and Olsson (2008) reviewed a number of potential hypotheses for multiple paternity in reptiles, including their own offering: because males benefit from multiple matings, as long as costs of mating are low for females, multiple paternity will occur or evolve without the need for direct (e.g., nuptial gifts) or indirect (genetic) benefits. But there may be other benefits given the well-documented behavioral variation among individual offspring in many reptiles (Waters et al., 2017). For example, gartersnakes (*Thamnophis*) from the same litter show variation in prey preferences (Burghardt, 1975) and antipredator behavior (Herzog et al., 1989). Given that environments can vary, a female may effectively hedge her bets by producing offspring with different survival tactics. This may be especially feasible when a mother produces large litters or clutches, only a few of whom will reach maturity. It may be that in many species the likelihood of multiple paternity is correlated with mate-encounter frequency. Indeed, we now know that in a wide variety of snakes and lizards, failure to find a mate can lead to parthenogenesis and all-male offspring (review in Mendelson et al., 2019). This could facilitate colonization of new habitats such as islands and help explain the remarkable diversity and speciation in some groups.

The above collectively suggest that, in fact, polygynandry (promiscuity) may be the predominant mating system in reptiles (Kamath & Losos, 2017; While et al., 2019); most species show polygyny with some degree of polyandry (Rivas & Burghardt, 2005; Whiting & While, 2017). There is evidence for monogamy in some reptiles, however. Although sea turtles show both polyandry and polygyny (reviewed in P. Lee, 2008), one population of green sea turtles (*Chelonia mydas*) potentially showed relatively high levels of (serial) monogamy (L. Wright et al., 2012). In lizards, monogamy is mainly associated with mate guarding (C. M. Bull, 2000). For example, a male may defend a territory that contains a single female (e.g., cleft lizard, *Sceloporus mucronatus*, Lemos-Espinal et al., 1997) or a male may track a female for an extended period before, during, or after mating to prevent access by rival males (e.g., broad-headed skink, *Plestiodon* [*Eumeces*] *laticeps*, Cooper & Vitt, 1997; the snow skink, *Niveoscincus microlepidotus*, Olsson & Shine, 1998). There can also be monogamous and polygynous individuals within a species, yet the extent of this is largely unknown (reviewed in C. M. Bull, 2000; Chapple & Keogh, 2005). Seasonal monogamy is common in some crocodylians such as the Nile crocodile (*Crocodylus niloticus*) (Pooley & Gans, 1976) or the Orinoco crocodile (*C. intermedius*) (Thorbjarnarson & Hernández, 1993a), and in the tuatara (J. A. Moore, Daugherty, Godfrey & Nelson, 2009).

Long-term genetic monogamy appears to be widespread in a lineage of group-living lizards (genus *Egernia*) (Gardner et al., 2002; reviewed in Chapple, 2003; D. O'Connor & Shine, 2003; Stow & Sunnucks, 2004; Chapple & Keogh, 2005, 2006; While, Sinn, et al., 2009). However, a mixture of polygyny, seasonal monogamy, and long-term monogamy may be common, as in White's skink (*Liopholis whitii*) (Chapple & Keogh, 2005). In an exceptional case, one species, the Australian sleepy lizard (*Tiliqua rugosa*) exhibits long-term social monogamy during the breeding season (Plate 3.1) (C. M. Bull, 2000). This nonterritorial species forms monogamous pair bonds for six to eight weeks before mating in late spring. The pairs split up after mating, but usually reform the next year. In two studies, 77% and 79% of females were paired with the same male partner across years (C. M. Bull, 1988; C. M. Bull et al., 1998). In 141 partnerships, 110 lasted more than 10 years, 31 lasted more than 15 years, and the longest recorded was a remarkable 27 years and counting (Leu et al., 2015).

C. M. Bull (2000) predicted that long-term monogamy would be found in other lizard species, noting the general difficulty of studying mating systems in reptiles. The high recapture rates and long-term nature of the sleepy lizard studies offered a rare insight into the social mating system of a reptile. Though quantifying genetic data for revealing paternity is important, data on multiple matings and social pair bonding in nature are also desperately needed (e.g., A. Harrison, 2013). Let us hope that the legacy of the late Mike Bull is emulated by new recruits who are passionate about revealing social behaviors associated with mating systems.

Although an exhaustive review of mating systems and associated behaviors is beyond our scope, there are studies on reptiles with novel findings in the area of mating systems (see Chapter 5). For example, researchers experimentally revealed that more aggressive females sired more offspring from extra-pair matings than did less aggressive females (While, Sinn, et al., 2009). A. Harrison (2013) found size assortative pairing in a socially monogamous population of the border anole (*Anolis limifrons*); larger males were paired with larger females, while smaller males were paired with smaller females. In the gopher tortoise (*Gopherus polyphemus*), a few resident males sired a high (46%) proportion of offspring (Tuberville et al., 2011), which has important implications for conservation efforts involving the translocation of individuals and populations.

3.2. Social Structure and Social Organization

According to Kappeler (2019), "social structure" is the content, quality, and patterning of social relationships emerging from repeated interactions between pairs of individuals belonging to the same social unit (e.g., dominance hierarchies and social bonds), while "social organization" refers to the size and composition of a social unit (solitary individuals, pairs, colonies, groups, etc.). Lizards have long been the focus of studies of territoriality and dominance hierarchies in animals (Stamps, 1977; Fox et al., 2003); territoriality appears to be less common in other reptilian groups, but most of those groups are more secretive (aquatic, nocturnal, inconspicuous). According to Whiting and While (2017), major areas of study of lizard social structure have included contest competition and rival recognition, male alternative reproductive tactics, mate preference and mate choice, and communication via signaling. Mate preference and mate choice are discussed in Chapter 5, and communication is covered in Chapter 4; the other behaviors making up social structure in reptiles follow here.

Territoriality is widespread in lizards, with males and females defending territories, and males possessing territories that overlap with those of one or more females (Stamps, 1977; Rodda, 1992; Fox et al., 2003). Males and females may both defend territories (Cooper, 1993; Halloy & Halloy,

1997) and Whiting and While (2017) predict that the social bonds formed during interactions between males and females, and between females that share space, are the most common form of social structure in lizards. They note, however, that most studies have not traditionally quantified these social bonds (but see Strickland et al., 2014). Instead, social structure is inferred from patterns of spatial overlap: males do not generally overlap with other males, but they overlap with one or multiple females (e.g., Panov & Zykova, 1993; Zykova & Panov, 1993; Lemos-Espinal et al., 1997); females tend to overlap with other females. Experiments have revealed that larger body size, resident status, and contest initiation are associated with the likelihood of winning contests between territorial male lizards (Olsson & Shine, 2000; Fox et al., 2003; Whiting et al., 2006; Sacchi et al., 2009; Umbers et al., 2012). Other factors may explain contest success, however, such as ultraviolet coloration (Stapley & Whiting, 2006; Whiting et al., 2006).

Some of the early pioneering studies of territorial and lek-like mating systems were those of Dugan (1982a, 1982b) in Panama on green iguanas, *Iguana iguana*. Dugan documented large males establishing territories they defended from other males. These territories were chosen based on visibility and defendability and had nothing to do with food resources or nest quality, as is typical in birds. At the beginning of the six-week mating season, females "shopped" around and chose a territory defended by a male, and males could have harems of up to 13 females (Rodda, 1992). Females in such a defended territory were protected from forced copulatory attempts by marginal males unable to secure an attractive territory, as well as by small female-mimicking males that quietly hung around the edges of large-male territories, waiting for opportunities for mating attempts when the male was preoccupied. After the mating season ended, the territories dissolved, and many weeks later females migrated to communal nesting sites where further dramatic social interactions took place. Wikelski and Bäurle (1996) described virtually the same type of social system in marine iguanas (*Amblyrhynchus cristatus*). Further information on iguana social behavior is in Chapter 9 and elsewhere in this book. A detailed analysis of the ecological factors intertwined with the varying social systems in iguanas is found in Dugan and Wiewandt (1982) and, for marine iguanas, in W. K. Hayes et al. (2004). Due to the threatened conservation status of many Caribbean iguanas, detailed analyses of many of these species are now available as well (A. C. Alberts et al., 2004; Iverson et al., 2016).

Are snakes territorial? Early researchers thought not, but then the snake "courtship dances" many researchers had observed were revealed to actually be male-to-male combat, which C. Lowe (1948) interpreted as territorial behavior. However, despite male-to-male combat and various forms of aggregations, social structure in the form of territoriality, as observed in lizards, is currently unknown in snakes (C. C. Carpenter, 1977; H. W. Greene, 1997; W.-S. Huang et al., 2011; but see below). Aggression in snakes appears to be restricted to male-to-male combat (aside from cannibalism); these contests have been interpreted as gaining access to females (Shine 1978; Shine & Fitzgerald, 1995), rather than defending a territory; indeed females have been seen lurking in the area where males are combating. However, it is worth noting that a female is rarely sighted during these contests; most combat observations have occurred in the absence of food, a mate, or a confined space (C. C. Carpenter, 1984; see Chapter 5). In a phylogenetic analysis, Senter et al. (2014) reconstructed the evolution of behaviors associated with male-to-male combat in snakes; their analysis suggests that early (Cretaceous) combat involved raising the head and neck in an attempt to topple one another, while later, in the Lampropeltini, toppling was replaced by coiling, and body-bridging was added. Similar male-to-male combat is also frequently observed in lizards; one of the most conspicuous of these is monitor lizards (*Varanus*) and beaded lizards (*Heloderma*) engaging in bipedal wrestling (Murphy & Mitchell, 1974; Auffenberg, 1981; Greer, 1989; Beck & Ramirez-Bautista, 1991; Phillips & Millar, 1998; Schuett et al., 2009).

Biology is rife with exceptions to broad patterns; in a remarkable one, females in an island population of Taiwanese kukri snakes (*Oligodon formosanus*) exhibit territoriality by actively defending sea turtle nests by repelling conspecifics for long periods (weeks) until the turtle eggs hatch or are consumed (W.-S. Huang et al., 2011). The authors hypothesized that the reason for this exceptional behavior was in the diet: a clutch of turtle eggs comprises a large, long-lasting food resource, unlike the prey types exploited by other species of snakes. They also concluded that the lack of territoriality in blindsnakes, which gorge on massive colonies of social insects, is due to the lack of weaponry capable of repelling intraspecific predators. Interestingly, territoriality in kukri snakes incurred a survival cost: males and nonterritorial females experienced lower mortality and lived longer than territorial females (C. Lee et al., 2019).

Turtles are not known to defend territories despite aggression between males in many species (Auffenberg, 1977; Bury, 1979; Aguirre et al., 1984; Kaufmann, 1992, 1995; Harless et al., 2009). The essential characteristics of a territory are a fixed location, defense against intruders, and exclusivity of the area (J. L. Brown & Orians, 1970). In studies and anecdotes of the common snapping turtle (*Chelydra serpentina*), there is evidence of the first two of these, but not the third; home-range cores overlapped between individual males (Galbraith et al., 1987). In contrast, Sonoran mud turtles (*Kinosternon sonoriense*) maintained nearly exclusive use of pools, but defense of these pools is uncertain (Emslie, 1982; Hall & Steidl, 2007). Male tuataras defend stable territories containing multiple burrows but allow some spatial overlap with neighboring territories of individuals of either sex (Gillingham et al., 1995; J. A. Moore, Daugherty & Nelson, 2009; review in Cree, 2014). Large males were more successful at monopolizing areas where females were most dense and more effective at guarding females by consistently winning conflicts with other males (J. A. Moore, Daugherty & Nelson, 2009). The degree of territoriality in crocodylians is poorly understood. It is generally understood that alligators and caimans (*Caiman*), which form mating aggregations, are less territorial than species living in groups with a dominant male, such as many *Crocodylus* crocodiles, or occurring at low densities and mostly solitarily, such as Sunda gharials (*Tomistoma*) and dwarf caimans (*Paleosuchus*). This is based on anecdotes, however, rather than focused research (Dinets, 2011c), and there is a suggestion that death roll behavior in all species is related to intraspecific competition (Drumheller et al., 2019).

Complicating matters in a fascinating way, sexual selection can generate phenotypic diversity, often manifest in alternate phenotypes in males

(Andersson, 1994; Taborsky & Brockmann, 2010). For example, in a classic example, male side-blotched lizards (*Uta stansburiana*) exhibit one of three discrete morphs (with orange, blue, or yellow throats; Sinervo & Lively 1996). Orange-throated males are hyperaggressive and defend large territories, blue-throated males defend smaller territories and are less aggressive, and yellow-throated males do not defend territories but sneak mates from orange males (Sinervo & Lively 1996). Paternity analysis (Zamudio & Sinervo, 2000) revealed that blue-throated males mate-guard their females and avoid cuckoldry by yellow-throated sneaker males, but mate guarding is ineffective against aggressive orange-throated males. The ultradominant orange-throated males are highly polygynous and maintain large territories, overpowering blue-throated males and co-siring offspring with their females, but are often cuckolded by yellow-throated males. Yellow-throated sneaker males sire offspring via secretive copulations and often share paternity of offspring within a female's clutch. Finally, sneaker males sire more offspring posthumously, indicating that sperm competition may be an important component of their strategy (Zamudio & Sinervo, 2000).

Another example involves males mimicking females in the Broadley's flat lizard (*Platysaurus broadleyi*). Older territorial males are aggressive toward male intruders, but young males have female-like coloration and can thus gain access to females from the territories of resident males (Whiting et al., 2009). The deceit of these female-mimicking males breaks down if they get too close to the older territorial males (due to olfaction giving away their gender); as a result, female-mimicking males keep their distance from dominant males (Whiting et al., 2009). Alternative reproductive tactics can also be behavioral rather than morphological; large male water skinks (*Eulamprus quoyii*) adopted subtly different reproductive tactics to achieve matings, despite having no obvious morphological differences among them (D. W. A. Noble et al., 2013). Sneaker males and morphological deception are also recorded in green iguanas (Dugan, 1982b; Weldon & Burghardt, 1984).

When reptiles such as lizards are at high densities, territoriality gives way to a dominance hierarchy (Brattstrom, 1974, and references therein). Dominance hierarchies in lizards, which range from complex fighting and submission (e.g., iguanids, agamids) to more subtle interactions (e.g., vocalizations in geckos) were reviewed by Brattstrom (1974), who noted that increases in population sizes cause increases in aggression, a generality common to other vertebrates. High population density can lead to either increased dispersal or decreased spacing, and in the latter a dominance hierarchy develops (L. T. Evans, 1951; C. C. Carpenter, 1966; Brattstrom, 1971, 1974). Experiments manipulating territories of lizards revealed that when a dominance hierarchy developed, larger lizards that signaled more often had higher ranking status (Brattstron, 1974). Dominants may gain their status mainly through the submissive behavior of subordinates, rather than by the dominant's displaying or fighting. Submission can take a diversity of forms ranging from flattening or adpressing to arm waving to flipping upside down (C. C. Carpenter, Badham & Kimble, 1970; Brattstrom, 1971, 1974). These submissive displays can allow a subordinate individual to move freely within the realm of a dominant. However, a predator deterrent function of limb waving has also been supported in wall lizards (*Podarcis*) (Font et al., 2012). Hierarchies are not restricted to lizards; tortoises and turtles exhibit dominance hierarchies and male fighting is common (Auffenberg, 1969; Boice, 1970; Froese & Burghardt, 1974). The same is apparently true for most crocodylians, although there is much diversity and the general opinion among researchers and keepers (based largely on unpublished experience) is that vertical hierarchy is much more "strict" in crocodiles than in alligators, caimans, and gharials (Dinets, 2013b).

3.3. Group Living

There are five recent reviews that cover this topic in reptiles. Gardner et al. (2016) defined aggregations and their contribution to complex social behaviors in squamates, Fox et al. (2003) and Whiting and While (2017) tackled sociality in lizards, While et al. (2019) focused on stable social groups in lizards, and Schuett et al. (2016) synthesized social behavior in rattlesnakes. Early snake research was reviewed in Gillingham (1987). To their theses we add our perspectives on the slim evidence, to date, for group living in turtles, crocodylians, and the tuatara.

3.3.1. Aggregations as Social Behavior

There is little doubt that complex social behavior must have evolved from solitary behavior by proceeding through intermediate steps of social tendencies (e.g., E. O. Wilson, 1975; Emlen, 1982a, 1982b; Graves & Duvall, 1995; Crespi, 2001; Gardner et al., 2016). One such cascade, for example, might proceed from solitary behavior to aggregations, based on a limited resource, with little or no important social interactions (e.g., communication) among individuals, to short-term aggregations based on conspecific attraction, to longer-term aggregations or other complex social behaviors in which individuals live and interact together permanently in groups. Here we discuss aggregations primarily involving adults or families. Aggregations in juveniles are discussed in Chapter 9.

We consider all forms of aggregation to be of potential importance in social behavior evolution. Here the critic would point out that some aggregations may simply reflect a limiting resource rather than any true social interaction (e.g., Graves & Duvall, 1995; Nieuwoudt et al., 2003; Visagie et al., 2005; Lancaster et al., 2006; Schutz et al., 2007; Schuett et al., 2016). For example, communal nesting, which is common in reptiles (see Chapter 6), has often been argued as the inevitable consequence of too many nesting females for available suitable nest sites, yet experiments have demonstrated clear conspecific attraction of mothers to eggs or other mothers (Stamps, 1988; reviewed in Doody, Freedberg & Keogh, 2009). Moreover, egg-laying reptilian mothers, which presumably find the eggs of conspecifics through chemosensory mechanisms, are typically secretive. While the onus is thus on empirical research to reveal conspecific attraction in aggregations (Visagie et al., 2005; Schutz et al., 2007; Doody, Freedberg & Keogh, 2009; see Chapter 6), we cannot assume that such social attraction is absent in any system. Also, multiple generations of limited nest sites would quite likely select for the evolution of conspecific attraction (Doody, Freedberg & Keogh, 2009). Thus, we do not agree with the strict definition of aggregations as "temporary assemblages of like (conspecific) or unlike (heterospecific) individuals brought together by physical forces in the environment or the attraction of many individuals to external stimuli or resources" (Wittenberger, 1981; Schuett et al., 2016). Aggregations will fall along a continuum of simple to complex social behavior—dichotomizing systems into aggregations versus social groups hinders interpretation. We are also inclusive in our consideration of mating behaviors or mate guarding as social aggregations (E. O. Wilson, 1975), especially since such behaviors could lead to other complex social behaviors (Boomsma, 2009; Gardner et al., 2016). Finally, we use social behavior and sociality interchangeably, despite some authors distinguishing sociality as "animals living together" (R. D. Alexander, 1974; E. O. Wilson, 1975; van Veelen et al., 2010). Clearly, most reptile species fall somewhere on a continuum of behaviors with solitary living at one end through territorial and overlapping home-range usage to obligatory permanent group living at the other.

There are some notable reptile aggregations. Perhaps their most remarkable feature is the number of individuals involved. For example, some 20,000 individual common gartersnakes (*Thamnophis sirtalis*) overwinter together in underground dens in Manitoba, Canada (Joy & Crews, 1985; Shine, LeMaster, et al., 2001). Spring emergence en masse involves intense mating before the snakes disperse to their summer habitats (P. T. Gregory, 1974; P. T. Gregory & Stewart, 1975) (Plate 5.4). The same species, when placed in enclosures, actively sought social interactions, preferred to remain in larger aggregates, and associated nonrandomly with certain individuals or groups (Skinner & Miller, 2020). In anacondas, *Eunectes murinus*, many males will simultaneously court a much larger female in a process than can go on for days. This remarkable behavior is recounted in detail in Rivas (2020).

The legendary "arribadas"—communal breeding aggregations of Kemp's and olive ridley sea turtles (*Lepidochelys kempii* and *L. olivaceus*)—once included hundreds of thousands of turtles per beach, and still reach tens of thousands per beach (see Chapter 6). These turtles migrate, sometimes through thousands of kilometers of ocean, then aggregate and mate as they approach the nesting beaches, disbanding after producing several clutches of eggs over weeks to months. In a freshwater example, the giant Amazon river turtle (*Podocnemis expansa*) migrates to river beaches to nest communally, but it may also escort hatchling turtles down river to their feeding grounds after nesting (see Chapter 6). However, as in most other turtles, mothers do not attend the nest and so they may not be expected to recognize or escort their own offspring down the river. Other notable breeding aggregations include those in iguanas and the tuatara (A. S. Rand, 1967; Wiewandt, 1982; Bock et al., 1985; Bock & Rand, 1989; Mora, 1989; Thompson et al., 1996; Iverson et al., 2004; Refsnider et al., 2013; Knapp et al., 2016; see also Chapter 6).

Although we deal with mating behavior in Chapter 5, mating aggregations in snakes deserve special mention here because they exhibit a potentially dichotomous grouping. Some species form aggregations termed "mating balls," which consist of one female and several males (Plate 5.4) (Starin & Burghardt, 1992; Rivas & Burghardt, 2001; Shuster & Wade, 2003). These mating clusters are seasonal, and the males attempt to court females simultaneously, without any overt aggressive behavior among the males. Species forming such aggregations may show female-biased sexual size dimorphism (Shine, 1994; Shine & Fitzgerald, 1995). At the other end of the spectrum is male-to-male combat, in which two males engage in wrestling (see Chapter 5). The males are said to be competing for a resident female, although she is rarely observed (Doody, Trembath, et al., 2010; but see G. Turner & James, 2010). In species engaging in male-to-male combat, males may be larger than females (Shine, 1994; Shine & Fitzgerald, 1995), a prediction advanced by Darwin (1871) himself. One species complex, the carpet python (*Morelia spilota*), apparently contains populations that form mating balls and other populations that engage in male-to-male combat (Pearson et al., 2002). A nagging problem in confirming the link between mating behavior (mating balls vs. male-to-male combat) and sexual size dimorphism is that these behaviors are rarely observed. We cannot assume that a species with no report of these behaviors does not possess them (Halliday & Verrell, 1986).

Reptiles can aggregate in great numbers for reasons other than breeding or overwintering. The yellow-bellied sea snake (*Hydrophis platurus*) aggregates on marine "slicks" (convergence zones creating smooth areas at the water's surface with floating organic material, such as foam, plant debris, and plankton) in numbers of up to 1,000 individuals per hectare off the Central American coast (W. Lowe, 1932; Dunson & Ehlert, 1971; Kropach, 1971, 1975; Tu, 1976; Lillywhite et al., 2010; Lillywhite, Iii, et al., 2015). Lillywhite et al. (2010) proposed three hypotheses for the aggregations: (1) increased feeding opportunities due to fishes seeking shelter beneath the floating debris; (2) passive convergence of snakes with currents, similar to floating debris; or (3) camouflage within the floating debris. Fascinatingly, the patchiness of sea snakes at different spatial scales might be related to the spatiotemporal distribution of rainfall because these snakes dehydrate in seawater, and so must rehydrate by drinking in the surface lens of freshwater created by rainfall at sea (Lillywhite et al., 2008, 2012; Lillywhite, Heatwole & Sheehy, 2015).

In some cases, aggregations are clearly related to food (see also Chapter 10). Although largely anecdotal, the Fernandina racer (*Pseudalsophis occidentalis*) of the Galápagos Islands aggregates to feed on both marine fish in tidal pools and on hatchling marine iguanas (Merlen & Thomas, 2013; B. Moss, 2017). Multiple snakes chasing single iguanas may reflect cooperative hunting. The approximately 20,000 individual Shedao pitvipers (*Gloydius shedaoensis*) on a very small (~0.73 km^2) island off the coast of China appear in the greatest numbers during bird migrations; indeed the snakes feed almost exclusively on birds (M. Huang, 1984, 1989; Li, 1995; Sun et al., 2001). The Broadley's flat lizard aggregates in large numbers to feed on black flies (Simuliidae) or fruiting fig trees (*Ficus sp.*) during the day, sheltering in large groups in rock crevices at night (Greeff & Whiting, 2000). Laboratory experiments showed that grouping was not due to social factors, leading to the hypothesis that aggregations were the result of spatial clumping of food resources and limited suitable overnight crevices (Schutz et al., 2007; see also Visagie et al., 2005, for examples with other cordylid lizards). The availability of free water has been implicated in large aggregations of tortoises (Medica et al., 1980; C. Peterson, 1996a, 1996b); for example, ~100 radiated tortoises (*Astrochelys radiata*) aggregated within a small area along a dry creek bottom in Madagascar, having anticipated rainfall at the beginning of the wet season (Doody, Castellano, Rakotondrainy, et al., 2011). In many cases, these aggregations involve conspecific attraction; indeed, even solitary foragers such as timber rattlesnakes

(*Crotalus horridus*) can use conspecific cues (public information) to find optimal or suitable foraging sites (R. W. Clark, 2007). Rattlesnakes in hibernacula develop social relationships among those they reside with, especially when grouping near the entrance when weather warms (Schuett et al., 2016). Green anoles (*Anolis carolinensis*) at the northern end of their range winter in small groups in crevices on south facing cliffs, emerging on warm days (Bishop & Echternacht, 2004). Little is known about what the relationships of the group members are to each other, but the selective small groupings may be based on attraction, even friendships based on prior interactions. Friendship was not studied in animal behavior until recently, but it is now commonly studied in primates, and it is certainly able to be objectively quantified with social network analysis, as is now being applied to reptiles (see Plate 11.1).

Gardner et al. (2016) found published documentation of aggregations for 22 of the 66 squamate families (33%), and their distribution was phylogenetically widespread. Of the 94 spp., 22 spp. (23%) exhibited "stable aggregations," defined as individuals residing together either all year or over a season with some common membership across years (Gardner et al., 2016). Of these, 18 spp. (19%) exhibited permanent aggregations, while 4 spp. (4%) displayed seasonal aggregations with common group members across seasons. These 22 spp. displayed a range of both aggregation duration (seasonal to permanent) and aggregation membership (stable to unstable). For example, armadillo lizards (*Ouroborus cataphractus*) live in groups year-round, but membership is transient (Mouton, 2011); the opposite occurs in the sleepy lizard (*Tiliqua rugosa*), in which pairs form long-term bonds across many years, but only for two months a year (C. M. Bull, 1988; Leu et al., 2015).

Fossil assemblages and trackways indicate aggregations in many extinct reptiles, but the social context of these assemblages (except for colonial nesting sites mentioned in other chapters) is difficult to elucidate. The best evidence for permanent groups is available for dinosaurs, particularly *Avimimus*, a Late Cretaceous feathered dinosaur that apparently formed mixed-age herds (Funston et al., 2019). Many extant or recently exterminated birds form, or formed, large stable aggregations, sometimes involving multiple species (up to 30 species in mixed flocks of forest birds in Colombia) (VD, pers. obs.) or, in other cases, millions or even billions of individuals (Møller & Laursen, 2019). In some cases, aggregations include only one age group—for example, subadults in the common raven (*Corvus corax*) (see Heinrich, 1988)—while others are sex-segregated, as in the great bustard (*Otis tarda*) (see Palacín et al., 2011); some aggregations can migrate together for thousands of kilometers, as in *Ardenna* shearwaters (Carey et al., 2014).

3.3.2. Stable Aggregations

While et al. (2019) distinguish between "egalitarian" social groups (unrelated individuals) and "fraternal" social groups of closely related individuals. While the former are taxonomically diverse, the latter are restricted to the Scincoidea (Gardner et al., 2016). Within the fraternal groupers, two features predominate: stable adult pair bonds and parent–offspring associations (Gardner et al., 2016; While et al., 2019). For example, kin aggregations have been reported for the desert night lizard (*Xantusia vigilis*) and the timber rattlesnake (A. R. Davis et al., 2011; R. W. Clark et al., 2012; A. R. Davis, 2012). To what extent kin relations reflect stable aggregations is not known for the vast majority of species, but it has been investigated in lizard species that live in conspicuous groups in nature. For example, molecular evidence has revealed 9–13 spp. of lizards that live in family groups, especially in the Australian skink genera *Egernia* and *Liopholis*. Three of the *Egernia* species (*E. cunninghami, E. saxatilis, E. stokesii*) commonly exhibit four or more overlapping generations (Plate 3.2) (Gardner et al., 2001; D. O'Connor & Shine, 2003, 2006; Stow & Sunnucks, 2004). Pair bonding across years has been demonstrated in *E. cunninghami, E. stokesii*, and the sleepy lizard, *Tiliqua rugosa* (C. M. Bull, 1988; C. M. Bull et al., 1998; Gardner et al., 2002; Stow & Sunnucks, 2004; Leu et al., 2015). Importantly, the extent of aggregations (permanence and membership), relatedness of members, and number of overlapping generations is unknown for most squamates (Gardner et al., 2016). For example, it is likely that we have underestimated the complexity of social behavior within *Egernia* and *Liopholis*, which contain 17 and 12 spp., respectively (Uetz et al., 2020).

What does a stable aggregation of lizards or snakes look like? Aggregations of lizards such as many geckos, skinks, girdled lizards, and spiny lizards are often observed in and around rock crevices, which are used as retreat sites (also termed shelter sites or overnight sites) (e.g., Kearney et al., 2001; Chapple, 2003). For example, in *Egernia* skinks, "family groups" consisting of related individuals of very different ages and sizes can be seen basking together on rock and boulder complexes that contain those crevices (Fig. 3.1). At least 11 of these species defecate in latrines or scat piles near their basking sites, crevices, or burrows (Chapple, 2003). Experiments revealed that scat-piling can be used to indicate resident status to a family group (C. M. Bull et al., 1999, 2001; C. M. Bull, 2000; see also Fenner & Bull, 2011a, 2011b; Wilgers & Horne, 2009). Such kin recognition is apparently advanced compared to latrine use by mammals (Chapple, 2003). In another fascinating example of a stable aggregation, the great desert skink, *Liopholis kintorei*, constructs an elaborate burrow system as a home for

Fig. 3.1. A mating pair of tree-crevice skinks, *Egernia striolata*, from Albury, New South Wales, Australia. Photograph by Martin Whiting.

Fig. 3.2. A great desert skink, *Liopholis kintorei*, near the entrance of an elaborate burrow system that houses family groups. Photograph by Adam Stow.

family members (Fig. 3.2) (McAlpin et al., 2011). The burrow system contains lateral tunnels with "pop" holes to the desert sand surface. Multiple generations contribute to burrow construction and maintenance, and burrows can be occupied for at least seven years (McAlpin et al., 2011).

In cooler climates snakes often overwinter together in dens—deep crevices, holes, or caves (see Manitoba gartersnakes example above). As temperatures warm in spring, rattlesnakes can be seen in piles around the hibernaculum opening (Sexton et al., 1992). This brings us to communal denning in snakes in general, which has attracted much attention for a very long time (Ditmars, 1907; Klauber, 1956, 1972; P. T. Gregory, 1984; P. T. Gregory et al., 1987; Sexton et al., 1992; W. Brown, 1993; Repp, 1998; Shine, O'Connor, LeMaster & Mason, 2001; Repp & Schuett, 2009; Amarello, 2012; R. W. Clark et al., 2012; Schuett et al., 2014). According to Schuett et al. (2016), these systems include bona fide examples of stable aggregations and related social behaviors. For example, using the term "cryptic sociality,"

R. W. Clark et al. (2012) revealed that pregnant timber rattlesnakes not only aggregate together in the spring but also preferentially aggregate with their (juvenile) kin for weeks after birth. Separation experiments revealed that these associations are important in further affiliative behavior between mothers and offspring and between siblings (Hoss et al., 2015). In fact, this form of parental care appears to be ubiquitous in North American pitvipers (Schuett et al., 2016).

Communal denning is not restricted to winter—dens are also used for birthing and used by pregnant mothers year-round, and some neonates and juveniles use dens for several years (Graves & Duvall, 1995; J. M. Parker et al., 2013; Schuett et al., 2016). R. W. Clark (2004) found that pairs of siblings associated more closely with each other than unrelated neonates, suggesting that timber rattlesnake females can recognize kin, even after being in strict isolation for more than two years. In fact, kin recognition was documented in several other snake species, suggesting some generality (reviewed in Schuett et al., 2016). In the related western diamondback rattlesnake (*Crotalus atrox*), communal dens are a mixture of kin groups and non-kin groups (Schuett et al., 2014, 2016, and references therein). Kin recognition in timber rattlesnakes was underpinned by the tendency of the animals to show high fidelity to their communal birthing dens across years and probably throughout their lifetime (W. Brown, 1993; see also Repp, 1998; Schuett et al., 2016); western diamondback rattlesnake dens, however, were not used by birthing mothers, neonates, or juveniles (Schuett et al., 2016). Again, communal denning is not all about overwintering, at least in some populations. Because the climate affecting western diamondback rattlesnake dens is often mild, Schuett et al. (2016) proposed that the function of their communal dens, and perhaps those of other rattlesnakes of southern latitudes, is fundamentally different from those in higher latitudes (in the latter, survival appears to be the primary and most important function of aggregation).

Rattlesnakes and other pitvipers exhibit other characteristics consistent with advanced sociality, including group defense, conspecific alarm signals, and maternal defense of young (Graves & Duvall, 1987, 1988; Graves, 1989; H. W. Greene et al., 2002; Schuett et al., 2016). Adult female Arizona black rattlesnakes (*C. cerberus*) were more likely to form preferred associations with one another than with juveniles or males; males, however, did not form preferred associations (Schuett et al., 2016). Other behaviors are intriguing and require more study to elucidate how complex the social interactions are; for example, newborn sibling sidewinder rattlesnakes (*C. cerastes*) form "balls" that stabilize temperatures in a highly fluctuating thermal environment (Reiserer et al., 2008). Remarkably, R. W. Clark (2007) experimentally demonstrated that rattlesnakes were more likely to select an ambush site in an area where there were chemical cues from conspecifics that had recently fed. This suggests the possibility that other solitary foragers also use public information.

In their review on squamates, Gardner et al. (2016) examined possible commonalities in species with stable aggregations. For example, all species with stable aggregations are viviparous, leading to the hypothesis that viviparity may lead to closer proximity of parents and offspring, which might facilitate the evolution of kin-based sociality (Chapple, 2003; A. R. Davis et al., 2011; Gardner et al., 2016; Halliwell, Uller, Holland & While, 2017). This association, analyzed by Gardner et al. (2016), was tested with a phylogenetic analysis by Halliwell, Uller, Holland and While (2017); both found a strong relationship between viviparity and social grouping in squamates. There were limitations to their analyses, however. For example, despite the viviparous genera *Egernia* and *Liopholis*, most skinks are oviparous, and so a finer resolution of social structure and mating system within the Scincidae would be required to explore the commonality between reproductive mode and aggregation behavior. Moreover, viviparity is phylogenetically conserved within the Egerniinae—the subfamily that includes *Egernia*, *Liopholis*, and *Tiliqua*. Thus, viviparity likely evolved once in the Egerniinae ancestor, resulting in one datapoint for the comparative analysis involving viviparity and aggregation behavior. We do not refute the association but rather join the call for more data and finer-scale phylogenetic-based comparisons to further test this association. According to Halliwell, Uller, Holland and While (2017), despite the strong relationship between viviparity and social grouping, the relative rarity of social groups among squamates, as well as the occurrence of social groups in oviparous species, indicates that viviparity alone is insufficient for social groups to form.

Another possibility suggested by Gardner et al. (2016) is that seasonal aggregations for thermoregulatory reasons may preadapt a species for more complex sociality—examples given were overwintering in snakes, diurnal aggregations in geckos, basking in iguanas, and communal egg-laying. This is a specific hypothesis of a broader notion that we introduced earlier in this chapter—the idea that simple aggregations for nonsocial reasons, such as food or thermal environment, could lead to complex sociality. We thus agree but note that communal egg-laying in squamates is generally *not* related to thermoregulation as the authors purport; moreover, experiments have revealed that conspecific attraction underpins communal egg-laying in lizards and snakes (Doody, Freedberg & Keogh, 2009; see Chapter 6). Longevity could be correlated with aggregation behavior because longevity can apparently suppress conflict in highly related societies (Port and Cant, 2013). Longevity data are sparse, but range from 8 to 50 years in the squamates with stable aggregations (Gardner et al., 2016). It would be interesting to compare the remarkably long-lived, and often small, skinks and geckos in New Zealand with those of more fast-paced living species in Australia, Africa, and other continents (Cree, 1994).

Stable aggregations may be more likely to evolve in habitats that select for philopatry of adults (Emlen, 1982a, 1982b; R. D. Alexander, 1991; Crespi, 2001). Gardner et al. (2016) concluded that several squamate species with stable aggregations that rely on clumped resources, such as home sites or food, may provide a context for the evolution of sociality. For example, species with long-term pair bonding and delayed dispersal tend to rely on permanent shelter sites such as rock crevices, tree hollows, or deep complex burrows (Chapple, 2003; McAlpin et al., 2011), which can be separated by distances that create barriers to dispersal (reviewed in While et al., 2019). Such barriers can affect mate-finding, and thus the degree of pair bonding, and dispersal ability, which can in theory cause delayed dispersal and prolonged parent–offspring associations (Halliwell, Uller, Chapple, et al., 2017; Halliwell, Uller, Wapstra & While, 2017; While et al., 2019). In contrast, other species that occupy simple burrows in soil tend to exhibit a more homogeneous spatial arrangement of shelter sites, which in turn may preclude barriers to dispersal that could lead to the evolution of pair bonding and delayed dispersal (While et al., 2019).

Monogamy, by reducing polyandry, could provide direct benefits for males and potentially lead to more complex social behaviors such as paternal care and stable social groups (Boomsma, 2007; While, Uller, & Wapstra, 2009a, 2009b). Indeed, high levels of monogamy occur in *Egernia* species with stable aggregations in which genetics have revealed mating systems (Gardner et al., 2002, 2016; Stow & Sunnucks, 2004). However, parental care is rare in reptiles (Shine, 1988; Somma, 2003), and long-term monogamy in the sleepy lizard has not led to aggregations of related individuals (Godfrey et al., 2014).

Whiting and While (2017), focusing mainly on the *Egernia* group of skinks, examined a number of factors associated with, or underlying, sociality. For example, stable aggregations in *Egernia* rely on permanent shelter sites such as hollow trees, rock outcrops with numerous crevices, or burrow systems. The most commonly utilized of these are rock outcrops, and their patchy distribution in a matrix of unsuitable habitat (Duffield & Bull, 2002) has been suggested to be the key feature influencing the emergence and diversification of social organization in the *Egernia* group (Whiting & While, 2017). Given a limited number of outcrops and crevices within those outcrops, crevice sharing is inevitable (Duffield & Bull, 2002; Chapple, 2003). In fact, lizard group size correlated with various attributes of rocky outcrops (Michael et al., 2010; but see Stow & Sunnucks, 2004; Gardner et al., 2007).

Another potential benefit of stable aggregations is reduced predation risk via factors such as enhanced vigilance. In fact, groups detected predators sooner than solitary individuals in two *Egernia* species (Eifler, 2001; Lanham & Bull, 2004). The thermal environment might also stimulate social aggregations such as huddling or the buffering of temperature oscillations in rocky outcrops or burrow systems (Henzell, 1972; Webber, 1979; Rabosky et al., 2012; Whiting & While, 2017). Mating can underpin social aggregations; male mating success has been shown to increase in areas of high rock crevice availability and in more social individuals in White's skink

(*Liopholis whitii*) and the desert night lizard (Chapple & Keogh, 2006; Rabosky et al., 2012). Offspring care in the form of delayed dispersal is prevalent in both the *Egernia* group and in the desert night lizard (Gardner et al., 2001; A. R. Davis et al., 2011); this would certainly increase the potential for juvenile lizards to copy adults or older individuals. Indeed, social learning has recently been demonstrated in the water skink (*Eulamprus quoyii*) and in the western bearded dragon, *Pogona vitticeps* (D. W. A. Noble et al., 2014; Kis et al., 2015; see also Chapter 11, section 11.6). Benefits from associating with parents in these systems include increased access to basking locations, foraging opportunities, retreat sites, inheritance of territories, and a reduction in the risk of conspecific aggression or infanticide (C. M. Bull & Baghurst, 1998; Lanham & Bull, 2000; Post, 2000; Gardner et al., 2001; D. O'Connor & Shine, 2004; Whiting & While, 2017). Parasite prevalence and the risk of infection could also underpin social grouping; for example, marine iguanas in groups remove ectoparasites from one another (see also Iverson et al., 2017, for the same in an iguana). In sleepy lizards, males that separated from their partners had higher parasite loads than males that retained their partners (C. M. Bull & Burzacott, 2006), and risk of infection from parasites in Gidgee skinks (*Egernia stokesii*) was affected by both sharing shelter sites and relatedness (Godfrey et al., 2006, 2009).

Interestingly, lizards with kin-based sociality or monogamy (*Egernia, Tiliqua rugosa, Xantusia*) are relatively long-lived compared to most lizards (Table 13.2 in Whiting & While, 2017). The former appear to live 10–50 years, while many other lizard species live for months, a year, or perhaps most often a few years (Pianka & Vitt, 2003). Longevity is correlated with delayed maturity, fewer offspring, and higher incidence of parental care (including delayed dispersal), making lizards a good model system to investigate the influences of habitat and life history on the evolution of complex social behavior (Whiting & While, 2017).

The presence or absence of aggregations has been reported for <1% of squamates (Gardner et al., 2016), and this is likely the case for turtles. In crocodylians, aggregations are known for many species, but their social structure is poorly understood (Dinets, 2011c). There is currently no solid evidence for stable aggregations with membership in turtles or the tuatara, and only anecdotal evidence for crocodylians (Dinets, 2011c). However, using network analysis from video recordings, K. M. Davis (2009) found stable interactions among individuals in a multispecies exhibit of New World emydid turtles.

4 |

Communication

Communication is a central aspect of social behavior studies, but communication can occur between predators and prey as well as between conspecifics such as mates, competitors, parents, and offspring. As discussed later, social play can involve communication across wide taxonomic chasms, as easily seen in YouTube videos of dogs playing with crows, parrots with humans, deer with kangaroos, and many other odd and unexpected, often unnatural, pairings.

But what is communication? The definitional issue has been batted around for years. After reviewing diverse early definitions, Burghardt (1970b) opted for communication involving a sender emitting a cue that, when responded to, would confer a fitness advantage to the sender or his or her group (thus accommodating kin selection, altruism, and cooperation). Of course, the recipient of the cue could respond, leading to further responses from the initial sender. This entrained a functional circle building on the Umwelt concept of Jakob von Uexküll (Burghardt, 1998b). The manipulation approach is merely a derivation from the sender-centered approach as are, ultimately, other approaches to communication, including more information processing terminology and explicit sender–receiver or Machiavellian language (Font & Carazo, 2010; Lucas et al., 2018), which often specifically involves intraspecific deception (Weldon & Burghardt, 1984, 2015). Deception from mimicry to crypticity is common in predator and prey interactions. Although the communication literature is vast, including on reptiles, and the underlying mechanisms often complex and obscure, some basic description of the great diversity of social communication in reptiles is necessary.

Reptiles use visual, acoustic, chemical, and tactile signals for intraspecific communication. Until recently, research has been mostly focused on visual signaling, because acoustic signaling was believed to be rare in reptiles, chemical signaling is usually difficult for humans to record and study, and tactile signaling was thought to be too simple and straightforward to be of interest. And since diurnal, easily observed lizards use visual signaling most often, almost all early research on reptile communication has been centered on them. As we are a highly visual species, anthropocentrism is also a source of our emphasis on this modality. We will discuss communication mechanisms in all groups of reptiles; while the modalities used by turtles, crocodylians, and the tuatara are rather uniform across these now depauperate taxa, among lizards, especially, the diversity in reliance on the various senses is great.

Pioneering work on complex and overall fascinating communication behavior in crocodylians by Garrick and associates (Garrick & Lang, 1977; Garrick et al., 1978; Vliet, 1989) was not followed up until the studies by Dinets (2010, 2011c, 2013a, 2013d). Discoveries described below made it clear that turtle and snake communication is also complex and worth studying, and numerous papers were published on chemical communication in snakes (R. T. Mason & Parker, 2010), although almost all of them are focused on a handful of "model" species (J. Martín & López, 2011). As for fossorial reptiles, little is known about their communication, just as little is known about all other aspects of their behavior: there are published studies on just one species of fossorial lizard and one fossorial snake (see below). Communication in marine reptiles is also very poorly understood.

Visual communication is usually by postures or expandable display structures; the latter are very common in some lizard families and in birds, and are believed to have been common in non-avian dinosaurs. Acoustic communication is mostly vocal, but there are exceptions. In some cases, the line separating acoustic and tactile communication is unclear: vibrations and particularly infrasound (used by crocodylians) are often perceived by tactile organs rather than ears (see below).

Pheromones are widely used for communication, and numerous taxa have specialized glands for producing them (overviewed in J. Martín & López, 2011). Usually these are granular glands; most reptiles have them in the anal region, and some lizards have up to four kinds of such glands around the vent (Kent, 1987). In addition to anal glands, crocodylians have a row of glands along the back and two large glands under the lower jaw (Kent, 1987; Grigg & Kirshner, 2015). Gopher tortoises (*Gopherus polyphemus*) have huge glands on the throat, which grow even larger during the breeding season; the secretions apparently serve as chemical signals in close-range interactions (Legler & Vogt, 2013). Using feces (independently of associated pheromones) for transmitting information has been demonstrated in western fence lizards (*Sceloporus occidentalis*) (Duvall et al., 1987) and various snakes (Chiszar et al., 1979). Fecal chemicals of adult male Carpetane rock lizards (*Iberolacerta cyreni*) allow for self-recognition, recognition of familiar males, and signaling of body size, suggesting that feces act as a multimodal signal (visual and chemical) in territorial marking (López et al., 1998; Aragón et al., 2000). Dalmatian wall lizards (*Podarcis melisellensis*) defecate on the largest rock within their territory, apparently to increase the detectability of the signal (Baeckens et al., 2019). Juvenile snakes are attracted to shelters previously used by other snakes, apparently by chemical cues that very well could include urine and feces (Burghardt, 1983). Remarkably, at least one species of mammals, the Siberian chipmunk (*Tamias sibiricus*), can hijack snake chemical signals by rubbing itself with skin fragments from dead snakes; this behavior likely protects it from predation by snakes (Kobayashi & Watanabe, 1986).

Mammals, a sister group of extant reptiles, are similar to snakes in relying heavily on chemical communication—a feature apparently inherited from early synapsids (Duvall et al., 1983). However, almost all species also use tactile, visual, and acoustic signals, described in detail in D. E. Wilson and Mittermeier (2009–2019); some taxa use those types of signals more than chemical communication. Extendable display structures (e.g., crests and dewlaps) that are so common in birds and lizards are all but absent in mammals (although many species erect their dorsal hair in threat displays); one notable exception is the nasal bubble of the hooded seal (*Cystophora cristata*). Acoustic signals can include infra- or ultrasound or nonvocal sounds, such as drumming in numerous taxa (reviewed in Randall, 2010) and breaching in cetaceans. A few lineages have developed very complex systems of acoustic signaling, known as "languages" in humans (*Homo sapiens*). There is an ongoing debate on the accuracy of using this term for other animals (see, for example, Bateson, 1966; Pinker & Jackendoff, 2005); it seems to be justified at least in cases when signaling systems are largely culturally acquired, such as in some cetaceans (Janik, 2014). The possibility that communication behavior in reptiles might to some extent be individually or culturally acquired has rarely if ever been tested, or even proposed; Dinets (2013a) failed to consider it, even though his study data hinted at such a possibility (see below and Chapter 11).

4.1. Archosaur Communication: The Pinnacle of Beauty

Elaborate communication behavior is particularly widespread in extant archosaurs and also seems to have been in extinct ones. Many pterosaurs, particularly the larger species, apparently had sexually dimorphic crests, likely used in displays (S. C. Bennett, 1992). Numerous dinosaurs had dazzlingly diverse structures, such as crests, frills, and horns, that were probably used for mating displays or ritualized combat (Knell et al., 2012; Hone & Naish, 2013). Some of these structures, like the huge crests with labyrinthine air passages found in *Parasaurolophus*, may have been used for producing species-specific calls (Wiman, 1931; Diegert & Williamson, 1998). Feathered taxa interpreted as bird-like dinosaurs or early birds had ornamental feathers or feather-like structures: *Jeholornis* had a frond-like array of tail feathers (J. K. O'Connor et al., 2012); *Epidexipteryx* possessed four ribbon-shaped, feather-like caudal structures (F. Zhang et al., 2008); some (possibly only male) *Confuciusornis* had long tail feathers (Chinsamy et al., 2013). Such structures in extant birds are mostly used in courtship displays, or at least play a role in sexual selection, so they could have had similar functions in extinct taxa (Naish, 2014). Growth changes seen in oviraptorosaur forelimb feathers (Xu et al., 2010) and the presence of feather iridescence in the deinonychosaur *Microraptor* (Q. Li et al., 2012) add further support to the idea that these dinosaurs were using feathers in display (Knell et al., 2012).

Extant birds have evolved a wonderful diversity of visual and acoustic signals, often used in combination; the examples below are described in detail in del Hoyo et al. (1992–2013) unless stated otherwise. The exquisitely beautiful displays by peacocks (*Pavo*) and many other Phasianidae, as well as by lyrebirds (*Menura*), birds-of-paradise (Paradisaeidae), nightingales (*Luscinia*), and many other songbirds (Passeri), are among nature's greatest spectacles. A few taxa use nonvocal acoustic communication: for example, snipes (*Gallinago*) and some hummingbirds (Trochilidae) produce sound with specialized tail feathers, while woodpeckers (Picidae) use species-specific drumming. Some birds use tools for communication; examples include elaborate structures built by bowerbirds (Ptilonorhynchidae) and nuptial gifts used by terns (Sternidae). Cassowaries (*Casuarius*) and peacocks are known to use infrasound (Mack & Jones 2003; Freeman & Hare, 2015), while some hummingbirds and songbirds use ultrasound (Narins et al., 2004; Pytte et al., 2004; Krull et al., 2009). Tactile signals are often used in later-stage courtship, by chicks of many species when begging for food, and by mated pairs in mutual preening—one remarkable example is talon-locking during courtship flights by some raptors (Accipitridae). Chemical communication is also present and possibly widespread but poorly understood (Hagelin & Jones, 2007).

Crocodylian communication, with early studies by Herzog (1974), Garrick and Lang (1977), and Vliet (1989), is remarkably well developed. It is also ancient: there is anatomical evidence of vocal capabilities in two species of *Caipirasuchus*, Early Cretaceous crocodylomorphs (Dias et al., 2020). Crocodylians begin to vocalize before hatching; young hatchlings already use diverse calls for various purposes, communicating with adults as well as with siblings (see Chapter 9). From that age onwards, the bottom frequency of certain vocal signals is an honest signal of size; mothers respond more strongly to calls of smaller offspring (Chabert et al., 2015).

Mating-season displays of crocodylians are probably the most complex displays in the animal kingdom: they include sounds, infrasound (a long-distant tactile signal), postures, odors, and probably object manipulation. The account below is summarized from Dinets (2013b, 2013c).

There are three major types of sound used by crocodylians in various combinations. The first type is a sound produced vocally above the water. It is traditionally called "bellowing" in alligators (*Alligator*) and "roaring" in crocodiles (Crocodylidae) and caimans (*Caiman, Melanosuchus, Paleosuchus*) (Plate 4.1). Roars and bellows of most species, except for particularly closely related ones (such as most species in the genus *Crocodylus*), can be distinguished by a human observer. Dwarf (*Osteolaemus* spp.) and narrow-snouted (*Mecistops* spp.) crocodiles produce at least four types of vocal signals during courtship (Staniewicz, 2020)

The second type of sound, produced only by males, is infrasound, or nonvocal vibrations, normally produced below the water surface at frequencies below the range of human hearing (~10 Hz). The infrasound is also outside the range of crocodylian hearing and is likely perceived by integumentary sense organs (Grigg & Kirshner, 2015) located in the center of certain scutes (see Chapter 2). The infrasound can be visually detected by the so-called "water dance" effect (Vliet, 1989), apparently created by Faraday waves; this effect has always been considered an accidental by-product of infrasound, but it might actually be a visual signal (Kofron & Farris, 2015). A recent study (Urra et al., 2019) found that the "water dance" (Fig. 4.1) can be used as an honest signal of size and physical power.

The third type of sound includes headslaps (sounds made by slapping the head against the water surface) and jawslaps (sounds made by slapping the jaws together at or below the water surface). These slaps have a very sharp onset, a feature known to make locating the source of the sound easier, and probably serve as location beacons. They are usually combined with infrasound, which is difficult to locate underwater due to its long wavelength.

Fig. 4.1. Male American alligator (*Alligator mississippiensis*) bellowing in HOTA posture. Note the "water dance" above his back. Big Cypress National Preserve, Florida. Photograph by Vladimir Dinets.

All these signals are produced in a particular posture that makes it easy to visually ascertain the size of the animal from afar. Garrick et al. (1978) called it "head oblique tail arched" (HOTA) posture. HOTA postures (Fig 4.1) are exhibited by all crocodylian species except the Indian gharial, in which it is replaced with a head-up posture (see below). HOTA posture is probably an honest signal of size, addressed to animals observing from above the water, while infrasound is addressed mostly to submerged or partially submerged animals. In some species, HOTA looks slightly different; for example, in black caimans (*Melanosuchus niger*) the head is held horizontally, making the animal look cobra-like (Dinets, 2011d).

There is evidence (Garrick & Lang, 1977) of odor being yet another component of displays. It is produced by gular glands (paired glands located on the underside of the lower jaw) and cloacal glands, and it possibly carries information about the animal's species, sex, and sexual maturity (Weldon & Wheeler, 2001). Gular glands are everted during some displays (Garrick & Lang, 1977).

Females occasionally produce roars or slaps, but without infrasound and usually not in HOTA posture. In some (possibly all) species of Crocodylidae, head slapping displays also serve as signals of dominance and can be performed by adult females kept in absence of males in captivity (Dinets, 2011b). Roars of mugger crocodiles (*Crocodylus palustris*) sound somewhat similar to roars of lions (*Panthera leo*), and the crocodiles would sometimes roar in response to lions (Dinets, 2011a). The first (and, for over a century, the only) description of the roar of the mugger crocodile was a poetic one, in a fictional story by Kipling (1895): "It was a thick voice—a muddy voice that would have made you shudder—a voice like something soft breaking in two. There was a quaver in it, a croak and a whine."

The courtship displays of the Indian gharial (*Gavialis gangeticus*) are different from those of other crocodylians. Instead of assuming HOTA posture in the water, gharials often assume a head-up posture on land, which is believed to be a territorial display and a signal of sex and maturity (Singh & Rao, 1990). Adult males have a huge bulbous growth called *ghara* (from a Hindi word for "cooking pot") on the tip of the snout; it becomes particularly visible in the head-up posture. This species is not known to produce infrasound. The only sounds associated with courtship are soft buzzes, given in close proximity to other animals, and incredibly loud "pong" sounds, given by both males and females (R. Whitaker & Basu, 1982). The mechanism used to produce those "pongs" is still a mystery, although it may involve the huge bullae (balloon-like structures) present in gharial skulls. Notably, the mechanism of infrasound production by other crocodylians is also a mystery, although photographic evidence suggests that the infrasound is somehow produced by vibrating walls of the chest cavity (Dinets, 2013b). Some extinct crocodyliforms, such as the giant *Sarcosuchus* of the Early Cretaceous, had snout-tip structures that were similar to gharas of Indian gharial but anatomically different; they were probably also used in displays (Buffetaut & Taquet, 1977; Sereno et al., 2001), although recent evidence suggests that, at least in the case of giant Cretaceous alligatorid *Deinosuchus* of North America, the structure at the snout tip was used for thermoregulation (Cossette & Brochu, 2020).

Females use hisses, growls, and sometimes bites to repel unwanted males, but if the courtship continues, it can last for up to an hour and include mutual touching (particularly with the underside of the lower jaw, which carries musk glands and might be an erogenous area), swimming in tight circles (accompanied with soft buzzing sounds in the Indian gharial), blowing bubbles, and sometimes play behavior (see Chapter 11).

4.2. Why Are Crocodylian Displays So Complex?

Why do crocodylian displays include so many different components (vocal sounds, slaps, infrasound, "water dance," HOTA or head-up posture, and odor)? The benefit of having multiple components may be in their differing ability to spread and carry information through air and water (Dinets, 2011c). Vocal sounds are optimal for carrying information about the location of the animal and its size through air, as the bottom frequency and formants of roars and bellows depend on the animal's size (Reber et al., 2015). Slaps are optimal for carrying information about the location of the animal through water. HOTA posture and infrasound are honest signals of the animal's size; infrasound and the accompanying "water dance" are also signals of sex as they are only produced by males. The head-up posture used by gharials is a signal of size, sex, and maturity. Together, all components form a robust and flexible signaling system.

If displays containing loud vocal signals are better adapted for carrying information through the air, while those containing headslaps are better adapted for carrying information through the water, then it can be expected that the former are used more in fragmented aquatic habitats such as overgrown marshes and small ponds, and the latter in continuous aquatic habitats such as rivers, large lakes, and coastal lagoons. Does this prediction hold true?

Indeed, a comparative study of extant crocodylians (Dinets, 2013c) found such a pattern. Species that are habitat generalists use both vocal signals and slaps; species inhabiting predominantly continuous aquatic habitats use slaps but few, if any, vocal signals; and species inhabiting predominantly fragmented aquatic habitats use vocal signals but few, if any, slaps.

In studies (Dinets, 2011c) of American alligators (*Alligator mississippiensis*) and Nile crocodiles (*Crocodylus niloticus*), such differences were also found between populations of each species inhabiting different habitats, but the observed pattern was complicated. Alligators in all populations use a lot of bellowing displays, while headslapping displays are used almost exclusively in continuous aquatic habitats. In Nile crocodiles it is the opposite: all populations headslap, but only those in fragmented aquatic habitats use a lot of roars; in continuous aquatic habitats, roars are seldom used in displays, and these roars are usually reduced to quiet "coughs."

One half of the prediction works only for alligators, and the other half only for crocodiles. The possible explanation for this is that social structures of the two species are different. Nile crocodiles have a strict hierarchy, in which headslaps, in addition to being part of pre-mating advertising displays, serve as signals of dominance. So, all populations need to use headslaps, while roars can be used only in those places where they are needed for individual advertising. American alligators have a different social structure; hierarchy and dominance do not play such an important role in their lives, but they have complex group courtship (see below). So, they can dispense with headslaps in places where they do not work well, but they cannot abandon bellowing, because, unlike crocodile roars, alligator bellows have a second function in addition to personal advertising: they attract other alligators to sites of group courtship.

How do those habitat-dependent differences in displays between populations and species appear? Do individual animals change their displays in reaction to the parameters of their habitat, or is it a long evolutionary process? To answer this question, Dinets (2013a) studied yacare caimans (*Caiman yacare*) living in savanna lakes in two areas near the Brazil-Bolivian border. In one of the study sites, water levels were dropping during this time, so lakes broke into numerous small ponds. On the other site, rains caused small ponds to rise and completely flood a river valley. But, despite the dramatic changes of the sound-carrying capacity of their habitats, caimans did not change the composition of their signals at either site. This was interpreted by the author as suggesting that the differences between populations that are described above result from differential evolution in different habitats, rather than from changes in the behavior of individual animals. But an alternative explanation is that caimans learn the use of various components of their signals from conspecifics and develop culturally acquired dialects at each location.

The fact that the signaling system of crocodylians, with its repertoire of physically different signals, can be easily adapted to diverse habitats simply by changing the usage of two signal components might account for the preservation of this system since the Late Cretaceous (Senter, 2008). Although the signal composition is highly variable and can differ between closely related species, and even between conspecific populations, the overall repertoire differs little between most species and appears to be extremely well-conserved. Despite having diverged more than 70 million years ago, crocodiles and alligators still understand each other's signals (Garrick & Lang, 1977). This allows the use of the differences in signaling repertoire to reconstruct the evolutionary history of the signaling system. The most parsimonious scenario is that the common ancestor of crocodiles, alligators, and caimans was a habitat generalist and used both vocal signals and slaps, as well as infrasound and HOTA posture. If that common ancestor had been a habitat specialist, inhabiting either only fragmented or only continuous aquatic habitats, one of the signal components (either slaps or vocal sounds) would have been lost instead of being inherited by all descendants of this common ancestor, as has evidently occurred in some descendant species with specialized habitat preferences (Dinets, 2011c).

Extant crocodylians are unique among major vertebrate lineages in one bizarre way: there are no truly small species among them. The smallest ones can reach sexual maturity at less than 1 m in length, but full-grown males exceed 1.2 m in all species. All these animals are apparently capable of producing infrasound when fully grown, and they use it in their displays. Males of larger species begin to accompany their bellows with infrasound at approximately the same length, so 1.2 m is possibly the minimum size at which producing infrasound becomes physically possible. The importance of producing underwater infrasound, which can only be emitted by animals of sufficient size, could be a limiting factor in the evolution of crown-group crocodylians (Dinets, 2013c). Males that were too small to produce infrasound were ignored by females, so no species could evolve small size. Some extinct crocodylian lineages were apparently free of this constraint, as they included small and fully terrestrial species.

4.3. Turtles and Tortoises: Not So Silent After All

Until recently it was believed that chelonian communication was rudimentary and, with an exception of an occasional threatening hiss or grunt, limited to mating-related behavior. We now know that turtles can communicate in many ways and in many situations. A study of painted wood turtles (*Rhinoclemmys pulcherrima*) found that at least 12 signals (visual, chemical, and tactile) are used during courtship (Hidalgo, 1982). That list did not include passive visual signals such as colorful stripes and patches, which are common in many turtles and are produced by carotenoid-based pigments (Steffen et al., 2015).

Tactile communication is particularly spectacular in Emydidae turtles. The following account is based on Legler and Vogt (2013). In 12 species of Emydidae, males have elongated foreclaws: the courting male swims backward facing the female and rapidly vibrates his foreclaws near her face, sometimes accompanying this so-called titillation with head bobbing. Remarkably, in some species such as the pond slider (*Trachemys scripta*), males in some populations have the elongated foreclaws and use titillation, while males in other populations do not. Males of some other species vibrate their heads in close proximity to the female's snout or blow water through their nostrils into her face. The degree to which those behaviors are exhibited can also vary between populations of the same species; males of some species and populations use only prolonged biting during courtship (Iverson, 1985). Why so many species exhibit such a profound difference between geographical areas is unknown, but it is a very interesting question.

Burmese roof turtles (*Batagur trivittata*) sometimes kick up large amounts of sand with their feet while basking; why they do this is unknown, but the researchers who discovered this behavior suspect that it is social, possibly territorial, and aimed at preventing conspecifics from approaching (S. G. Platt et al., 2019).

Chemical communication is widespread among freshwater turtles; in spiny softshell turtles (*Apalone spinifera*), even hatchlings can use chemical signals to locate conspecifics and tell relatives from nonrelatives (Whitear et al., 2016). Males of most species sniff the female's cloaca while pursuing her (Legler & Vogt, 2013). In contrast, a study by Belzer (2002) found that male eastern box turtles (*Terrapene ornata*) were able to detect females only if the latter were visible and moving; nonvisual cues apparently were not used. Most turtles (freshwater and marine) have specialized glands apparently used for chemical communication; these glands (axillary and/or inguinal) have apparently evolved multiple times and are collectively known as Rathke's glands (Ehrenfeld & Ehrenfeld, 1973). Visual signals are common: in some species a courting male bobs his head while pursuing the female, then stops her using biting and ramming; the pair then moves in diminishing concentric circles until the female presents herself by raising the rear of her shell (Legler & Vogt, 2013).

Recently some turtles have been found to be surprisingly vocal underwater. Long-necked freshwater turtles (*Chelodina oblonga*) use at least 17 different sounds (Giles et al., 2009), while giant Amazon river turtles (*Podocnemis expansa*) use at least six (Ferrara, Vogt, Harfush, et al., 2014). Many of these sounds play a role in courtship and mating, but not all: giant Amazon river turtles apparently communicate with their offspring (see Chapter 9). Acoustic communication might yet prove to be rather common, as it has been found in species belonging to many different clades (Ferrara et al., 2017). About 50 species of turtles and tortoises are now known to emit sounds (see overview by Ferrara, Vogt, Giles & Kuchling, 2014). Phylogenetic mapping suggests that use of sound is an ancestral trait of S-necked turtles (Cryptodira); analysis of these sounds clearly shows that they have evolved for communication, although the mechanism of producing them is still unknown (Galeotti et al., 2005).

Embryos of at least one freshwater turtle (the pig-nosed turtle, *Caretochelys insculpta*) respond to vibrations of sibling embryos to expedite hatching; rapid hatching and emergence is likely critical during flooding (Doody, Stewart, et al., 2012). Although embryo–embryo vocalizations have not been ruled out, mechanical stimulation will induce rapid hatching in the species (see Chapter 8).

Very little is known about communication in sea turtles. Males use tactile stimulation during courtship (see Chapter 5). It is strongly suspected that chemical (Bernardo & Plotkin, 2007, and references therein) and acoustic (Ferrara, Vogt, Giles & Kuchling, 2014) communication is used by ridleys (*Lepidochelys*) to synchronize arribadas (mass nesting events). Sea turtles begin to produce sounds before hatching (McKenna, 2016); these sounds are used to synchronize hatching by developing leatherback sea turtles (*Dermochelys coriacea*) and possibly other species (Ferrara, Vogt, Harfush, et al., 2014). Adult leatherback sea turtles also produce a variety of sounds, but their role is unknown (Mrosovsky, 1972).

Tortoises use visual, acoustic, chemical, and tactile signals during courtship and during aggressive encounters between males; these signals are species-specific and include ritualized walking, stereotypical head movements ("bobbing"), nose touching, leg biting, shell ramming, olfactory communication ("sniffing"), and sounds (Eglis, 1962; Auffenberg, 1966; Ruby & Niblick, 1994). Male and female gopher tortoises rub their forearms against their chin glands, and both sexes wave their forearms only at males during courtship, possibly as a sign of not being receptive to their courting (Weaver, 1970); receptive females instead perform semicircular movements (Auffenberg, 1966). In closely related Texas tortoises (*Gopherus berlandieri*), males respond to the mental gland secretions of conspecific males with combat behavior, while females respond with head bobbing, a courtship behavior (Rose, 1970).

Tortoise mating sounds are of particular interest. They are very diverse and have been variously described as "hisses," "grunts," "bellows," "clicks," "chirruping," "moans," and so on. An experienced human observer can recognize sounds of different species and even individual animals. Males of some species use one type of sound when approaching a female and another one during mating (Auffenberg, 1977). Sacchi et al. (2003) and Pellitteri-Rosa et al. (2011) have shown that sound parameters of mating hisses (particularly the call rate) are honest signals correlating with the male's size, condition, and reproductive success in various *Testudo* species. Chemical signaling is also important: in Hermann's tortoises (*T. hermanni*), loss of olfactory function disrupts reproductive behavior (Chelazzi & Delfino, 1986).

4.4. Lizards and the Tuatara: Colorful Displays

The tuatara (*Sphenodon punctatus*) and many lizards use very similar signaling, suggesting that some features of their communication predate the split between rhynchocephalians and squamates and might be ancestral for lepidosaurs. A courting male tuatara inflates his body, raises the crests on his head and torso, changes the color of his shoulders and head crest, and approaches the female in a strange walk called "proud walk"; he then bites the female until she stops moving and assumes a receptive posture (Gans et al., 1984). Similar displays and croaking sounds are used in aggressive encounters between males, until dominance is established (Gans et al., 1984). It is suspected that pheromones also play a role (J. Martín & López, 2011). Some of the earliest studies using models, including moving ones, were carried out by Gillingham and colleagues (see Fig. 5.3) (reviewed in Gillingham et al., 1995).

Inflating or flattening the body, raising crest, changing color, and walking in a peculiar way are also used by numerous lizard species in threatening and courtship displays. Of course, not all lizards have crests, but countless species have other extendable parts, such as frills or triangular flaps of skin on the throat called dewlaps (Plate 4.2). Dewlaps are often brightly colored; the most striking example is the enormous dewlap of the fan-throated lizard (*Sitana ponticeriana*), which is brilliant-blue with black and scarlet bands. Extending the dewlap is an anatomically complex process, at least in anoles (Lovern et al., 2004; Losos, 2009). Sympatric anole species tend to have different dewlap colors (red, yellow, or white), which play a role in species recognition (Losos, 1985) and have been used in studies of taxonomy for discovering cryptic species (Webster & Bums, 1973). Size and brightness of dewlaps also correlate with habitat parameters (Fitch & Hillis, 1984; Losos & Chu, 1998). Some studies (i.e., B. Greenberg & Noble, 1944) have failed to show that females prefer males with dewlaps of a particular color, but these studies have been mostly conducted on the Carolina anole, which was not allopatric with other anoles until recent introductions of a few Caribbean species within its range. More recent studies using video playbacks and robotic models suggest that species-specific dewlap color and head-bobbing frequency in anoles serve species recognition between animals of the same sex, rather than between males and females (Losos, 2009).

If there is no dewlap, the entire throat area can be extended and/or flushed with color. Visual signaling is not limited to adults: hatchling green iguanas (*Iguana iguana*), which are highly social, use their black eyes or eye-like spots on closed eyelids as passive signals of position or deceptive signals of being awake and alert (Burghardt, 1977a) (see Chapter 9).

Also, very common among lizards are movements called headbobs or push-ups (Martins, 1993); they can be remarkably similar in unrelated lineages. C. C. Carpenter (1986) pioneered the filming and analysis of these displays in many species and compared them using a display action pattern (DAP) graph. Although these movements seem simple, at least some species can change parameters such as frequency and amplitude to produce different signals for different situations and convey lots of information, including the signaler's sex, intentions, and species identity (Vicente, 2018). Individual male green iguanas in the same population have individually distinct signature displays (Dugan, 1982a). Some species have an impressive arsenal of display behaviors: in just one species, the garden lizard (*Calotes versicolor*), 25 distinctive postures and gestures have been documented (Pandav et al., 2007).

In many anoles, both sexes engage in displays. For example, in Carolina (*Anolis carolinensis*) and Puerto Rican brown (*A. sagrei*) anoles, both sexes perform head bobbing, but males make it much more conspicuous by flashing their brightly colored dewlaps. Courtship and aggressive displays can be very similar but lead to very different outcomes. Darwin (1871, p. 354) described aggressive displays by the Puerto Rican crested anole (*A. cristatellus*):

During the spring and early part of the summer, two adult males rarely meet without a contest. On first seeing one another, they nod their heads up and down three or four times, and at the same time expanding the frill or pouch beneath the throat; their eyes glisten with rage, and after waving their tails from side to side for a few seconds, as if to gather energy, they dart at each other furiously, rolling over and over, and holding firmly with their teeth. The conflict generally ends in one of the combatants losing his tail, which is often devoured by the victor.

One lizard group in which display behavior is well known is chameleons (Chamaeleonidae). They are diurnal, their displays are highly visual (Plate 4.3), and many species are now bred in captivity, so there is much accumulated information on their social behavior in captivity, although much of it has not been scientifically published, and some behaviors observed in captivity might look very different in the wild. Most chameleon species are social and live in loose colonies, with densities of up to one individual per square meter (Nečos, 1999). Some species form permanent pairs, for example, the helmeted chameleon (*Trioceros hoehnelii*) (Toxopeus et al., 1988). Nečos (1999) lists as many as 33 different visual signals used by various chameleon species during male-to-male, female-to-female, and courtship displays; these signals include changing body shape, exposing teeth or tongue, changing color and/or pattern, and performing repeated movements of eyes, head, tail, or entire body. Color changes can be particularly diverse and complex. But in some species, highly territorial males make limited, if any, use of those signals and engage in combat almost immediately, using bites or horn strikes. There are also passive signals: prominent structures such as casques, horns, and crests are more common in chameleons living in places with many chameleon species, and apparently aid in species recognition (Nečos, 1999).

The following color changes are used (in addition to body postures) by the veiled chameleon (*C. calyptratus*) during courtship (Nečos, 1999):

- Male approaching female: lime-yellow stripes on turquoise-green background
- Female rejecting the male: orange and green spots on black background

- Female being receptive: bright yellow stripes on black and green background
- Male responding to rejection by female: dull colors
- Female responding to male response: orange spots on green background

On top of that, numerous chameleon species, particularly those living in shady habitats with a high proportion of ultraviolet (UV) in the ambient light, have been recently found to use fluorescence for signaling: they have protruding bone crests that are visible through transparent "windows" in the skin and emit blue light when exposed to UV light (Prötzel et al., 2018.). UV reflectivity of the throat (see Plate 10.2) serves as an honest signal of fighting ability in Broadley's flat lizards (*Platysaurus broadleyi*) and differs between territorial and floater males (Whiting et al., 2006). UV reflectance may be an important aspect of social communication in many lizards (c.f., Font et al., 2009).

Male Indian rock agamas (*Psammophilus dorsalis*) can change color within seconds; the color change is more dramatic when the male begins to court a female, but faster when he meets another male (Batabyal & Thaker, 2017). The same study found that agamas living in urban environments have much duller colors and change them slower, possibly as a result of chronic stress.

Coloration is also an important signal of sex and status in many species that cannot voluntarily change color. Studies using models have shown, for example, that male side-blotched lizards (*Uta stansburiana*) begin courtship behavior in response to visual, rather than olfactory, information (Ferguson, 1966). Ord et al. (2002) successfully used video playbacks to elicit reciprocal display behavior in the jacky dragon (*Amphibolurus muricatus*). A few lineages with complex social systems have different types of males that behave differently and develop different coloration (see Chapter 3). Bright coloration developed by males of some species appears to increase predation risk, which explains why it is often present only during the mating season (Amdekar & Thaker, 2019).

Lizards can also use dominance and assertion displays with nonconspecifics as threat displays, communicating intent to attack or defend oneself, as most zookeepers know. Lizards can also use other visual cues to alert or distract predators or competitors from approaching or attacking them. Varanid lizards such as Komodo dragons will tense and partially raise and stiffen their tails when irritated or threatened. Hatchling green iguanas, which also have very long tails, will perform slowly undulating tail movements minutes after emerging from their nests, as repeatedly seen and filmed (GMB, unpubl. data). Curly-tailed lizards (*Leiocephalus* sp.) even get their common name from such displays. Some tail signals might actually be addressed at interspecific potential predators, warning them that they have been spotted and an attack would be futile; such signals are known as pursuit-deterrent signals (PDS) (see Woodland et al., 1980; Hasson, 1991). Doody, Schembri, and Sweet (2015) observed two species of monitor lizards apparently using their tails to make PDS addressed to the observers.

4.5. Chemical Communication in Lizards

Over the last 30 years it became clear that chemical communication is much more widespread in lizards than previously thought. The following account is summarized from an overview by J. Martín and López (2014), unless noted otherwise. Lizards produce a broad variety of chemical compounds from specialized glands, most commonly femoral, precloacal, or cloacal. Some of these compounds serve as signals, while others increase or decrease the volatility of the secretions, thus making them more efficient for airborne communication or more durable, particularly in arid climate (A. C. Alberts, 1990; Bruinjé et al., 2020). In many lizard species, males use the secretions of femoral or precloacal glands to scent-mark their territory; species living in environments where scent-marking is less practical also produce lots of volatile compounds such as fatty acids (A. C. Alberts et al., 1992). It has been experimentally shown that common wall lizards (*Podarcis muralis*) react differently to the odor of gland secretions of familiar versus unfamiliar individuals (Mangiacotti et al., 2019). Female Algerian sand racers (*Psammodromus algirus*) can tell the age of males by the scent of their femoral gland secretions and feces (Nisa Ramiro et al., 2019). Olfactory signals are often more important than visual ones: this has been shown for male western fence lizards (Duvall, 1982), Iberian wall lizards (*Podarcis hispanica*) (López & Martín, 2001), and Pacha tree iguanas (*Liolaemus pacha*) (Vicente & Halloy, 2017). In species with "sneaky males" (see Chapter 3), those males can successfully mimic females by developing similar coloration, but the deception only works as long as the territorial male is not sufficiently close to receive chemical signals, which apparently cannot be faked. Male Erhard's wall lizards (*Podarcis erhardii*) translocated to predator-free islands increased the complexity of chemical signals and proportions of octadecanoic acid, oleic acid, and α-tocopherol (the three compounds that are known to be associated with lizard territoriality and mate choice) within just four years, showing that, just as with visual and acoustic signaling, chemical communication can be limited by predation risk (Donihue et al., 2020).

Fossorial lizards might rely entirely on chemical communication to locate conspecifics: in the only such species studied to date, the Iberian worm lizard (*Blanus cinereus*), Cooper et al. (1994) found some evidence that males use olfaction to quickly determine presence and sex of conspecifics, although vomeronasal perception is also used. Males of this species produce squalene (a chemical) and respond to it aggressively (López & Martín, 2009); this signal of sex is used by males of many lizard and some snake species. Various alcohols are found in the secretions of some male lizards, and are thought to be signals of dominance.

Male leopard geckos (*Eublepharis macularius*) often mistake females in the process of shedding skin for males, likely because the odor of their skin changes; skin secretions of females normally contain long-chain methyl ketones, which are lost after shedding the skin (R. T. Mason & Gutzke, 1990). Male Puerto Rican dwarf geckos (*Sphaerodactylus nicholsi*) apparently recognize females by taste, as they lick them (Regalado, 2003). Males of most species can follow females' scent trails, sometimes for very long distances, as in the desert monitor (*Varanus griseus*) (Tsellarius & Tsellarius, 1996).

Compounds secreted by cloacal glands are sometimes added to copulatory plugs, which are used by males to seal the female's cloaca after mating. This possibly allows males to scent-mark females they mate with: male Iberian rock lizards (*Iberolacerta monticola*) can distinguish their own copulatory plugs from those of other males and even assess the dominance status of other males by chemical cues from copulatory plugs (Moreira et al., 2006).

In addition to sex-specific compounds, lizard secretions commonly contain steroids in species-specific combinations, possibly used for species recognition. Many of these steroids are of plant or microbial origin and have to be obtained from food, so the quality of a lizard's diet affects the characteristics of its gland secretions. This might explain why, in some lacertid lizards, females prefer the scent of males with high proportions of such compounds in their secretions. Another compound of dietary origin believed to be used as an honest signal of male quality is linoleic acid, found in the secretions of male Argentine black and white tegus (*Salvator merianae*). In Algerian sand lizards (*Psammodromus algirus*), females show higher responses to femoral gland secretions of males with low blood parasite loads and stronger immune responses, which are apparently signaled by higher proportions of two alcohols (octadecanol and eicosanol) and lower proportions of their correspondent carboxylic acids. Use of chemical information to avoid potential inbreeding has been reported in many species.

Different compounds found in the same secretions may carry different messages for males and females. When presented with male secretions, female Carpetane rock lizards have higher responses to cholesta-5,7-dien-3-ol and to ergosterol, while males respond mostly to cholesterol. It is possible that the levels of cholesta-5,7-dien-3-ol and ergosterol in male secretions signal the quality of a potential mating partner, while cholesterol level signals the body size of a potential opponent. Experimentally adding ergosterol to territorial markings of males made their territories more attractive to females, while adding cholesterol allowed the males to win more fights.

4.6. Lizard Calls

Geckos (Gekkonidae sensu lato) and closely related snake-lizards (Pygopodidae) are almost unique among lizards in that many species use sound for communication. Some will call frequently throughout their active time;

the calls are usually considered to be territorial proclamations as well as mating calls (Brillet & Pailette, 1991), although in most cases their role is actually unknown. Barks, squeaks, whistles, and other calls by geckos are a characteristic component of nighttime soundscapes in many subtropical and tropical locations, from the deserts of Turkmenistan and Namibia to the rain forests of Borneo and Congo and the city streets of Miami, Nairobi, and Singapore. The number of known call types per species varies from one in parthenogenetic Indo-Pacific geckos (*Hemidactylus garnotii*) to 10 in Mediterranean house geckos (*H. turcicus*); in the latter species, males have different calls for repelling other males and for attracting females (Frankenberg, 1982a, 1982b). Some species use ultrasound signals (A. M. Brown, 1984), while others compose their signals of various sounds combined in a particular sequence (Brillet & Pailette, 1991). Males of different species of genus *Gekko* are either "loud callers" that use sounds for long-range communication or "soft callers" that only use them when approaching a female (J. Chen et al., 2016). Using acoustic signals does not mean that geckos do not rely on visual signals; despite being nocturnal, they can use a complex system of postures (Regalado, 2003). However, visual courtship displays are mostly used by diurnal geckos (Marcellini, 1977).

A recent study by Brumm and Zollinger (2017) found that tokay geckos (*Gekko gecko*) can change their calls in response to a noisy environment. However, unlike most birds and mammals, which respond to noise primarily by calling louder, geckos change the composition of their signals, increasing the share of high-frequency syllables.

Some chameleons use low-frequency sounds (described as growls) to find mates at dusk or at night (Schmidt, 1993). The veiled chameleon is the only lizard (so far) known to use vibration for signaling: males produce it by shaking the branches that both they and the female perch on (Barnett et al., 1999).

Winck and Cechin (2008) found that adult male Argentine giant tegus (*Tupinambis merianae*), but not females or juveniles, exchange low-frequency vocal signals while basking communally, and the sounds lead to other males changing position; acoustic communication starts after the animals emerge from winter hibernation sites and ceases when they establish territories and stop basking communally. Monitor lizards hiss during aggressive interactions (Pianka & Vitt, 2003), while Paraguay caiman lizards (*Dracaena paraguayensis*) and Canary Islands wall lizards (*Gallotia*) produce sounds during courtship (Strüssmann, 1997; Pianka & Vitt, 2003).

Besides emitting sounds, lizards can also eavesdrop on the alarms and other sounds from birds and other animals. Thus, even though they may not use sound for their communication, they can recognize and respond appropriately to the communication signals of other species. Using playback experiments, Ito and Mori (2010) demonstrated that the Madagascan spiny-tailed iguana (*Oplurus cuvieri*) eavesdrops on a flycatcher, distinguishing between alarm calls and songs, with no apparent benefit for the flycatcher. In Florida, the invasive Puerto Rican brown anole discriminates between calls of predatory and nonpredatory birds (Cantwell & Forrest, 2013). Similarly, the weeping lizard (*Liolaemus chiliensis*) eavesdrops on the alarm calls of conspecifics, and there is evidence for local population adaptation to stress call variation (Labra et al., 2016).

4.7. Snake Communication: Following the Scent

Unlike lizards, snakes appear to largely communicate with conspecifics by way of chemical and tactile signals. Visual signals are also used (many snakes have huge eyes in relation to their head size, even diurnal species). After conspecifics are located or recognized, body size may be very important in courtship and male-to-male combat, but other visual signals may be subtle and difficult for human observers to notice (Jellen & Aldridge, 2014). This is a bit surprising because many snakes have specialized structures for producing acoustic and visual displays, and they use hard-to-miss acoustic and visual signals to warn predators and repel pesky primates. Well-known examples include rattling sound by rattlesnakes (*Crotalus*, *Sistrurus*), "sizzling" sound produced by specialized scales of saw-scaled vipers (*Echis*), a loud hiss made by the resonating air sac of the puff adder (*Bitis arietans*), open-mouth display by cottonmouths (*Agkistrodon palustris*, *A. conanti*), and "hood" display by cobras (*Naja* and related genera), as well as various cobra mimics. Spitting cobras (rinkhals *Hemachatus haemachatus* and some species of *Naja*) have aposematically colored hoods (with contrasting black and white bands), and H. W. Greene (2013) convincingly argued that these striking visual displays and the ability to spit venom have evolved specifically as a way to avoid harassment and killing by monkeys and apes, including various extinct and extant species of humans (*Homo*). Rattling displays in rattlesnakes probably evolved from tail-wagging displays used by many species of pitvipers and the nonvenomous snakes that mimic them, for example, by Ussuri mamushi (*Gloydius ussuriensis*) and its mimic, steppe ratsnake (*Elaphe dione*), in Ussuriland (VD, pers. obs.). Tail-wagging is also used by many snakes to lure small birds and lizards, and the tail is often brightly colored or banded to look like a wiggling worm or arthropod; the most spectacular example is the recently described spider-tailed viper (*Pseudocerastes urarachnoides*), with its tail scales modified to look like moving spider legs (Fathinia et al., 2009, 2015).

Young (1997) suggested that intraspecific acoustic communication is absent among snakes for two reasons: (1) the imbalance between the frequency of sound produced (typically over 3,000 Hz) and the auditory range (which rarely extends over 1,000 Hz), and (2) the absence of frequency and amplitude modulation and temporal patterning in snake sounds that limits the potential information content. He noted, however, that there are exceptions to both rules: some degree of patterning can occur, and some species, such as the king cobra (*Ophiophagus hannah*), produce sounds that are at least partially within their auditory range. That led Young to predict that intraspecific acoustic communication might eventually be discovered in some snake species. This has not happened so far.

Intraspecific visual signals are so far known to be used only by male snakes; the general pattern seems to be that male-to-male combat (see Chapter 5) is preceded or substituted mostly by visual contact and postures, while courtship behavior is initiated after exchanging chemical signals. Here is a description by R. Repp (in Schuett et al., 2016) of visual signaling by male western diamondback rattlesnakes (*Crotalus atrox*) at a large communal den, unexpectedly noticed while watching documentary footage (Fig. 4.2):

Through it all, the male *atrox* we named Tyson commanded center stage. He was omnipresent, in nearly every frame, and always on the move. He was alert and aggressively took note of every nook, cranny, and snake on his turf. He appeared to have complete command of his world. . . . To be sure, there was mating and fighting, but that is not what got me out of my chair. What truly piqued my attention was the intense tail waving that transpired. In one sequence, Tyson was investigating (rapid tongue flicking) a cluster of about 15 adult conspecifics. As he crawled on top of the pile, four tails immediately rose out and began waving back-and-forth in sinusoidal fashion, not unlike caudal luring [a tail movement used by many snakes to attract small prey, particularly birds]. Similar behavior occurs in male copperheads during dominant-subordinate episodes or bouts. . . . Tyson zeroed in on one of those tails and used his snout to push and extract the snake that owned it. It turns out this other snake was a male, but not quite as big as Tyson. The excellent footage clearly showed that his rattles were tapered and complete—a younger snake, perhaps an "upstart." . . . Once Upstart caught wind he was noticed by Tyson, he rapidly fled the pile. . . . Male dominance was clearly exhibited without sparring in outright combat with neck-to-neck vertical postures.

Snakes often show great color and pattern variation and even polymorphism, but unlike in some lizards, this does not seem to play much role in intraspecific communication. One reason may be that there is no experimental evidence yet of light wavelength or color discrimination in snakes (Terrick et al., 1995), although many have a trichromatic cone retina (Simões et al., 2016). It may eventually be shown that the anatomical and physiological substrates for color vision can translate into behavioral discrimination uncontaminated by brightness cues; in that case there would likely be numerous reliable anecdotal reports of such discrimination, but we are not aware of any.

Tail waving is now believed (Schuett et al., 2016) to signal submission and sometimes sexual identity (by males trying to avoid mating attempts by another male). Raising the head and up to a third of the body vertically and swaying back and forth is apparently a challenge to combat used by many species that practice male-to-male combat in "vertical stance" (see Chapter 5); it results in the adversary either fleeing or accepting the chal-

Fig. 4.2. Large male western diamondback rattlesnake (*Crotalus atrox*) nicknamed "Tyson" on top of a group of conspecifics in front of a communal den. Note the head of an adult female poking out from beneath the lower body of another rattlesnake. Photograph by Roger A. Repp from Schuett et al., 2016.

lenge (C. C. Carpenter, 1977). This vertical stance behavior in pitvipers and other snake taxa may also be used by females in sexual identification, as well as in mate choice and male superiority (Schuett & Duvall, 1996).

Male snakes courting females use a broad variety of tactile signals, overviewed by C. C. Carpenter (1977). Different species use various combinations of chin rubbing, body jerking, caudocephalic or cephalocaudal waves (running from tail to head or vice versa), tail searching, pushing, nudging, and biting; boas and other species possessing spurs can use them to stroke or scratch the female's cloacal region, apparently to pacify her and induce her to twist her cloaca laterally and open it. Successful insertion of a hemipenis requires elaborate tactile signaling, so that the female opens her cloaca just as the male is ready to penetrate her. It is possible that species-specific sequences of these signals sometimes act to prevent hybridization.

Chemical signaling, both passive and active, appears to be very widespread, possibly universal, in snakes (see overviews by N. B. Ford &

Burghardt, 1993; R. T. Mason & Parker, 2010; J. Martín & López, 2011). Even hatchlings can follow scent trails of conspecifics (Constanzo, 1989). Males often have to trail the female for hours before getting within visual range (Kurbanov, 1985). Cloacal gland secretions can be used as alarm signals: prairie rattlesnakes (*Crotalus viridis*) exposed to a threatening stimulus experience a larger rise in their heart rate in the presence of cloacal gland materials from conspecifics (Graves & Duvall, 1988). In brown tree snakes (*Boiga irregularis*), females use anal gland secretions to repel unwanted males (M. J. Greene & Mason, 2000). But skin secretions seem to be the most-used chemical signals in mating-related communication: skin secretions of females of multiple species were found to elicit stronger response in conspecific males than cloacal gland secretions (G. K. Noble & Clausen, 1936; Andrén, 1986).

In a study of brown snakes (*Storeria dekayi*), G. K. Noble and Clausen (1936) showed that outside the mating period, males and females follow chemical cues of conspecifics regardless of their sex, but during the mating period, males exclusively follow the trails of females, while females still follow the trails of both males and females. This led them to suggest that receptive females switch from passive to active signaling by producing pheromones. This theory was later confirmed by numerous studies (i.e., N. B. Ford, 1986; N. B. Ford & O'Bleness, 1986; LeMaster & Mason, 2002), almost all of them on the gartersnakes, primarily the common gartersnake (*Thamnophis sirtalis*), which may be the most studied serpent.

It is now known that, just as in many lizards (see above), methyl ketones are found in the skin of female gartersnakes, where they serve as sex attractiveness pheromones (R. T. Mason et al., 1990). However, females with cloacal plugs inserted by males after mating do not attract other males' attention, suggesting that a different chemical signal is present in the plug (O'Donnell et al., 2004). Nevertheless, such supposed anti-aphrodisiac pheromonal plugs are clearly not sufficient to prevent multiple mating (Schwartz et al., 1989). Male common gartersnakes can assess the body size and body condition of females based solely on pheromone cues: larger and healthier females have higher content of unsaturated methyl ketones in their pheromonal blend (LeMaster & Mason, 2002; Shine, Phillips, et al.,

2003). In species that practice male-to-male combat, males sometimes trail not just females but also males, possibly looking for a chance to obtain access to a female by fighting off the male who has found her first (M. J. Greene et al., 2001).

One interesting discovery made in the course of research on common gartersnakes is that when males from some populations in the northern part of the species' range emerge from hibernation, they produce female sex pheromones and are courted as if they were females. These sexually attractive males were termed "she-males" (R. T. Mason & Crews, 1985); note that this term has a different meaning in lizards: in snakes males go through this phase temporarily, while in lizards it is a permanent morph. Shine, Phillips, et al. (2001) suggested that males mimicking females may benefit because large mating balls of warmer males form around them, warming them up and protecting them from predators.

In sea kraits (*Laticauda*), the least sea-adapted of sea snakes, mating occurs on land, and communication is typical for elapids: males use tactile stimulation to court females, while the latter's skin produces pheromones that stimulate male behavior; these pheromones are species-specific and possibly play a role in sympatric speciation (Shine, Reed, et al., 2002). In the turtle-headed sea snake (*Emydocephalus annulatus*), which is completely aquatic, males utilize visual cues to locate potential mates, but reception of female skin lipid pheromones by tongue flicking is necessary for them to initiate courtship (Shine, 2005).

Nothing is known about communication in fully fossorial snakes, except that gland secretions of Texas blind snakes (*Rina dulcis*) attract conspecifics and repel ant- and termite-eating snakes of other species (Gehlbach et al., 1968).

Aggregation in snakes seems to be facilitated by chemical cues. Ringnecked snakes *(Diadophis)* and natricine snakes have been shown in experimental studies to preferentially use refuges previous snakes have used. The cues are chemical and are derived from feces or skin chemicals (Dundee & Miller, 1968; Burghardt, 1980, 1983; B. A. Allen et al., 1984; Lyman-Henley & Burghardt, 1994). More recent studies have shown this may be widespread among snakes.

5 |

Courtship and Mating

Mating-related rituals performed by various animals, from mollusks to humans, are among the most fascinating spectacles to see on our planet, and those practiced by reptiles are no exception. A peculiar aspect making reptile courtship behavior different from that of other vertebrates is that so much of it remains unknown to us. Some of the most impressive, colorful, and weird displays have only been discovered recently, and hundreds more can be presumed to remain unobserved and undescribed. Courtship and mating have been seen in the wild in some species of all the major groups, including some crocodylians, lizards, snakes, turtles, and the tuatara. Often these are species that are readily seen by humans, such as iguanas, alligators, tortoises, and gartersnakes emerging from brumation at hibernacula. However, details of courtship are better known for species in captivity, especially those bred in captivity for conservation, ranching, or the pet industry. Even in these cases, the lucky breeders often miss the event (that is particularly likely to happen with nocturnal species) or fail to publish their observations, as evidenced by many YouTube videos not followed by scientific publications. Since the available data on courtship and mating are often based on observations of captive pairs, many aspects of these behaviors, such as group courtship or male-to-male aggression, remain unknown even for better-studied species. In many reptiles, pre-mating interactions are based on exchanging chemical signals; these signals are not usually noticed by a casual observer, if at all. For some groups, such as worm lizards and many fossorial families of lizards and snakes, there is still no information whatsoever on pre-mating or mating behavior, although some observations hint at the possibility of a complex social life in worm lizards (J. Martín et al., 2011; Borteiro et al., 2013).

Three scientific trends, in particular, have had an unfortunate effect on the studies of reptile breeding behavior. One is the concept of "model species": a species that is exceptionally easy to breed and manipulate in captivity is chosen to represent an entire major taxon, and studied in great depth while thousands of other species are largely ignored (Fields & Johnston, 2005). The other is the idea that the only "truly scientific" way to study animal behavior is by conducting manipulative experiments in controlled lab settings (Kennedy et al., 1990); some proponents of this dogma have derogatorily termed observational studies in the wild "Boy Scout science" and "mere natural history," as if that is nonessential to scientific progress. The third is the claim that observation and experimentation is useless unless carried out to support or test some preconceived theory or concept (S. J. Arnold, 2003). H. W. Greene (2005) has reviewed and strongly countered these views. However, the trend in scientific research generally is heading toward preregistration of studies with almost all details specified in advance and a single subject being tested (cf., Burghardt, 2020), and even ethological journals are eliminating descriptive studies (e.g., *Ethology*; see Chapter 1).

Over time it became clear (see examples below) that extrapolating results from model species to other taxa, as well as replacing observations in the wild with studies in captivity, is more often misleading than informative. Moreover, serendipitous observations are at least as likely to result in transformative discoveries as are preplanned studies. Of course, testing preconceived theories, using model species, and using captive animals can be useful, but these methods should never be treated as a valid substitute for broad studies of natural history (Futumya, 1998; H. W. Greene, 2005; McCallum & McCallum, 2006). Even for the four most-studied species of reptiles among each of the major groups—crocodylians, the American alligator (*Alligator mississippiensis*); turtles, the red-eared slider (*Trachemys scripta*); snakes, the common gartersnake (*Thamnophis sirtalis*); and lizards, the Carolina anole (*Anolis carolinensis*)—the results of studies in captivity have been shown to often create an inadequate and oversimplified picture of their behavior (see, for example, Larsen et al., 1993; Tucker et al., 1998; Tokarz, 2002; Dinets, 2010).

Lack of basic data hampers broader understanding: for example, it has been proposed (Wade, 2002) that the mechanisms of hormonal control of behavior in birds and reptiles are substantially different from those in mammals, but our understanding is severely hampered by the fact that most such studies in birds have been limited to just one "model" species, the zebra finch (*Taeniopygia guttata*), and in reptiles also to one species, the already mentioned Carolina anole (e.g., N. Greenberg & Crews, 1990), although the common gartersnake has also been studied (e.g., R. T. Mason & Parker, 2010).

5.1. Fossilized Courtship

We know that mating displays and courtship rituals exist in all major reptile lineages; we can also assume that numerous extinct taxa had them. Among the earliest reptiles (in the broad sense) were animals sporting dorsal crests (commonly called "sails"), such as synapsids *Dimetrodon* and *Edaphosaurus*. These sails were once thought to be thermoregulatory organs, but the currently favored explanation is that they were used in displays of some sort. This explanation is supported by evidence that in *Dimetrodon* sails were possibly sexually dimorphic and did not fully develop until maturity, while in *Edaphosaurus* they were decorated by elaborate bone growth with no known function (Tomkins et al., 2010; Huttenlocker et al., 2011). Notably, the sails in the extant lizards *Hydrosaurus* and *Basiliscus* are restricted to males (Pianka & Vitt, 2003).

Interestingly, pronounced sexual dimorphism and complex displays, as well as structures evolved specifically for display or intraspecific combat, are relatively rare in extant synapsids (mammals). They are absent in marsupials, monotremes (except for venomous spurs possessed by male platypuses, *Ornithorhynchus anatinus*), and most orders of placentals, and they are present only in some carnivores, odd-toed ungulates, cetaceans, elephants, and primates. In most mammals (including all members of the two largest groups, rodents and bats) pre-mating social behavior appears limited to simple exchanges of chemical and sometimes acoustic signals and occasional male-to-male fights, and it is generally less elaborate than in many reptiles and most birds (Nowak, 2018).

Fossil evidence (indirect, of course) of complex pre-mating behavior is particularly abundant for Archosauria. This is hardly surprising, considering how ubiquitous courtship displays are in modern birds and how complex pre-mating behavior is in crocodylians (see below). A peculiar type of trace fossils found in Colorado has been interpreted as leks or display arenas of theropod dinosaurs, with scraping marks resembling those left by the "nest scrape displays" common in extant ground-nesting birds; these arenas can reach 50 m in diameter (Lockley et al., 2016). Many pterosaurs had large, sometimes apparently sexually dimorphic, bony crests that were fully developed only in adults, suggesting that they were used in mating displays or other reproduction-related behaviors (Naish & Martill, 2003; Martill & Naish, 2006; Hone et al., 2011; Manzig et al., 2014; X. Wang et al., 2014). Numerous dinosaurs had dazzlingly diverse structures, such as crests, frills, and horns, that were probably used for mating displays or ritualized combat (Knell et al., 2012; Hone & Naish, 2013). Some of these structures, like the huge crests with labyrinthine air passages found in *Parasaurolophus*, may have been used for producing species-specific calls (Wiman, 1931).

Feathered taxa interpreted as bird-like dinosaurs or early birds had ornamental feathers or feather-like structures that were probably used for courtship displays (see Chapter 4, section 4.1). It has even been suggested (Senter, 2007) that ultra-long necks of giant sauropod dinosaurs evolved under sexual selection, probably for use in ritualized combat or pre-mating displays. Unfortunately, this imagination-provoking theory is unlikely to be correct (Taylor et al., 2011).

5.2. Displays of Living Archosaurs

Modern birds have all kinds of display structures, sometimes remarkably similar to those of extinct reptiles (for example, the crests of cassowaries are superficially almost identical to those of *Corythosaurus* dinosaurs). Extant crocodylians do not have such structures (with the exception of the Indian gharial, *Gavialis gangeticus*), but it does not mean their courtship is simple or uninteresting. Indeed, crocodylian courtship and the displays preceding it can be outstandingly spectacular. The following account is abbreviated from the review by Dinets (2013c), unless stated otherwise.

Crocodylian displays include acoustic, tactile (infrasound vibrations), visual, and olfactory components (see Chapter 4, sections 4.1–2). They are performed most often in morning hours during the mating season, but can occasionally be seen at other times of the day and the year (Garrick & Lang, 1977; Vliet, 1989, 2001), and some species repeat them in the evening or conduct them at night.

Alligators and caimans (Alligatoridae) have two distinct types of displays: a bellowing/roaring display and a headslapping display, both of which are usually performed with raised head and tail, a posture termed "head oblique tail arched" (HOTA) by Garrick et al. (1978), and discussed further in Chapter 4 (see Fig. 4.1). Both types include infrasound in males but not in females. Alligators (*Alligator* spp.) usually display in the morning; black caiman and dwarf caimans (*Paleosuchus* spp.) do so at night (the former also at dawn). Common caimans (*Caiman* spp.) mostly display in the morning but may have a second peak in the evening. Both displays are highly contagious and performed by both sexes in *Alligator* and *Caiman*, but headslapping displays are rarely performed by females. There is no solid evidence of contagion or of females displaying in dwarf and black caimans, which are generally more solitary and territorial.

Alligator and *Caiman* bellowing displays are often performed in choruses (group displays joined by most adults within hearing range). The most impressive are bellowing choruses by American alligators; they can involve hundreds of animals and last for up to an hour. Here is a description of one such chorus, observed in Everglades National Park, Florida (Dinets, 2013b, p. 8):

The lake was barely visible in pink fog. The forest was eerily quiet after the crickets-filled night. The purple sky was criss-crossed with golden jet contrails and lines of high cirrus clouds. The sun was just about to come up. The alligators were all in the water, floating like black rotten logs. Suddenly the largest one, a beast almost as long as my car, lifted its massive head and heavy rudder-like tail high above the water surface. He (such huge individuals are usually males) froze in this awkward position for at least a minute, while others around him were also raising their heads and tails one by one, until there were twenty odd-looking arched silhouettes floating in the mist.

Then the giant male began vibrating. His back shook so violently that the water covering it seemed to boil in a bizarre, regular pattern, with jets of droplets thrown about a foot into the air. He was emitting infrasound, acoustic vibrations too low for human ears to hear. I was standing on the shore at least fifty feet away, but I could feel the waves of infrasound within every bone in my body. A second later, he rolled a bit backwards and bellowed—a deep roar, terrifying and beautiful at the same time. His voice was immensely powerful. It was hard to believe that a living creature could produce what sounded more like a heavy tank accelerating up a steep rampart. He kept rocking back and forth, emitting a bellow every time his head was at the highest point and a pulse of infrasound every time it was at the lowest. All around the lake, others joined him. They were all smaller, so their voices were higher-pitched and less powerful, but still impressive. Clouds of steam shot up from their nostrils (weren't they supposed to be cold-blooded?). Trees around the lake—huge bald-cypresses— were shaking, dropping twigs and dry leaves on churning waters. I stood there, frozen, fascinated, hearing alligators in other lakes, near and far, as they joined this unbelievable show of strength and endurance. For about an hour, waves of bellows and infrasound rolled through forests and swamps all across southern Florida.

Then, gradually, they stopped. It was quiet again. The alligators were floating silently in the black water of the lake as if nothing had happened. I waited for two hours, and not a single one of them moved. Nothing moved there, except the rising sun and flocks of snowy egrets that sailed across the sky on their way from their night roosts to some fish-filled ponds.

Unlike alligators and some caimans, crocodiles (Crocodylidae) are not known to display in choruses. In most species, displays are performed mostly by territorial (i.e., dominant) males. They are usually given in HOTA posture and include infrasound. Another difference in crocodiles is that roars and slaps can be combined within the same display, although headslaps are rare or absent in some species, while roars are rare or absent in others. Slender-snouted crocodiles (*Mecistops*) vocalize particularly frequently. Crocodiles display in the morning, but some species also do so at night.

The courtship displays of the Indian gharial are different from those of other crocodylians. Instead of assuming HOTA posture in the water, gharials often assume a head-up posture on land, which is believed to be a territorial display and a signal of sex and maturity (Singh & Rao, 1990). Their acoustical signals are unique among living crocodylians (see Chapter 4, section 4.1).

5.3. Crocodylians: Tender, Caring, and Intimate

What happens if all those displays work and animals get together? The answer for crocodiles, caimans, and gharials is relatively simple. Both sexes can initiate the next stage of the courtship: sometimes a male approaches a female, and sometimes one or a few females immediately approach a male (Fig. 5.1) after it roars or headslaps (Dinets, 2011a). If the approached animal is not interested, it would signal its displeasure by hissing, growling, or snapping at the intruder (Garrick & Lang, 1977; Vliet, 2001). If the courtship continues past this point, it can last for up to an hour and include mutual touching (particularly with the underside of the lower jaw, which carries musk glands and might be an erogenous area), swimming in tight circles (accompanied with soft buzzing sounds in the Indian gharial), blowing bubbles, and sometimes play behavior (see Chapter 11). Mating is also a tender affair; it can last for up to half an hour, although usually it is only a few minutes. Typically, the male is on top of the submerged female, with the caudal part of his body wrapped around her lizard-style, but matings in which both animals were lying on their sides have also been observed (Dinets, 2013b). They do not move much, except the female sometimes raises her snout to breathe.

In alligators, mating looks the same, but courtship is only simple in places where just a few animals are present. In areas with high population

Fig. 5.1. Female mugger crocodiles (*Crocodylus palustris*) courting a large male after he roared. Sasan Gir National Park, India. Photograph by Vladimir Dinets.

density, alligators gather for courtship in large numbers. In Florida, their mating season normally falls in late April to early May, the last month of the dry season, and dozens or even hundreds of animals can gather in remaining ponds. In the morning all present adults bellow in chorus, and at night they engage in spectacular group courtship known as "alligator dances" (Plate 5.1). Astonishingly, despite being common and relatively easy to observe, this behavior was not noted and recorded until 2006 (Dinets, 2010). Countless naturalists who have studied the mating biology of the American alligator, starting with Bartram (1791), have never noticed the "dances," probably because they all have conducted their observations during daylight hours or on captive animals. Here is the description of the first observation of a "dance," made in Everglades National Park (Dinets, 2013b, p. 13):

Two hours after sunset all tourists were gone. I switched my tiny headlamp to red light and sat on a wooden bench, watching alligator eyes circle below the boardwalk. Soon I noticed that they were all gathering in one part of the lake, and becoming more and more active. Eventually about thirty alligators gathered in an area less than sixty feet across, swimming like crazy, splashing, hissing, slapping their heads and tails, occasionally getting into brief but violent fights. Some of them would form pairs, then break up again. New ones were arriving every few minutes, alone or already in pairs, smaller females following their males. Others were leaving, but many remained in that small area until dawn. . . . What were they doing? To me it looked like village dancing parties, where people would come, alone or with their spouses, to socialize, have fun, and, in the case of singles, search for mates.

The exact role of these "dances" can only be guessed, because it is extremely difficult to follow individual animals during these nighttime melees, and the social organization of alligators is still poorly understood (see Chapter 4, section 4.1).

5.4. Courtship in Turtles: Love in the Fast Lane

Like crocodylians and some primitive birds, turtles and tortoises have one penis; its anatomy is remarkably similar to mammalian penises, almost certainly as a result of convergence (Kelly, 2004). Relative penis size and structural complexity differs from large, relatively simple penises in sea turtles to enormous, highly complex ones in tortoises; the latter can be half as long as the shell and have structures such as lobes, multiple openings, and independently controlled grooves (Zug, 1966), but nothing is known of their function or of corresponding structures in female cloacae. Multiple paternity and the ability of females to store sperm in their oviducts (possibly for up to seven years) have been reported for many turtles; nests of giant Amazon river turtles (*Podocnemis expansa*) can contain eggs fathered by up to 10 males (Fantin et al., 2017). In contrast, in some species, such as the leatherback turtle (*Dermochelys coriacea*), the females are monogamous within each breeding season (Pearse & Avise, 2002). In loggerhead sea turtles (*Caretta caretta*), a female might resist mating with one male but immediately accept another (Kawazu et al., 2017). There have been no experimental studies of female mate choice, but male northern map turtles (*Graptemys geographica*) prefer larger females, as shown by a study using decoys (Bulté et al., 2018).

Unlike tender and patient crocodylians, turtles and tortoises can get rough and even violent during courtship and mating. In some species, copulation looks like a serious fight with multiple bites (Fig. 5.2). Turtles and tortoises are also very fast, often accomplishing the whole behavioral sequence in just a few minutes (Mahmoud, 1967); although in some species, such as painted wood turtles (*Rhinoclemmys pulcherrima*), the male might have to trail the female for days or even weeks before mating (Hidalgo, 1982).

In sea turtles, copulation takes much longer (up to six hours), but it can still be stressful for females (see below). Based on very limited sampling, Berry and Shine (1980) suggested that there is the following general pattern:

- In most terrestrial species, males engage in combat with each other, and typically grow larger than females.
- In semiaquatic and "bottom-walking" aquatic species, male combat is less common, but males often forcibly inseminate females; they also grow larger than females.
- In truly aquatic species, male combat and forcible insemination are rare; instead, males utilize elaborate precoital displays, and female choice is highly important; males are usually smaller than females.

However, there are many exceptions to this pattern, and the third point might be wrong in the majority of cases, and thus Berry and Shine's (1980) approach and analysis may have been premature. For example, although sea turtles are truly aquatic, they do not have "elaborate precoital displays." According to J. Booth and Peters (1972), female green sea turtles (*Chelonia mydas*) have to spend many days being chased by groups of five to six males, and they use a variety of behaviors to avoid unwanted matings: covering the cloacal opening with hind limbs, swimming away, crawling onto a beach, facing and biting the pursuers, swimming in a vertical position, or hiding in mysterious "female sanctuaries" (areas on the bottom that are avoided by males for unknown reasons). Males in search of females would attempt mating with almost any floating object of roughly appropriate size, including other males and human observers; they sometimes bite non-receptive females violently.

Mating behavior of softshell turtles *Apalone* (described by Plummer, 1977) is somewhat similar. During the mating season, adult males patrol their home ranges and actively investigate all conspecifics, sometimes chasing them or swimming in parallel. Nonreceptive females aggressively repel males by spinning around, chasing, and biting, often drawing blood and leaving distinct bite marks on the edge of the male's carapace. Receptive females allow males to mount but keep moving slowly, and since softshell males cannot grasp females effectively, they have to swim at the same speed to remain attached. Other turtles are much better at holding on to females: they use their tail and legs for grasping, and some are even capable of standing up vertically while the female swims horizontally (Pope, 1939).

Observing mating behavior in many turtle species is difficult because they mate underwater, often in turbid waters with near-zero visibility. Perhaps as an adaptation to poor visibility, courtship rituals in many freshwater turtles involve a lot of tactile signaling. Males of the Indian softshell turtle (*Lissemys punctata*) flip over and press their back to the back of the female (Duda & Gupta, 1981). Males of some Austral side-necked turtles (Chelidae) stimulate the female by cloacal touching, stroking the female's head and neck with front claws, and touching the sensitive barbules on the female's neck with similar barbules on their own necks (Murphy & Lamoreaux, 1978). A male yellow mud turtle (*Kinosternon flavescens*) touches the female's tail and cloaca, taps on her shell with his, and rubs or gently bites her head and neck (Lardie, 1975). Males of a few genera of pond turtles in North America (Emydidae) have greatly elongated foreclaws and either caress the female's chin with the back of these claws (Eglis, 1962) or, more commonly, rapidly vibrate them back and forth in front of the female's face to create water movement (this is called "titillating"). Males of some pond turtles also position themselves above the female and titillate her neck or face from above, or they press their neck to the female's head and vibrate it; they might also use their tubular nostrils to blow water in her face or vertically rub noses with her (Pope, 1939; Eglis, 1962;

Jackson & Davis, 1972; Jenkins, 1979; K. M. Davis, 2009). The courtship goes through a sequence of stages, sometimes rather choreographic: for example, in Florida redbelly turtles (*Pseudemys nelsoni*), the male titillates the female while swimming on top of her (Kramer & Fritz, 1989). Interestingly, juvenile males of this species often "practice" courtship behavior, preferring particular individuals, always conspecific (Kramer & Burghardt, 1998).

Courtship rituals tend to differ not just between species, but also between subspecies and populations. For example, males of various races of the red-eared slider use either biting, blowing water, or titillating, and in one race, females also court males by chasing, titillating (although they lack modified claws), blinking (they have white nictitating membrane that flashes dramatically), and nosing the cloacal region (Lovich et al., 1990). In painted turtles (*Chrysemys picta*), males of some populations court females, while males of others force them into submission by delivering painful bites (Moldowan, 2014). Titillation displays, formerly thought to only occur in adult males, do occur not only in juveniles but also in other contexts in both sexes (K. M. Davis, 2009).

Mating rituals of Asian river turtles (Geoemydidae) are less known, but males and females of furrowed wood turtle (*Rhinoclemmys areolata*) have been reported to engage in mutual nose rubbing and sniffing of each other's mouth and cloacal region (Pérez-Higareda & Smith, 1988), while males of the four-eyed spotted turtle (*Sacalia quadriocellata*) repeatedly strike the female's nape with their chin (Liu et al., 2008). In Mexican spotted wood turtles (*Rhinoclemmys rubida*), the male lies side by side with the female, then forces himself underneath her and repeatedly (15 times in 33 seconds) "wipes" the top of his rostrum against the female's chin and throat, bobbing his head between wipes (Legler & Vogt, 2013).

Interestingly, in some freshwater turtles there is little pre-mating courtship, but much ritualized touching, rubbing, and biting during copulation (Bels & Crama, 1994; Brito et al., 2009). The purpose of this "belated" behavior is unknown. For example, two particularly large alligator snapping turtles (*Macrochelys temminckii*) were observed mating in a small area of transparent water flowing from an underwater spring into an otherwise murky river, at a depth of ~3 m; the copulation (Fig. 5.2) lasted for about five

Fig. 5.2. Final stage of the violent courtship of alligator snapping turtles (*Macroclemys temminckii*) from White Lake, Louisiana. Photograph by Vladimir Dinets.

minutes, with both animals almost continuously exchanging what looked like painful bites to the neck, although no blood was drawn (VD pers. obs.). In some cases, such as in Blanding's turtles (*Emydoidea blandingii*), biting and other stimulation take place after the male mounts the female, but before the intromission, and apparently serve to persuade the female to extend her tail or otherwise make intromission possible (R. E. Baker & Gillingham, 1983). Males of some mud turtles (*Kinosternon*) have patches of rough scales on the hindlimbs that serve to hold the female's tail to the side during mating (Berry & Shine, 1980).

5.5. Tortoises, Wild and Emotional

Tortoises (Testudinidae), being terrestrial, do not have to court under conditions of poor visibility, so their rituals involve more visual displays, such as ritualized walks, before proceeding to the usual menu of touching, biting, and ramming; chemical signals are also used (Auffenberg, 1977). Courtship in some terrestrial turtles, such as the common box turtle (*Terrapene carolina*), is remarkably similar to that of tortoises (L. T. Evans, 1953; Belzer, 2002). Since box turtles are derived from aquatic pond turtles, this is an example (one of many in reptile behavior) of convergence, or homoplasy, in behavior.

Males of many tortoise species engage in ramming contests during the mating season; they often get flipped by high-speed collisions with other males, and they have evolved an improved (compared to females) ability to right themselves (Bonnet et al., 2001). Courtship in tortoises tends to be complex and diverse; it has been described as "wedding dance" in the Mediterranean spur-thighed tortoise (*Testudo graeca*). An overview of tortoise courtship by Eglis (1962) lists numerous species-specific behaviors, such as ritualized walking, stereotypical head movements ("bobbing"), nose touching, leg biting (probably to force the female to retract her legs and thus prevent her from attempting to walk away), olfactory communication ("sniffing"), and sounds. Head bobbing, biting, and sniffing can also be used in aggressive encounters between males (Ruby & Niblick, 1994), and ramming is sometimes a part of courtship.

5.6. Tuatara and Lizards: Beautiful Aggression and Ugly Romance

Many common courtship behaviors of lizards are probably very ancient, because they are also found in the tuatara (*Sphenodon punctatus*), the only surviving representative of a lineage that separated from lizard ancestors about 220 million years ago. According to Gans et al. (1984) and Gillingham et al. (1995), a courting male inflates his body, raises his crests (one on the head and one on the torso; Fig. 5.3), changes the color of its head crest and shoulders to a darker tone, and approaches the female in an unusual walk, with legs moved in a circular manner. He then bites her, causing her to move. The sequence is repeated many times until the male stops biting and repeatedly attempts to mount the female. Eventually the female stops moving completely, and the male, while perched on top of her, lowers

Fig. 5.3. Male displays in the tuatara (*Sphenodon punctatus*): *top,* male displaying to a rival male; *middle,* male displaying to a smaller female; *bottom,* model tuatara with detachable crests used to study behavioral responses to display. Photographs by Jennifer Moore (*top*), Alison Cree (*middle*), and James Gillingham (*bottom*).

his tail and lower body underneath hers to achieve cloacal contact. Although that contact lasts only about 15 seconds, the whole courtship and mating takes almost an hour. Similar displays and croaking sounds are used in aggressive encounters between males until dominance is established.

Unlike all other extant reptiles, the tuatara lacks a penis. Male lizards and snakes have two so-called hemipenes, of which one is used at a time during mating; some males prefer to use them in alternating order during multiple matings (Tokarz, 1988). The complex anatomy of hemipenes is best studied in anoles (*Anolis* spp.); see Lovern et al. (2004) for an overview. Despite this difference, lizards use the same mating position as tuatara and crocodylians, although the exact position of the male varies between species. The male would often use his jaws to hold the female by her neck or leg. Repeated pelvic thrusts can be present or absent; sometimes even closely related species differ in this aspect (G. K. Noble & Bradley, 1933). Legless lizards for which information is available, such as the slow-worm (*Anguis fragilis*), mate in typical lizard manner, with bite-holds and the male on top, rather than in the way common in snakes (G. K. Noble & Bradley, 1933). In blind lizards (Dibamidae) females are legless but males have tiny front legs, used exclusively for holding females during mating (Bauer et al., 1998).

All components of tuatara courtship (ritualized walk, body inflation, crest raising, color changing, and biting) can be found in many lizard species. In fact, some iguanid displays are almost identical. There is a spectacular diversity of crests, dewlaps, and frills, as well as repeated movements often designed to show off the bright display structures; color changes are used by many species but are particularly complex in chameleons (Chamaeleonidae) (see Chapter 4, section 4.4). In many (probably most) species, all displays are performed by males; the only thing the female has to do is stay still for a while, or at least allow herself to be caught and held in place. Interestingly, some racerunner species of hybrid origin that do not have males and reproduce by parthenogenesis (i.e., without mating) exhibit typical male courtship behavior (Crews & Fitzgerald, 1980). One theory suggests that it has changed function and is used for aggressive displays (Y. L. Werner, 1980). Another theory (G. K. Noble &

Bradley, 1933) is that virtually all male displays in all lizards are aggressive to begin with and are addressed at other males rather than females (Plate 5.2), while male–female interactions can be best described as rape. This appears to be true for many species, but certainly not for all, as examples given below illustrate.

Male behaviors known for certain to be parts of courtship and mating, rather than aggressive or territorial displays, include tail-wagging, licking (sometimes mutual), leg-rubbing, rubbing the cloaca against the ground, and demonstrating bright-colored gape. In some species, the courtship is very simple and consists of the male following the female until she stops and allows herself to be held by the neck. In the striped forest whiptail (*Kentropyx calcarata*) such following lasts for only a few minutes (Costa et al., 2013), but in the desert monitor (*Varanus griseus*) it can go on for a few days (Tsellarius & Tsellarius, 1996). Females of some lizards, such as the Namib rock agama (*Agama planiceps*), have been observed soliciting sex from males (Heideman, 1993).

In some species, including the common ameiva (*Ameiva chrysolaema*), *Plestiodon* skinks, and many anoles, homosexual matings are common, and sometimes two males would penetrate each other's cloacae in turn. G. K. Noble and Bradley (1933) were the first to report such matings and suggested that they can be explained by the inability of males to distinguish between other males and females.

Lizard mating behavior has been the subject of much research, starting with the work of Gachet (1833). Most of that research was on the dynamics of mating systems (see Chapters 3 and 4) and on hormonal regulatory mechanisms. Still, pre-mating and mating behaviors are well documented for only a small percentage of the known species (their total number now over 6,000), and no recent overview is available. Hormonal studies have mostly been conducted on anoles and, to much lesser extent, on racerunners (*Aspidoscelis*); they have shown that high levels of testosterone are required for male courtship behavior (P. Mason & Adkins, 1976; Lindzey & Crews, 1986). However, this does not appear to be true for some other species, such as the Yucatan banded gecko (*Coleonyx elegans*) (Golinski et al., 2011). Female anoles show consistently elevated levels of progesterone throughout the mating season (which is a few months long), while their levels of estradiol-17β and testosterone fluctuate with the ovulatory cycle (Lovern et al., 2004).

Multiple paternity and sperm storage are common in lizards (Tolley et al., 2014; see also Chapter 3). It has been suggested (Passek, 2002) that females that have mated (voluntarily or not) with multiple males can exercise "cryptic female choice" by using only the sperm of the male(s) they like for fertilization; supporting evidence is so far only available for some anoles (Calsbeek & Bonneaud, 2008).

An overview of anole natural history by Losos (2009) lists many interesting facts about anole mating behavior, illustrating its surprising diversity. The length of courtship diminishes if the male and the female are well familiar with each other; in some anoles this has evolved into long-term pair bonding. The length of intromission varies more by species, from just a second in the Webster's anole (*A. websteri*) to more than half an hour in the clouded anole (*A. nebulosus*). Females of some species become non-receptive after one mating and remain that way until the next ovulatory cycle, while others can mate many times per day. Some species do not seem to be choosy about the location of mating, but others would preferentially mate in either conspicuous places or secluded ones.

Despite many decades of research (see Losos, 2009, for bibliography), there is still a lot of controversy surrounding anole mating behavior, particularly the existence of female choice and the roles of particular display elements in courtship. A study by L. T. Evans (1938) found that females choose the male that extends his dewlap more often. However, the role of dewlap display in courtship has been questioned (Tokarz, 2002). This controversy might be a result of the shortcomings of some experimental settings, or of inherent limitations of studies in captivity in general. The suggested role of dewlap color in species recognition has also been questioned (see Chapter 4, section 4.4).

Many lizards can have trouble telling apart their own species and closely related ones, even if there are obvious color differences. For example, male Broadley's flat lizards (*Platysaurus broadleyi*) fall for particularly large female Cape flat lizards (*P. capensis*), even if females of their own species are present (Wymann & Whiting, 2003). Perhaps for this reason, some species groups have evolved mechanical barriers to intraspecific mating (Richmond et al., 2011). Male common chameleons (*Chamaeleo chamaeleon*) will attempt mating with skinks, agamas, and even turtles (G. K. Noble & Bradley, 1933).

Monitor lizards (*Varanus*) are known for spectacular ritualized fights between males (Fig. 5.4), and these have been summarized by Earley et al. (2002) from a game theory perspective. Details vary between species, but in the most spectacular version used by larger monitors, two animals embrace each other with their front legs while standing upright and try to push each other out of balance. Smaller species often forgo the upright pushing, instead engaging in circular chases and embraces that often are mistakenly taken for mating. According to Earley et al. (2002), the winner then mounts his adversary in a simulation of mating; however, first-hand observations of such behavior seem to be lacking. Similar fights have also been observed between females and between a male and a female, but much more rarely. Combat in the Kimberley rock monitor (*Varanus glauerti*) can last nearly two hours (JSD pers. obs.). Phylogenetic mapping (Schuett et al., 2009) shows that bipedal fighting has most likely evolved once in the ancestor of *Varanus* and was later lost in some lineages. As these "fights" are often non-damaging and do not involve overt biting and injury, the possibility that they may represent play fighting, as found in so many mammals, should not be ruled out; certainly, we know that monitors engage in object play (Burghardt et al., 2002; Burghardt, 2005). Courtship displays in monitors, if present, are also a bit unusual: for example, male rock monitors (*V. albogularis*) shudder spasmodically, to which the female responds by flattening her body (Phillips & Millar, 1998). Fights between males of related beaded lizards (*Heloderma* spp.) are similar to those of smaller monitors, with back-arching and leg-pushing but no bipedal standing; however, their movements are much slower, and bouts of fighting can last for over an hour (Beck & Ramirez-Bautista, 1991). Male earless monitors (*Lanthanotus borneensis*) often have scarred faces, presumably from fights (Langner, 2017).

There is some (limited) evidence for mate guarding in lizards. In the lek system of green iguanas, male territories are often arboreal and in trees

Fig. 5.4. Male-to-male combat in squamates: *left,* lace monitors (*Varanus varius*); *right,* Northern Pacific rattlesnakes (*Crotalus oreganus*). Photographs by (*left*) John E. Hill (Creative Commons Attribution–Share Alike License) and (*right*) Dawn Endico (Creative Commons Attribution–Share Alike License).

that can be defended and where competitors seen. In Panama and Venezuela, these can be barren trees in which females congregate with a dominant male. The leafless tree clearly does not offer food or defense against aerial predators, but it does provide a safe harbor from marauding low-ranking males (Dugan, 1982b; Rodda, 1992). The geometry of the trees often facilitates male mate guarding with female escape routes limited or blocked (Fig. 5.5).

In an intensive study of >2,000 copulations of 60 pairs of Rosenberg's monitor lizards (*Varanus rosenbergi*), Rismiller et al. (2010) found that the male would enlarge the burrow of the female and roost with her. After two to seven days the pair would initiate copulation and would cohabitate in the nuptial burrow while copulating for an additional 7–17 days. In most cases the primary male would chase any intruding males away from the female and nuptial burrow, but he rarely engaged in physical contact. In several cases, however, two males shared a female with no aggression, and

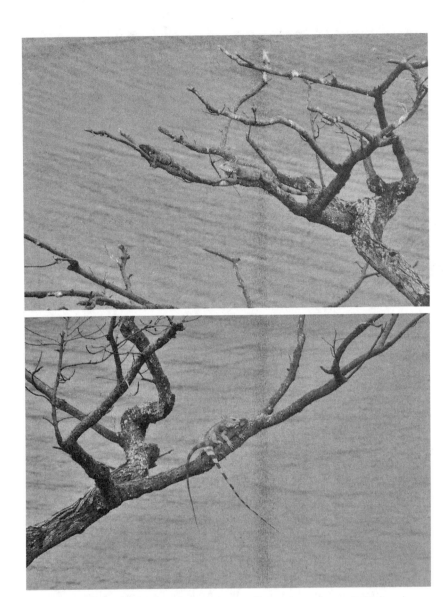

Fig. 5.5. Top: A harem-holding male green iguana (*Iguana iguana*) guarding a female from leaving his treetop mating territory. Females often move among territories, as a male can only mate once a day, even if he has a territory with 10 or more females. Females not in a dominant-male territory suffer from harassment and forced copulations by both small female-mimicking males and large peripheral males without territories. *Bottom*: Male green iguana in same territory mating in a tree both exposed and high above the water on Flamenco Island in the Pacific Ocean outside Panama City, Panama. Photographs by Beverly Dugan.

in one case two males sharing a female tolerated a third male to mate with her (monogamy within a season ranged from 40% to 80%). Two to three weeks after copulation males began to forage but regularly visited females. Females were never seen actively foraging or hunting before egg-laying, but on one occasion a male dragged a fresh wallaby carcass to within 5 m of the nuptial burrow, and over a five-day period both male and female fed from the carcass. Three to five weeks later females nested in termite mounds and stayed with the nest and defended it for up to three weeks, and the primary male stayed with the female in 8 of 85 cases. Both primary and secondary males defended nest sites against intruding males, which maraud the eggs. In six instances pairs spent the winter together in the same burrow (Rismiller et al., 2010). An online video reveals another fascinating case of potential mate guarding; a flying snake (*Chrysopelea*) had several coils around a tokay gecko (*Gekko gecko*) in a house in Thailand when a second gecko foiled the predation attempt by circling and biting the snake (https://www.youtube.com/watch?v=n4JBeUV6uvk). The attacking gecko was clearly a male (swollen area near the cloaca), but the sex of the attacked gecko cannot be determined from the video. The most parsimonious explanations for the behavior are mate guarding (if the geckos are a mating pair), kin altruism (if they are relatives), or reciprocal altruism (if they are unrelated). All three are virtually unknown in reptiles (but see the next section for potential examples of mate guarding in snakes).

5.7. Snakes: Fistless Fights and Armless Embraces

Males of many snake species engage in one-to-one fights. Such combat behavior has been known to humans for a long time; there are Hopi and Yomud Turkmen dances imitating male-to-male combat in rattlesnakes (*Crotalus*) and the Turkestan cobra (*Naja oxiana*), respectively (G. W. James, 1913; Kurbanov, 1985), and a remarkable number of amateur videos of wild and captive snakes engaged in combat can be found on YouTube. However, the literature on these fights is very limited compared to that on lizards. The following account is mostly based on overviews by C. C. Carpenter (1977) and Rivas and Burghardt (2005). At the time the first of these overviews was published, its author estimated that "combat ritualistic behavior [had] been noted for only 5.5% of the genera and just under 2% of the known species of snakes, and courtship and mating for 8% and 3% respectively" (C. C. Carpenter, 1977, p. 218), with almost all of these observations conducted in captivity. This was a gross underestimation, but not surprisingly: as the authors of the second overview have noted, although more than 3,000 species of snakes were known at the time of Carpenter's observation, almost all detailed research on mating behavior was still focused on just one outlying population of one species, the common gartersnake.

Lack of limbs and streamlined body shape apparently had a profound influence on the evolution of snake mating behavior. Embraces and bite-holds used by males of other reptiles to subdue the females became all but impossible (although a few colubrid snakes do use bite-holds). Only pythons (*Pythonidae*) and boas (*Boidae*) still have spur-like hind leg remnants; these are more developed in males and are sometimes used during mating for grasping and stimulating females, but they are too small to hold an unwilling female in place. All snake sex appears to be consensual (although a study by Shine, Langkilde & Mason, 2003a, claims that this is not so; see below). This difference from many lizards and turtles, in which females are often forced into mating (for example, by repeated biting), might explain why in many snakes females are larger than males, while in other reptiles, males are usually (but not always) larger. Snakes also are rarely territorial or have dominance-based social systems. The exceptions, predictably, are snake species that practice male-to-male combat, apparently putting males under selection for larger size (Shine 1978, 2003). As larger female reptiles often have larger litters and clutches than smaller females, female snakes freed from male patriarchic dominance may be more able to exploit this reproductive advantage (Rivas & Burghardt, 2005).

Male-to-male combat is known in snakes belonging to three major lineages (Schuett et al., 2001). In many boas and pythons, the males intertwine the posterior parts of their bodies. The anterior parts are sometimes kept separate and used to gain better support but can be intertwined as well. Then the males tighten their grip on each other as they wiggle and rotate, trying to scratch the opponent with their spurs, constrict, or bite him. The loser is chased as he tries to escape and can be attacked again (C. C. Carpenter et al., 1978; D. G. Barker et al., 1979; Pizzatto et al., 2006; Pizzatto & Marques, 2007). In snakes belonging to the group Caenophidia, which is often considered the advanced lineage and includes the vast majority of extant taxa, the combat is more ritualized. Two males usually intertwine the middle portions of their bodies and raise the anterior parts at a certain angle above ground. That angle differs between taxa—the anterior parts could be almost horizontal, like in kingsnakes (*Lampropeltis*), bronzebacks (*Dendrelaphis*), and the puff adder (*Bitis arietans*); close to 45°, like in Turkestan cobras and asp vipers (*Vipera aspis*); or almost vertical, like in many rat snakes (*Elaphe, Pantherophis, Ptyas*), as well as in rattlesnakes (*Crotalus, Sistrurus*) and other pitvipers (see Fig. 5.4). In the upright posture the combatants sway back and forth, each trying to position his head above the head of his opponent and press him down. This might end the fight or lead to another round; some males "dance" in this manner for an hour or more. In vipers (Viperidae), the fight is often preceded by one male displaying: he rises the anterior portion of his body vertically, as if inviting the opponent to join the fight (C. C. Carpenter, 1977; Schuett, 1997; VD, pers. obs.). Studies on copperheads (*Agkistrodon contortrix*) have shown that losing a fight has a long-term effect on the male. The loser is rendered incapable of further fighting and, just as significantly, courtship for many hours or even days, possibly due to stress-induced suppression of circulating hormone levels (particularly of testosterone), elevated levels of circulating corticosterone, or both (Schuett, 1996).

Senter et al. (2014) proposed the following scenario of male-to-male combat evolution in snakes. Raising heads and attempting to topple an adversary evolved in the Late Cretaceous in the common ancestor of boas, pythons, and Caenophidia snakes. Boas then added poking with spurs, while Lampropeltini (kingsnakes and relatives) switched to coiling without neck-raising.

Snake combat is somewhat similar to that of monitors and beaded lizards (see above). Both monitors and beaded lizards belong to the lizard clade most closely related to snakes, and it is tempting to suspect that snakes have inherited their combat style from their lizard ancestors. However, such combat behavior is not known in snake lineages currently considered basal (see analysis by Schuett et al., 2001), so Rivas and Burghardt (2005) suggested that snake combat must have evolved independently. As virtually nothing is known about the behavior of basal snake lineages, combat behavior might still be discovered, especially considering that some of these snakes, such as the American pipe snake (*Anilius scytale*), have spur-like protrusions of vestigial pelvic bones that could be used in combat (Harrington & Reeder, 2017).

Male-to-male combat is relatively common in some groups of snakes (Shine, 1978; Schuett et al., 2001). In an overview of mating behavior of Italian snakes (all but one belonging to Caenophidia), Capula and Luiselli (1997) found that all species for which sufficient data were available exhibited this behavior, except for grass snakes (*Natrix*), in which males only occasionally pushed each other while pursuing a female. Elsewhere, many snakes do not engage in one-to-one fights, but instead chase females in groups and eventually form so-called mating balls. Shine (1978, 2003) found that in species practicing combat, males tend to be larger than females, while in those forming mating balls, males are either smaller or the same size as females. Male-to male combat in snakes appears to be more ritualized than in other reptiles: there is hardly any evidence of biting or injuries.

Courtship in all snakes for which information is available begins with the male aligning his body with that of the female. Male boas and pythons use their spurs to stroke, scratch, or vibrate the female's cloacal area. This causes the female to stop moving, turn the subcaudal part of her body sideways, and open the cloaca, allowing the male to insert one of his hemipenes. In the Colubridae and Elapidae families (recently merged, see Chapter 2), a male will reach the same effect by caressing the female's back with his chin, embracing her with his tail, "searching" her body with his tail tip, slightly biting her, or pressing to her with waves or loops running along his body, either from head to tail or vice versa. Shine, Langkilde, and Mason (2003a), who studied common gartersnakes, suggested that those waves running along the male's body interfere with the work of the female's lung (snakes have only one), causing her to begin suffocating and open her cloaca in distress, thus allowing the male to forcibly penetrate her. Male vipers chase or approach the female with their head raised, then entwine with her or climb on top of her and "search" her body with their tail. Male pitvipers (Crotalinae) add jerking movements of their body and slight prodding of the female with their snout.

Senter et al. (2014) proposed that the ancestral form of courtship in boas involved the male rubbing the female with his spurs (still observed in some species), while in Caenophidia, courtship ancestrally involved chin rubbing and head or body jerking. Later various groups added other behaviors, such as moving undulations in gartersnakes, kingsnakes, and related taxa, coital neck biting in Old World rat snakes (*Elaphe*), and tail quivering in New World rat snakes (*Pantherophis*).

It is believed that male-to-male combat is preceded mostly by visual contact and postures, while courtship behavior is initiated after exchanging chemical signals. Indeed, males often have to track the female for hours before getting within visual range (Kurbanov, 1985). Successful insertion of a hemipenis requires elaborate tactile signaling, so that the female opens her cloaca just as the male is ready to penetrate her. It is possible that species-specific sequences of these signals sometimes act to prevent hybridization.

The most elaborate courtship behavior observed to date in snakes takes place between male and female Montpellier's snakes (*Malpolon monspessulanus*). According to De Haan (1999 and unpubl. data), a male (which in this species is usually much larger than the female) establishes a long-term bond with a female he has copulated with and guards her continuously until a few days prior to oviposition (Plate 5.3), aggressively protecting her (even from large predators such as humans) and sometimes helping with prey capture. There is anecdotal evidence that in some populations of this species, the guarding male provides the female with freshly killed prey (small rodents); this behavior has reportedly been filmed for a documentary (Schmedes, 2002). De Haan (1999) also noticed that large males become territorial during the mating season, but they tolerate the presence of a few small males that act submissively, suggesting possible existence of "sneaker males." *Malpolon* and related genera (tribe Psammophini) are unique among snakes in having smooth, slender hemipenes (De Haan,

1999), as well as unusual glands, possibly used for scent-marking themselves and their mates (see Chapter 4, section 4.7); the significance of these traits is unknown.

Once inserted, the hemipenis in many species of snakes apparently locks into position (being held in place by backward-facing spines), ensuring successful copulation (Friesen et al., 2014); the male then remains motionless on top of or alongside the female, although in some species there are slight movements of the tail and adjacent portion of the body (Schuett & Gillingham, 1988). The locking mechanism appears to be particularly strong in the yellow-bellied sea snake (*Hydrophis platurus*), which mates in the open ocean, with the male wrapped around the female: a pair captured during copulation spent 1.5 hours trying unsuccessfully to swim apart (Vallarino & Weldon, 1996). Copulation in snakes can last for a long time: calico snakes (*Oxyrhopus petola*) can remain attached for four hours (Zacariotti & del Rio do Valle, 2010), and Indian pythons (*Python molurus*) up to six hours (D. G. Barker et al., 1979). Evidence suggests that this is a form of mate guarding (Wilmes et al., 2011). Some species, such as the common gartersnake, leave a gelatinous mass called copulatory plug in the female's cloaca (see Chapter 4, section 4.7). This plug is often said to be another form of mate guarding, but recent data suggests that it is a kind of spermatophore, holding sperm and slowly releasing it as the gelatinous matrix gradually dissolves (Friesen et al., 2013). The discovery of high rates of multiple paternity in snakes with plugs suggests that the plug is not an effective deterrent (Schwartz et al., 1989). Female big-eyed pitvipers (*Trimeresurus macrops*), the only truly arboreal snakes in which mating has been observed in the wild, sometimes drag males from the ground into the trees (Strine et al., 2018).

The idyllic picture of the male patiently following the female and seducing her by tender stimulation breaks down into apparent chaos in species that form mating aggregations, such as anacondas (*Eunectes*), Nearctic water snakes (*Nerodia*), and the common gartersnake (Plate 5.4). The female at the center of an anaconda mating ball spends weeks being surrounded by a mass of males that frequently try to stimulate her by tickling her with their spurs, fight for proximity to her cloaca, and tighten their collective embrace defensively every time yet another male tries to join them. If a male manages to achieve penetration, he remains attached to the female for more than an hour. However, Rivas et al. (2007) have found that there might be some order in the chaos: their observations suggest that the female somehow manages to choose mates and helps the ones she likes to maintain intromission longer (see also Rivas, 2020). Very long, thin snakes sometimes form mating braids rather than mating balls (Kaiser et al., 2012).

Sperm storage is common in snakes (Sever & Hamlett, 2002; Aldridge & Sever, 2016), but it is not known if females can opt to use sperm only from some of the males they have mated with. Some species are capable of parthenogenesis (see Chapter 2, section 2.10), but only one, the Brahminy blind snake (*Indotyphlops braminus*), is known to be obligatory parthenogenetic.

6 |

Communal Egg-Laying

Habitat Saturation or Conspecific Attraction?

Conspicuous communally laying reptiles have attracted the attention of herpetologists and ecologists for a very long time. Familiar examples include the massive arribadas of sea turtles, island-nesting iguanas, and the tuatara. But the majority of reptiles are small lizards that can fit in one's hand, with natural histories that are anything but conspicuous, including their egg-laying habits. Finding the nests of these diminutive species is challenging, and as a result the gaps in our knowledge are massive and widespread. However, there has been a paradigm shift in our understanding of the social complexity that is thought to underpin communal egg-laying in reptiles. Conventional wisdom has evoked habitat saturation, or limited nest sites, as the primary cause, but in the last 10–15 years evidence has mounted for conspecific attraction to eggs, nests, or mothers, forecasting the revelation of more complex social behaviors than ever imagined, though anticipated over 50 years ago by Stan Rand (A. S. Rand, 1968). Suddenly, small, plain, secretive lizards (e.g., skinks, geckos), largely ignored by herpetologists, offer exciting new systems for studying social behaviors associated with reproduction. Although arguments have been made that viviparity is a major reason for the evolution of complex sociality in reptiles, it cannot explain everything. Birds are all egg-layers and have social and family lives far more complex than many viviparous mammals. Many snakes are viviparous as well. While viviparity may present opportunities for parental care and family bonding, care of nests and eggs can do likewise, as in crocodylians and birds. Communal nesting may also be a cause, as well as a result, of social evolution. Here we take a look into the secret lives of communally laying reptiles.

6.1. What Constitutes Communal Egg-Laying?

Communal egg-laying involves mothers laying eggs in a common area or adding their eggs to those of conspecifics. It is widespread in animals, occurring in all vertebrate classes and in several classes of invertebrates (see references in Doody, Freedberg & Keogh, 2009). The antiquity and persistence of communal egg-laying is evident in its occurrence in dinosaurs (Horner, 1982; K. Carpenter & Alf, 1994; K. Carpenter, 1999; Horner, 2000; Varricchio et al., 2008).

Communal egg-laying can occur at different spatial scales; conspecific eggs can be touching in the same "nest" under cover objects or in a crevice or hole, or individual nests can be clumped at a larger scale (i.e., buried nests within a small area such as a beach). Distinguishing between these scales can be nominally useful in interpreting the causes and consequences of communal egg-laying (Espinoza & Lobo, 1996; Doody, Freedberg & Keogh, 2009). Annobon dwarf geckos (*Lygodactylus thomensis*) laid communally in the mud-nests of wasps, but each chamber included a clutch from only one mother (Rodriguez-Prieto et al., 2010). We use "communal egg-laying" throughout this chapter because many reptile species do not construct a nest in the strict sense (Doody, Freedberg & Keogh, 2009). There are other terms used interchangeably with communal egg-laying, such as communal nesting, communal oviposition, colonial nesting, and egg dumping (reviewed in Doody, Freedberg & Keogh, 2009). Live-bearing animals can also give birth communally (Manning et al., 1995; L. D. Hayes, 2000); indeed, many reptiles are viviparous, and the causes, cues, and consequences of communal gestation and communal birthing may be similar to those of communal egg-laying (see review in Graves & Duvall, 1995). However, we mainly focus on oviparous species in which multiple clutches have been found in a single location.

Our knowledge of nesting biology varies across reptile groups because some leave clues and others do not. The most obvious nests are those of beach-nesting turtles, whose tracks in the sand lead directly to the nest, and crocodylians that create large conspicuous mounds. We thus know more about nesting in turtles and crocodylians than we do about nesting in rhynchocephalians, lizards, and snakes, which often nest underground, in rock crevices, or under cover. Although there are exceptions, we know very little about nesting in snakes in general (Shine, 1991; Braz & Manço, 2011), and what we do know comes largely from captive breeding. There are other interesting differences in nesting among reptilian groups, but it remains to be seen if these differences play a role in the evolution or maintenance of communal egg-laying. In contrast, communal nesting in birds has been studied extensively (above references).

Although all turtles, crocodylians, and the tuatara (*Sphenodon punctatus*) lay eggs, many lizards and snakes give birth to live young. Do they show anything comparable to communal nesting? We touch on this interesting point below, but it is known that pregnant females are sometimes found in aggregations during gestation (Reichenbach, 1983; Graves & Duvall, 1995).

6.2. Why Do Animals Lay Eggs Communally?

Despite the prevalence of communal egg-laying, and despite considerable research effort, a consensus on an explanation for the evolution of communal egg-laying in vertebrates remains elusive (Doody, Freedberg & Keogh, 2009). Although hypotheses have been raised as to why communal egg-laying occurs in some groups (e.g., Ward & Kukuk, 1998; Tallamy, 2005), the adaptive value of this behavior in the vast majority of cases remains unclear (reviewed in Siegel-Causey & Kharitonov, 1990; Danchin & Wagner, 1997). The remarkable diversity in life histories and natural histories of communally egg-laying groups presents a great challenge to discovering any generality for communal egg-laying (Doody, Freedberg & Keogh, 2009). For example, while communally nesting birds might be expected to benefit from cooperative defense, these benefits would not apply to animals where mothers abandon their offspring shortly after oviposition. In another example, in some communally egg-laying insects, mothers are faced with balancing trade-offs such as leaving the nest to forage versus eating the egg of a conspecific and then replacing the eaten egg with one of their own (Ward & Kukuk, 1998). The selective regimes associated with communal egg-laying are thus likely to differ among groups with diverse attributes, such as parental care (mammals, birds), brood parasitism (snails, beetles, bugs, bees, fishes, birds), communal spawning (fishes), male interference (frogs, dragonflies), and host plant specificity (butterflies). Despite this diversity, some benefits of communal egg-laying may be pervasive, such as a reduction in predation risk (Foster & Treherne, 1981; Sweeney & Vannote, 1982) or savings in time or energy (Wiewandt, 1982; Danforth, 1991). Adaptive benefits of the behavior are often assumed for traditionally more "socially complex" animals such as birds, but in other groups such as reptiles authors often invoke limited suitable nest sites as the cause (reviewed in Graves & Duvall, 1995). However, recent laboratory experiments, along with a field experiment (JSD, unpubl. data), demonstrate conspecific attraction to eggs in reptilian mothers, further revealing the potential for advanced social behavior and decision-making in animals that are often branded as "asocial" (Doody, Freedberg & Keogh, 2009; Doody, Burghardt & Dinets, 2013).

Reptiles can serve as models for investigating the causes, costs, and benefits of communal egg-laying, without many of the confounding factors found in other groups. For example, postnatal parental care is currently known to occur in only 3% of snakes and 1% of lizards and turtles (Shine, 1988; Iverson, 1990; Crump, 1995). Other factors, such as brood parasitism, broadcast spawning, male interference, and host plant specificity, have not been documented in extant reptiles (Doody, Freedberg & Keogh, 2009). Moreover, where mating and egg-laying are simultaneous (e.g., frogs), mate choice can confound nest site choice, and thus the decision to oviposit communally. In reptiles, the father plays no role in the choice of a nest site because mating and egg-laying are typically temporally dissociated, and pair bonding is very rare in egg-layers (some species of crocodiles being an exception, see Chapter 7, section 7.1). In reptiles, the decision to lay communally is entirely up to the mother, reducing the number of potential ecological and evolutionary explanations. Finally, in some groups of birds, communal egg-laying is obligatory (Seigel-Causey & Kharitonov, 1990). In reptiles, there are both solitarily and communally nesting individuals within species and within populations (Hirth, 1980; Eckrich & Owens, 1995; Bernardo & Plotkin, 2007; reviewed in Doody, Freedberg & Keogh, 2009), providing scope for revealing comparative costs and benefits of solitary versus communal egg-laying within a population. Moreover, social interactions between nesting reptilian mothers may be representative of early stages in the evolution of more complex social systems (Graves & Duvall, 1995; Doody, Burghardt & Dinets, 2013). Thus, reptiles are theoretically ideal for discovering any fundamental explanation for the evolution of communal egg-laying in animals.

6.3. Taxonomic Distribution of Communal Egg-Laying in Reptiles

How widespread is communal egg-laying in reptiles? Doody, Freedberg, and Keogh (2009) reviewed the literature for the prevalence of communal laying in extant reptiles and amphibians, finding it in 345 species of reptiles (Table 6.1). The proportion of reptiles laying communally was 6%, and averaged 21% by major group (range <1%–100%). Although the proportion of communally laying species is modest for each group (Table 6.1), this proportion may greatly underestimate its prevalence because the eggs and nests are unknown for a great number of species (Test & Heatwole, 1962; Shine, 1991; Perry & Dmi'el, 1994; Doody, Freedberg & Keogh, 2009). Although the difficulty in locating nests hampers our ability to determine the actual frequency of communal egg-laying, Doody, Freedberg, and Keogh (2009) pointed out that we can better estimate this proportion by dividing the number of known communally laying species by the total number of species, *excluding those for which eggs have not been found*. Indeed, in their analysis of three families of Australian lizards (Gekkonidae, Pygopodidae, and Scincidae), the proportions of species in each group rose from 4%–9%

to 73%–100% when excluding species for which nests were unknown (Table 6.2). This analysis had two important implications: (1) it strongly suggests that communal laying is much more common than previously realized; and (2) it highlights our inadequate knowledge of the eggs and nests of reptiles. Since that review, communal laying continues to be documented for additional reptile species (e.g., de Sousa & Freire, 2010; Pike et al., 2010; Rodriguez-Prieto et al., 2010; Lima et al., 2011; Luiselli et al., 2011; Montgomery et al., 2011; Mateo & Cuadrado, 2012; Doody, Castellano, Rhind & Green, 2013; Filadelfo et al., 2013; Ramos-Pallares et al., 2013; Rangel & Lopez-Parilla, 2014; Robinson et al., 2014; Doody, James, Colyvas, et al., 2015; García-Roa et al., 2015; Peñalver-Alcázar et al., 2015; Bernstein et al., 2016; Rojas et al., 2016; Doody, Coleman, et al., 2017; Calderon-Espinosa et al., 2018; Doody, McHenry, Brown, et al., 2018; Doody, McHenry, Durkin, et al., 2018; Doody, Zavala, et al., 2018; Carvajal-Orcampo et al., 2019).

Among extant reptiles, communal laying was reported in 255 species of lizards, 52 species of snakes, 30 species of turtles, 6 species of crocodylians, and the tuatara (Table 6.1) (Doody, Freedberg & Keogh, 2009). Communal laying is not known in amphisbaenians (worm lizards), but their eggs in the wild are virtually unknown, and the ~190 species of amphisbaenians are believed to be oviparous (Zug et al., 2001; Andrade et al., 2006). The incidence of communal laying in lizards and snakes did not correspond with the total relative numbers of those species worldwide: lizards = over 6,680 spp.; snakes = over 3,600 spp. (Uetz et al., 2020). However, communal laying in snakes may be much more common than is indicated by Table 6.1, and this is supported by many accounts of communal snake clutches in subterranean sites or other cryptic sites such as within hollow trees, ant nests, or bromeliads (e.g., Plummer, 1989; P. B. Whitaker & Shine, 2002, 2003). In many cases, these nests were found only by radio tracking gravid snakes (e.g., Madsen, 1984; Plummer, 1989, 1990; P. B. Whitaker & Shine, 2002, 2003; Blouin-Demers et al., 2004; Cunnington & Cebek, 2005). In contrast, many communal lizard clutches are found under rocks, logs, or other objects that are easily moved by an observer (the space under such objects is often too small for larger snakes).

Table 6.1. Family-level proportions of reptiles known to lay communally

Taxa with known communal nesting	Number of species	Number of oviparous species	Number known to nest communally (% of oviparous)
Reptilia			
Sauria			
Agamidae	381	375	3 (<1)
Anguidae	112	34	1 (3)
Chamaeleonidae	161	134	1 (1)
Cordylidae	54	15	2 (13)
Gekkonidae	1,076	1,057	129 (12)
Gerrhosauridae	32	32	2 (6)
Gymnophthalmidae	193	193	11 (6)
Iguanidae	36	36	13 (36)
Lacertidae	279	278	9 (3)
Phrynosomatidae	125	125	5 (4)
Polychrotidae	393	393	15 (4)
Pygopodidae	36	36	4 (11)
Scincidae	1,305	~744	41 (6)
Teiidae	121	121	7 (6)
Tropiduridae	309	224	12 (5)
Total lizards	4,232	3,422	255 (7)
Serpentes			
Colubridae	1,827	~1,663	40 (2)
Elapidae	138	137	6 (4)
Hydrophiidae	177	53	2 (4)
Leptotyphlopidae	93	93	3 (3)
Pythonidae	74	35	1 (3)
Total snakes	2,309	1,981	52 (3)
Chelonia			
Carettochelydidae	1	1	1 (100)
Cheloniidae	6	6	6 (100)
Chelydridae	2	2	1 (50)
Dermochelydae	1	1	1 (100)
Emydidae	41	41	10 (5)
Geoemydidae	69	69	3 (4)
Kinosternidae	25	25	2 (4)
Podocnemididae	8	8	4 (38)
Trionychidae	30	30	2 (7)
Total turtles	183	183	30 (16)

Table 6.1. (continued)

Taxa with known communal nesting	Number of species	Number of oviparous species	Number known to nest communally (% of oviparous)
Crocodylia			
Alligatoridae	8	8	1 (11)
Crocodylidae	16	16	4 (31)
Gavialidae	2	2	1 (100)
Total crocodylians	23	23	6 (26)
Rhynchocephalia			
Sphenodontidae	1	1	1 (100)
Total tuatara	1	1	1 (100)
Total reptiles	6,746	5,608	332 (6)

SOURCE: Adapted from Doody, Freedberg, and Keogh (2009), with more recent findings added.

Among lizards, communal laying was reported for 15 of the 26 recognized families (Table 6.1) (Doody, Freedberg & Keogh, 2009). According to their review, 51% ($N = 129$ spp.) of all communally laying lizards were geckos (Doody, Freedberg & Keogh, 2009). Geckos generally do not construct a nest but instead typically lay eggs under bark, in crevices in tree trunks or rocks, under rocks or logs, under cap rocks of granitic boulders, or in protected areas of caves and human dwellings (e.g., Krysko et al., 2003; Oda, 2004). Skinks were second to geckos in the incidence of communal laying (16%; $N = 41$ spp.). In contrast to geckos, skinks often construct a nest—generally a slight depression in the substrate, usually beneath some cover object, such as a rock or log, or beneath the leaf litter (e.g., Greer, 1989; Couper & Ingram, 1992)—although some species do not (Ota et al., 1989; Shea and Sadlier, 2000; Rodriguez-Prieto et al., 2010). The remaining 33% of communally laying lizards are spread across the other 13 families (<1%–6%).

Among snakes, communal laying is known in 5 of 18 families (Table 6.1) (Doody, Freedberg & Keogh, 2009). However, at least two of the snake families not known to lay communally contain only viviparous species (Acrochordidae, Aniliidae), and still others contain mostly viviparous species (Hydrophiidae, Tropidophiidae, Viperidae) (Uetz et al., 2020). Most (77%)

Table 6.2. Prevalence of communal egg-laying in three families of Australian lizards

Group (N)	Proportion known to lay communally (N)	Proportion known to lay communally, excluding species for which eggs/nests unknown
Gekkonidae	9% (10)	100% (3)
Pygopodidae	8% (3)	100% (10)
Scincidae	4% (11)	73% (15)
Combined	6% (24)	86% (28)

SOURCE: Adapted from Doody, Freedberg, and Keogh (2009).

communal snake clutches reported were of colubrids ($N = 40$ spp.). Only two snake species, the pine snake (*Pituophis melanoleucus*) and the eastern hognose snake (*Heterodon platirhinos*), are known to excavate their own burrows explicitly for nesting, and both nest communally (Burger & Zappalorti, 1991, 1992; Cunnington & Cebek, 2005).

There is no evidence of communal egg-laying among extinct lizards or snakes, probably because most lay eggs with soft shells that are seldom found as fossils (Fernández et al., 2015).

Much like the iguanas, the tuatara (*Sphenodon punctatus*) is a communal nester and excavates and backfills nest chambers (Cree et al., 1995).

Communal laying is evident in at least 9 of 13 turtle families ($N = 30$ spp.; Table 6.1) (Doody, Freedberg & Keogh, 2009). E. O. Moll (1979) suggested that communal egg-laying is associated with large aquatic species that lay eggs in well-defined ancestral nesting areas. Sea turtles are a classic example, with all seven recognized species nesting communally (Carr, 1967; Bjorndal, 1979). The highest number of species nesting communally in any turtle family are the emydids or basking turtles ($N = 10$ spp.), which often nest on sandbanks or beaches. In at least three freshwater species that lay communally, gravid females migrate to or move together among nesting beaches, indicating that information is shared among conspecific mothers (Mosqueira Manso, 1960; Doody, Georges & Young, 2003; Doody, Sims & Georges, 2003; Vogt, 2008). Most turtles excavate and backfill a nest in the ground (e.g., Doody, Georges & Young, 2003; Horne et al., 2003). Turtle eggs have variably hard shells, but (uniquely among vertebrates) these shells are made of aragonite rather than calcite, so they do not fossilize well and are rare in the fossil record (Lawver & Jackson, 2014). The only known evidence of communal egg-laying in extinct turtles is one suspected Miocene communal nesting site possibly belonging to *Bairdemys venezuelensis*, a large freshwater or marine turtle from Venezuela (Winkler & Sánchez-Villagra, 2006).

Six species of extant crocodylians, representing all three families, are reported to nest communally (Doody, Freedberg & Keogh, 2009; Table 6.1); although in one species multiple clutches in a nest may have been consecutive clutches from the same female (G. J. W. Webb et al., 1983), and another species might no longer nest communally due to lower population densities (Cott, 1961; Pooley, 1982). There is fossil evidence of communal egg-laying in Cretaceous crocodylomorphs (Marsola et al., 2016).

Evidence of communal egg-laying sites, sometimes with thousands of eggs, exists for numerous non-avian dinosaurs (but it is important to remember that multiple fossil nests found at the same location might not have been used at the same time). Interestingly, although the oldest such site is from the Early Jurassic, almost all others are from the Late Cretaceous (Reisz et al., 2012), suggesting that some environmental change might have caused such behavior to become more common. Just like many modern seabirds, some non-avian dinosaurs might have formed interspecific colonies (Sellés et al., 2013), sometimes possibly including crocodylomorphs and mammals (Srivastava et al., 2015; Weaver et al., 2020); similar behavior is known in extant reptiles (see below). There is also evidence of communal egg-laying in pterosaurs (X. Wang et al., 2014).

Among extant birds, about 13% of all species breed in colonies, ranging from loose colonies of a few breeding individuals in white storks (*Ciconia alba*) to dense aggregations of over 4 million burrow-nesting sooty shearwaters (*Puffinus griseus*) (Gill, 2007; Reyes-Arriagada et al., 2007). There is also evidence of colonial breeding in extinct birds, including some from the Late Cretaceous (Dyke et al., 2012).

In mammals, no extinct or extant egg-laying species is known to breed communally. As for viviparous species, giving birth communally is common only in bats and pinnipeds, with breeding aggregations reaching

20 million individuals in Mexican free-tailed bat (*Tadarida brasiliensis*) (Nowak, 2018). The percentage of extant mammalian species giving birth communally is difficult to estimate because breeding biology of many tropical bats is poorly known.

6.4. Factors Affecting the Taxonomic Distribution of Communal Laying in Reptiles

Doody, Freedberg, and Keogh (2009) reviewed factors affecting communal laying, asking why communal laying is prevalent in some families and not others, and, in families where it occurs, why it is more common in some groups than in others. Groups with a low proportion of oviparous species might be expected to include fewer communal layers, as would monotypic groups or those with low species richness. In the latter, the likelihood that communal laying would occur or be detected might be expected to be low. Indeed, (oviparous) species richness in communally laying lizard families (253 ± 76.3 SE) was significantly higher than that in families in which communal laying had not been reported (15 ± 6.6 SE). For example, the Lanthanotidae is monotypic, and there are only two species in the Anniellidae and Helodermatidae. Communal laying had not been reported in Opluridae (7 spp.), Xenosauridae (7 spp.), Corytophanidae (9 spp.), Crotaphytidae (10 spp.), or Hoplocercidae (11 spp.). The difference persisted when the Agamidae and Chamaeleonidae, for which only a combined four species were known to nest communally, were excluded. The only lizard family with >19 species without known communal layers was the Varanidae ($N = 59$ spp.) (Doody, Freedberg & Keogh, 2009). However, communal laying was later reported in two species of varanids (Doody, James, Ellis, et al., 2014; Doody, James, Colyvas, et al., 2015; Doody et al., unpubl. data).

Another potential influence is body size, which affects egg size and the type of nest site (Doody, Freedberg & Keogh, 2009). Mothers of smaller lizards such as skinks and geckos often deposit their small eggs in exposed areas (i.e., they do not bury them), which means they are often easily detected by conspecific mothers and by researchers. In contrast, larger lizards, such as varanids, iguanids, agamids, and chamaeleonids, generally

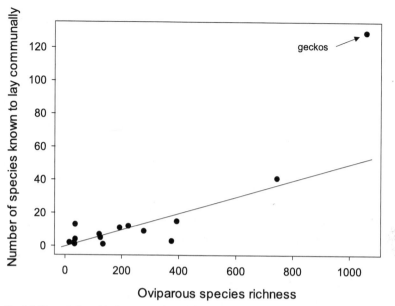

Fig. 6.1. The relationship between the number of communally laying species and (oviparous) species richness for 15 lizard families. Adopted from Doody, Freedberg, & Keogh. (2009).

excavate and backfill individual nest chambers, leaving their larger eggs hidden and potentially undetectable by conspecific mothers and researchers. This pattern is also seen across genera within at least one family—while most geckos do not deposit eggs in a nest, some larger species excavate nest chambers (Delean and Harvey, 1984; Couper, 1996; B. Branch, 1998).

Finally, after demonstrating a strong positive linear relationship between the number of species known to lay communally and species richness in lizard families, Doody, Freedberg, and Keogh (2009) noted that gekkonids exhibited a much higher than expected proportion of communal layers (Fig. 6.1). This finding may be related to the conspicuousness of gekkonid eggs, which could increase the likelihood of detection by both conspecific mothers and by researchers. In their review, 122 of the 129 (95%) geckos known to lay communally were gekkonines (5% were diplodactylines). This is significant because gekkonine eggs are usually deposited in somewhat exposed sites (e.g., rock walls, under bark, and around human dwellings) (see Bustard, 1969), as their rigid shells make them essentially independent of the hydric environment (Dunson, 1982; Thompson et al., 2004). In contrast, diplodactylines, like all other extant lizards, possess flexible, parchment-shelled eggs, which lose water quickly in dry conditions (Dunson & Bramham, 1981; Dunson, 1982; Ackerman et al., 1985), and these eggs are generally deposited in less-exposed sites (Bustard, 1965). As a result, gekkonines, which are often gregarious (e.g., B. Greenberg, 1943; Cooper et al., 1985; Kearney et al., 2001), lay eggs in areas where they forage and sleep, potentially increasing the likelihood that the eggs will be discovered by conspecific mothers. Doody, Freedberg, and Keogh (2009) concluded by hypothesizing that the high incidence of communal egg-laying in geckos relative to other lizards is attributable to oviposition site choice and egg physiology, which, in turn, lead to eggs that are more conspicuous than those of other lizard groups.

6.5. Macroevolution of Communal Egg-Laying in Reptiles

In order to better understand under which contexts communal egg-laying might evolve, a phylogenetically based analysis is desirable. Although a fine-scale analysis is precluded by the inability to confidently assign "no communal egg-laying" to a large proportion of species for which eggs or nests are unknown, a higher-order comparison can be made. The phylogenetic distribution of communal egg-laying was analyzed at the family level in squamate reptiles (Doody, Freedberg & Keogh, 2009). Communal egg-laying occurred in 16 of 29 families, and its presence–absence was mapped on a topology comprising a squamate phylogeny (Townsend et al., 2004) combined with several recognized but missing iguanid families (after Schulte et al., 2003). The conclusion was that communal egg-laying was widespread in reptiles and comprised multiple independent origins, and possibly reversals (Doody, Freedberg, & Keogh, 2009). This conclusion is unchanged despite recent changes to higher-order squamate relationships (Pyron et al., 2013).

We proceed by elucidating some noteworthy examples of communal egg-laying in reptiles, covering each major group. We then address proximate and ultimate causes of communal egg-laying in reptiles, including cues, costs and benefits, and the evidence for each.

6.6. Some Dramatic Examples of Communal Egg-Laying in Reptiles

6.6.1. The Ridley Sea Turtle Arribadas

Each year, tens to hundreds of thousands of olive (*Lepidochelys olivacea*) and Kemp's (*L. kempii*) ridley sea turtles nest together en masse over a few days on tropical beaches in the Atlantic and Indo-Pacific (Carr, 1967; Pritchard, 1969; D. A. Hughes & Richard, 1974; Valverde et al., 2012). For example, up to 500,000 individual turtles nest annually at Ostional Beach, Costa Rica (Plate 6.1; Valverde et al., 2012). During the Costa Rican dry season, female *L. olivacea*, and some males, migrate together from their oceanic feeding grounds to coastal waters, where the females emerge from the surf onto beaches to lay. The proximate cues are unknown but may include winds, moon phase, olfactory cues, or social facilitation (reviewed in Bernardo & Plotkin, 2007). More than one clutch may be laid at monthly intervals before the turtles migrate back to their oceanic feeding grounds, and so many clutches are deposited in such close proximity that mothers often inadvertently excavate the nests of other mothers (Bernardo & Plotkin, 2007). The millions of eggs draw a plethora of predators, including jaguars, coyotes, coatis, raccoons, and vultures, yet there are too many eggs for the predators to handle. That is, all except one predator—humans (Plate 6.2). Commercial harvesting began in Mexico in the 1960s, and in just five years 2 million olive ridleys were taken from coastal waters and beaches (Spotila, 2004). By the time the Mexican government had banned the harvest in 1990, the turtles had disappeared from three of the four major arribada beaches, and from 15 other beaches with smaller arribadas or solitary nesting (Spotila, 2004).

A similar fate has plagued Kemp's ridley, another arribada species. For decades the nesting grounds of this turtle were unknown to science. Finally, amid rumors of a mass nesting beach, in 1947, Andres Herrera, a young Mexican engineer and pilot, flew along the Mexican coast searching for evidence of mass nesting (Hildebrand, 1963; Bevan et al., 2016). On the 24th and last day of his survey, he discovered and filmed an estimated 40,000 nesting Kemp's ridleys on a single beach. The population then crashed, declining by 99% by 1985, and is currently at about 10% of its 1947 population size (Bevan et al., 2016, see also references therein). The turtles and eggs were being harvested for human consumption, and many more adults were being drowned in shrimp trawls (Spotila, 2004).

The plight of the arribada turtles reminds us that communal egg-laying cannot occur if population sizes or densities are too low. For example, communal egg-laying was once common in the Nile crocodile (*Crocodylus niloticus*) and in the painted terrapin (*Batagur baska*), but reductions in population sizes have essentially forced these species to nest solitarily (Cott, 1961; Pooley, 1982; references in D. Moll & Moll, 2004). These cases also remind us that, while it is an effective reproductive strategy against most predators, predator dilution is futile against humans armed with technology. Fortunately, conservation-minded humans, perhaps spurned by the "paradoxical losses" of superabundant species such as the communally roosting passenger pigeon (*Ectopistes migratorius*), have intervened to conserve these species and their mass nesting habits.

6.6.2. Underwater Vocalization in the Giant Amazon River Turtle

As if not to be outdone by their cousins of the sea, some freshwater turtles also lay communally, and the champion of communally nesting freshwater turtles is the giant Amazon river turtle (*Podocnemis expansa*) (Plate 6.3; Pritchard & Trebbau, 1984; Vogt, 2008). Numbers of nesting turtles today do not generally rival arribadas, but they commonly range from hundreds to a few thousand, and in 1945 an estimated 123,000 nested on one beach in one season (Pritchard & Trebbau, 1984).

Along the Rio Trombetas in Amazonia, as water levels begin to fall in the early dry season, females migrate upstream to the nesting beaches, where they bask for two to three weeks prior to nesting, when they emerge in great numbers (Plate 6.3) (Alho & Pádua, 1982; Vogt, 2008). As in sea turtles, mass nesting takes its toll on egg survival through the destruction of other nests by nesting mothers. But the social story does not end there. After nesting in large groups, the females remain in deep pools near the nesting areas (Vogt, 2008). Remarkably, late-term embryos vocalize while still in the egg, possibly to facilitate synchronous hatching within nests (Vogt, 2008; Ferrara et al., 2013). Perhaps even more remarkably, upon hatching the adult females and hatchlings appear to migrate downstream together, probably communicating by underwater vocalizations (Vogt, 2008; Ferrara et al., 2013).

Prior to European colonization, the harvest of mothers and eggs was apparently sustainable (Pritchard & Trebbau, 1984; Vogt, 2008). Between 1700 and 1900, millions of *P. expansa* were slaughtered for food and oil— estimates suggest that during the nineteenth century more than 200 million eggs were taken for oil products in the upper Amazon alone, and similar impacts occurred elsewhere (Vogt, 2008). Declines were evident in most *P. expansa* populations (e.g., see Pritchard & Trebbau, 1984). More recently, although exploitation continues, the protection of nesting beaches and subsequent hatchling releases throughout the Brazilian Amazon have resulted in increases of nesting females at most sites (Vogt, 2008).

6.6.3. The Motivated Mothers of the Green Iguana

Although communal laying is not that rare in reptiles, close behavioral observations of what actually transpires at these sites have rarely been conducted. For example, do the females arrive sequentially to lay eggs and thus not interact with each other, or do they arrive at about the same time? Also, does the communal laying occur in a small area with individual depositions, or do females share the same nest hole or burrow systems?

Communal egg-laying in the green iguana (*Iguana iguana*) was thoroughly revealed in the now classic studies pioneered by the late Stan Rand and colleagues in Panama beginning in 1968. Up to 100 females would, over a period of several weeks, swim across a freshwater channel from Barro Colorado Island (BCI) to a small islet called Slothia, on which a small clearing existed and which had the advantage of being too small to sustain populations of mammalian egg predators. The clearing had been used for many years and the hard soil was thus loosened and relatively easy to excavate. The initially arriving iguanas, quite nervous, would begin to excavate burrows that could be a meter or more deep. The female would excavate a chamber at the end large enough to hold the clutch, which could contain many dozens of eggs. The female would then back out of the nest, turn around, and then go back in tail-first and deposit the eggs. Several females may be doing this individually at the same time, but more females would continually arrive. The setting would become crowded and space limited. Agonistic interactions would ensue between females competing for a spot to dig. If they fail to secure a spot, they may dig into a burrow already used. If so, they could expand it in a different direction underground or actually dig up and scatter on the surface the previously deposited eggs of another female. More direct interactions would also occur as when one female would attempt to dislodge another already digging a burrow and perhaps almost done. A variety of lunges, bites, and wrestling might ensue.

It is important to remember that digging a nest burrow is perhaps the hardest and most energetically costly work an adult iguana female ever does. After this ordeal and laying the eggs, she is quite spent and physically weak and thus vulnerable. As the large clutch develops inside the female, she stops eating as there is no room in the stomach for the bulky leafy diet of the mother; she and her clutch are provisioned from fat reserves. If a female can avoid the costly digging exercise and usurp another animal's burrow, it probably aids her own survival.

Besides observing, filming, and describing interactions at close range from a blind on the islet (the setting was particularly fortuitous), as well as telescopic monitoring of the daily numbers of nesting females from a hill on BCI overlooking Slothia, in later years animals were captured for marking and radio tracking. Detailed quantitative observations allowed energy costs of nesting and fighting to be calculated and the dynamics of the agonistic competitions to be modeled stochastically in some of the first such studies on any animal (Rand & Rand, 1976, 1978). It was documented that females return to the island in subsequent years and that hatchlings will preferentially nest where they themselves were hatched. The typical explanations for the dense aggregations involved the paucity of areas with sun-exposed soil that facilitated egg development in the rain forest and riverine environment preferred by green iguanas, as well as egg predator avoidance. Later studies by the team documented the behavior and social interactions of the hatchlings as they emerged from nest burrows and dispersed across and from the islet (see Chapter 10). Slothia was not the only islet where iguanas nested, and they migrated from many disparate locations on BCI to nesting areas on these islets, as well as sites on the main island (Bock et al., 1989).

6.6.4. Deep Nesting in the Yellow-Spotted and Gould's Monitor Lizards

From as early as 255 million years ago, a diversity of vertebrate species have excavated mysterious, deep helical burrows called *Daimonelix* ("demonic corkscrews") (L. D. Martin & Bennett, 1977; A. M. A. Smith, 1987). Rarity of such burrows in extant animals has frustrated their adaptive explanation, but recently the first contemporary helical reptile burrows were discovered, created by the yellow-spotted monitor (*Varanus panoptes*), a 1.5 m long apex predator inhabiting the tropical savannas and deserts of northern Australia (Doody, James, Ellis, et al., 2014; Doody, James, Colyvas, et al., 2015; Doody, McHenry, Durkin, et al., 2018). The nesting burrows terminated in chambers that were the deepest known vertebrate nests in the world (up to 3.6 m deep, average = 2.7 m) (Plate 6.4), far exceeding the previously known deepest reptile nest (0.9 m, on average), which belonged to the leatherback sea turtle (*Dermochelys coriacea*) (Doody, James, Ellis, et al., 2014). Intriguingly, the nests were communal and excavated in open riparian areas. At one site researchers found 97 nests in a 4×4 m area, including 21 fresh nests and 76 clutches of eggshells from successful nests in previous years (Doody, James, Colyvas, et al., 2015). The decimation of yellow-spotted monitor populations by the cane toad invasion of northern Australia exposed the lizard as a top predator and made urgent the need to study its communal nesting behavior ahead of the invasion front (Doody, Freedberg & Keogh, 2009; Doody, Castellano, Rhind & Green, 2013; Doody, Soanes, et al., 2015; Doody, Rhind, et al., 2017).

Seeking more context in order to explain the behavior, the researchers then excavated a nesting warren of the species' closest relative, Gould's monitor lizard (*V. gouldii*), at a much drier site in the northern Australian desert (Plate 6.5) (Doody, McHenry, Brown, et al., 2018). The results were similar: Gould's monitors nested communally, creating helical burrows up to 3.7 m deep (average = 3.0 m; JSD, unpubl. data). At this site, researchers discovered 125 nests in a 4 × 4 m area, including 11 fresh clutches and 114 clutches of eggshells (JSD, unpubl. data). Surprisingly, hatchlings were discovered to excavate emergence burrows separate from their mother's burrow (Doody, McHenry, Brown, et al., 2018).

The benefits of deep communal helical nesting in these large lizards are uncertain, but it appears that large monitor lizard species that possess very long egg incubation periods (~7–10 months; Horn & Visser, 1989, 1997; Birchard & Marcellini, 1996) and inhabit areas with long dry seasons are forced to excavate very deep nests to provide adequate moisture for developing embryos. A recent excavation of a *V. panoptes* nest in a desert climate supported this idea (Doody, McHenry, Durkin, et al., 2018). Nest excavation by mothers can take up to two weeks (JSD, unpubl. data) and thus use considerable energy; mothers can save energy and time by nesting communally (and traditionally) in suitable areas due to an annual loosening of the substrate by nesting mothers. Regardless of costs and benefits, it is clear that many individual mothers of both species are congregating to nest in the same warrens each year, raising the possibility of more complex social interactions than previously appreciated. For example, how might these interactions influence the mating system—would males not use the communal egg-laying warrens as a means of accessing multiple females? We predict that the other large monitor lizard with a long incubation period that must span the long dry season (the perentie, *V. giganteus*) will also nest communally at great depths, and a current study has recently followed a gravid radio-tracked mother to a warren created by burrowing bettongs (*Bettongia lesueur*)—marsupial mammals that have been extinct for more than 150 years.

6.6.5. Traditional Communal Nesting in the Sri Lankan Golden Gecko

Deep nesting is unusual for reptiles—in fact most nest within 20 cm of the surface. The gekkonid geckos have hard-shelled eggs that are essentially independent of the hydric environment (Dunson, 1982), allowing them to nest in sites with lower humidity (e.g., open air in shaded places). The nesting strategy and nest sites of the Sri Lankan golden gecko (*Calodactylodes illingworthorum*) reveal the importance of tradition in communally nesting reptiles. This species glues its two eggs to cave walls (Plate 6.6a), and tens to hundreds of eggs can be found together (Wickramasinghe & Somaweera, 2002; Goonewardene et al., 2003; de Silva, Bauer, Austin, Goonewardene, et al., 2004; Karunarathna & Amarasinghe, 2011). But what makes this nesting unique is that, beneath the eggs, the "scars" of eggshells from previous years can be found together by the thousands, indicating traditional nesting that may span hundreds of years (Plate 6.6b). The dry environment and eggshell composition combine to preserve the egg remains for a very long time; this raises the question as to whether traditional communal egg-laying is widespread in other reptiles. The degradation of eggshells in typical environments (e.g., soil nests) and paucity of long-term nesting studies has prevented us from answering this question. In the Australian water dragon (*Intellagama lesueurii*) and in the yellow-spotted monitor, eggshells in shallow soil nests decompose in approximately four to five years, based on a multiyear study in the former and discrete categories of egg decomposition and color in the latter (JSD, unpubl. data; see also Burger & Zappalorti, 1992). The benefits of communal egg-laying in the Sri Lankan golden gecko are unknown, but the eggs are eaten by the Bengal monitor lizard (*V. bengalensis*), Dumeril's kukri snake (*Oligodon sublineatus*), ants, and rodents (Sreekar et al., 2010; Karunarathna & Amarasinghe, 2011). Interestingly, this gecko is depicted in rock art in some caves (Goonatilake & Peries, 2001; de Silva, Bauer, Austin, Goonewardene, et al., 2004; de Silva, Bauer, Austin, Perera, et al., 2004; Karunarathna & Amarasinghe, 2011). In at least one area, the tribal people are afraid of the geckos, believing that touching the geckos brings about disease (Karunarathna & Amarasinghe,

2011). In the closely related Indian golden gecko (*C. aureus*), 97% of eggs in public bathrooms are routinely destroyed by humans because they do not like the appearance of the eggs (Sreekar et al., 2010).

6.6.6. Cool Climate Communal Egg-Laying in Snakes

Snakes are generally more secretive than most other reptiles (Shine, 1991; Seigel & Collins, 1993; H. W. Greene, 1997), and their reproductive habits are no exception. Turtles often leave tracks to their nest sites, crocodylians typically guard their nests, and lizards often nest conspicuously in open areas (e.g., iguanas) or in conspicuous places in and around caves and human dwellings (e.g., geckos). Snakes tend to lay their eggs in out-of-sight places, such as in underground burrows or in logs, stumps, or trees. As such, by the mid-twentieth century there was little evidence of communal egg-laying in snakes. Thanks in part to the advent of radiotelemetry, researchers discovered communal egg-laying in snakes, which they could now follow closely (Brodie et al., 1969; W. S. Parker & Brown, 1972; Madsen, 1984; P. B. Whitaker & Shine, 2002, 2003; Blouin-Demers et al., 2004; Cunnington & Cebek, 2005). Two North American snakes have particularly interesting stories.

Turtles excavate their nests with their hind limbs (and incidentally never see their own eggs), while the lizards and crocodylians that excavate nests use their front limbs. Snakes generally do not excavate their own nesting burrows, presumably because they are legless. The northern pine snake (*Pituophis melanoleucus*), however, excavates its own nesting burrows in sandy sites in New Jersey, USA. Mothers excavate by pushing forward into the sand with their head and snout, crooking the neck, capturing the sand in the crook, and bringing sand out of the burrow to the "dump pile" (Burger & Zappalorti, 1991). In one study, 39% of 74 nests contained more than one clutch of eggs (Burger & Zappalorti, 1991). Pine snake mothers showed both high synchronicity (within two weeks) of nesting and high fidelity to their individual nest sites across years (traditional nesting) (Burger & Zappalorti, 1992).

Another snake that is fond of communal egg-laying is the black rat snake (*Pantherophis obsoletus*), which nests ~75 cm deep in decaying organic matter in hollow trees, logs, leaf piles, or compost piles (Blouin-Demers et al.,

2004). Research on Canadian populations demonstrated that eggs in communal nests were warmer than those in solitary nests, and lab experiments showed that mothers preferred to nest at temperatures more similar to those found in communal nests (Blouin-Demers et al., 2004). So, why do some females nest solitarily? The possible answer is intriguing. Burying beetles of the genus *Nicrophorus* are carrion feeders, but one species (*N. pustulatus*) has apparently undergone a remarkable dietary shift to exploit snake eggs (Blouin-Demers & Weatherhead, 2000; Keller and Heske, 2001; G. Smith et al., 2007). Communal nests may be more likely to be parasitized by burying beetles than are solitary nests (Blouin-Demers et al., 2004).

Remarkably, a landowner found and removed ~340 eggs and hatchlings of *P. obsoletus* from a single tree stump in Poolesville, Maryland, USA (M. Hilegis, unpubl. obs.) (Fig. 6.2). At an average clutch size of about 15, and considering one clutch per female per season, that is about 23 mothers nesting in the same stump. In snakes, the size of this nest is surpassed only by a communal nest of the yellow-faced whipsnake (*Demansia psammophis*); a nest of 500–600 eggs and freshly hatched eggshells within a rock crevice

Fig. 6.2. A communal nest of the black rat snake (*Pantherophis obsoletus*) in a recently cut stump from near Poolesville, Maryland. Based on the numbers of eggs and hatchlings recovered, the nest comprised at least 22 mothers. Photograph by Mike Hillegas.

was excavated by a bulldozer in eastern Australia (Covacevich & Limpus, 1972). At an average clutch size <10, this nest comprised >50–60 clutches.

6.6.7. Crocodile Mothers Are Gentle to Their Offspring

Crocodile mothers are known for their maternal care, protecting their nests, excavating their hatchlings and carrying them to the water, and even opening unhatched eggs (see Chapter 8). But nest site choice can also be thought of as parental care, and crocodylians are good at it. Some crocodylians build mound nests of vegetation for nesting, while others excavate holes in the sand, and communal egg-laying has been reported for several crocodylian species (Doody, Freedberg & Keogh, 2009, supplement). One species, the American crocodile (*Crocodylus acutus*), can utilize either method, and at one nesting beach in Cuba, as is also common in green iguanas, 20% of *C. acutus* eggs were destroyed by conspecific nesting mothers (Alonso et al., 2002).

Communal egg-laying can occur on sandy banks in the hole-nesting Johnston's crocodile (*C. johnstoni*), the only crocodile species that does not defend its nests. Communal egg-laying may have contributed to the loss of nest defense because several crocodiles may not be able to cohabit such a small area (A. M. A. Smith, 1987). Other hole-nesting, communally nesting species, such as American and Nile (*C. acutus* and *C. niloticus*) crocodiles, defend their nests, however.

Communal egg-laying was once common in the Nile crocodile, but it may no longer occur due to a reduction in population sizes (Cott, 1961; Pooley, 1982). The last two recorded significant Nile crocodile communal egg-laying sites, two small sandy beaches (~0.6 hectares each), contained 17 and 24 nests, respectively (Cott, 1961).

6.6.8. Tuatara Communal Egg-Laying: Insight into the Biology of a Long-Lost Group

The unique tuatara is the only surviving member of a once widespread group that flourished alongside the dinosaurs (Benton, 2000; Cree, 2014), and it retains many ancient features. Among extant amniotes, it is considered to be the most primitive, with features resembling those of many amphibians (e.g., ear, heart, and spine morphology and low thermal preference) (Thompson & Daugherty, 1998; Lutz, 2005; Besson & Cree, 2010, 2011). The tuatara is confined to New Zealand, where its stronghold is on the 150-hectare Stephens Island, which boasts some 40,000 animals (J. A. Moore et al., 2007). It is an atypical reptile in many ways, including its persistence in a relatively cold climate, where it is more active at night despite warmer daytime temperatures (Thompson & Daugherty, 1998). Relatedly, the cool temperatures prolong egg incubation (J. T. Nelson et al., 2006), which averages 365 days.

On Stephens Island, these lizard look-a-likes lay communally in north-facing slopes of open areas such as rocky outcrops and old sheep paddocks (Thompson et al., 1996; Refsnider et al., 2013). Mothers travel up to hundreds of meters from their roosting burrows to nesting rookeries (Refsnider et al., 2013), where they spend several nights excavating a shallow tunnel up to 1 m long and backfill the nest with soil after laying (Thompson et al., 1996). Up to ~20 nests can be found in a small (10 × 10 m) area (JSD, pers. obs.). The sometimes-hard substrate can require considerable effort to excavate, so mothers often search for the softer areas already excavated by other mothers; in so doing, they inadvertently excavate conspecific eggs (Refsnider et al., 2009). As a result, mothers have evolved nest defense: they curl up in the plugged nest entrance for a period of 1–10 days (mean = 3.0 ± 2.4 days) (Refsnider et al., 2009). Not all mothers guard—44% of 73 mothers abandoned their nest after laying—but the number of guarding mothers increased as the overall activity of conspecific mothers increased on the oviposition date (Refsnider et al., 2009).

Communal egg-laying in the tuatara is not particularly unique—for example, nest defense against communally nesting conspecific mothers has also evolved in iguanid lizards in central and South America, occurring in at least 12 species (reviewed in Iverson et al., 2004). However, the tuatara is nothing short of a taxonomic jewel, because it is the sole survivor of a once widespread group, and it requires conservation attention. The tuatara was once widely distributed throughout New Zealand, but disappeared from the mainland shortly after the arrival of humans and land mammals around 700 years ago (Wilmshurst et al., 2008). It now inhabits only 32 islands, mostly in low numbers (Gaze, 2001; Cree, 2014). Its survival as a species with impending climate change may depend on captive rearing and translocations (e.g., N. J. Nelson et al., 2002), along with relevant research and monitoring (e.g., Jarvie et al., 2016). It will be interesting to see if communal egg-laying, as occurs on Stephens Island, becomes a vital part of translocated populations' natural history.

6.7. Why Do Reptiles Lay Communally?

6.7.1. Habitat Saturation versus Conspecific Attraction

Now that we can accept that communal nesting is common and widespread in reptiles, we will explore why it may have evolved. Historically, communal egg-laying in reptiles was mainly attributed to habitat saturation—or the lack of suitable nest sites (see reviews in Graves & Duvall, 1995, and Doody, Freedberg & Keogh, 2009). Simply put, if there are more mothers than nest sites, nesting will be communal; this is irrespective of benefits (precluding the need to invoke adaptation) or social factors such as conspecific attraction to eggs, eggshells, or mothers. Although anecdotal, many authors have noted the apparent lack of available nest sites surrounding a communal reptile nest (see references in Graves & Duvall, 1995, and in Doody, Freedberg & Keogh, 2009). For example, some researchers concluded that lizards, snakes, and turtles concentrated their nests in small areas because these were the only sites available (Fowler, 1966; Carr, 1967; A. S. Rand, 1967, 1968; Brodie et al., 1969; Covacevich & Limpus, 1972; W. S. Parker & Brown, 1972; Swain & Smith, 1978; A. S. Rand & Dugan, 1983; Pritchard & Trebbau, 1984; Plummer, 1990; Burger & Zappalorti, 1991). Given some of these examples and the branding of reptiles as "asocial," the habitat saturation hypothesis became the favored explanation for communal egg-laying in reptiles, precluding the need for an adaptive explanation for communal egg-laying. Indeed, some authors continue to support the habitat saturation hypothesis (Limpus et al., 2003; Pianka & Vitt, 2003; Bernardo & Plotkin, 2007; Pike et al., 2010).

In sharp contrast, but also mainly based on anecdotes, other authors have noted the abundance of suitable nesting areas with seemingly identical conditions surrounding communal reptile nests (Swain & Smith, 1978; reviewed in Graves & Duvall, 1995, and in Doody, Freedberg & Keogh, 2009). For example, microhabitats used by lizards and snakes as communal nest sites were abundant in nearby non-nest sites (Magnusson & Lima, 1984; Mora, 1989; Brown & Duffy, 1992). A few authorities were more cautious, noting that the evidence was insufficient to discern between habitat saturation and conspecific attraction (P. T. Gregory, 1975; Swain & Smith, 1978; Plummer, 1981). If the habitat saturation hypothesis was insufficient to explain communal egg-laying alone, there were two additional possibilities: (1) reptile mothers are attracted to conspecific eggs, and (2) communal egg-laying is adaptive in reptiles. Impressively, G. K. Noble and Mason (1933, p. 13) were the first to suggest conspecific attraction in reptile mothers: "Many reptiles lay their eggs together in the most suitable sites, but it is difficult to account for the colonial nesting habit . . . without assuming that the gravid females are in some way attracted by the eggs of their own species."

Do reptile communal nests reflect forced communal use of a scarce microhabitat or complex social behavior? Reductions in available nest sites (via rock removal and fires) did not cause an increase in communal egg-laying in the eastern three-lined skink (*Acritoscincus duperreyi*), prompting Radder and Shine (2007) to argue against the habitat saturation hypothesis. Field data generally cannot disentangle these explanations, however. For example, how can we know that the reduction in nest sites in their study was sufficient to cause nest sites to be limiting? Fortunately, laboratory experiments can provide answers.

Research in the laboratory has now demonstrated conspecific attraction of reptile mothers to the eggs and eggshells of conspecifics, at least in squamates. In the first experimental evidence for conspecific attraction in communal egg-laying reptiles, 29 of 35 captive gravid rough green snakes (*Opheodrys aestivus*) nested in a single nest site, and four nested in a second nest site, when offered 10 identical nest sites in the laboratory (Plummer, 1981). Also in the laboratory, 20 of 26 gravid eastern three-lined skinks oviposited in nest sites with cotton-filled eggshells, compared to six that oviposited in control sites with no eggshells (Radder & Shine, 2007). In the laboratory, common keelback snakes (*Tropidonophis mairii*) and velvet geckos (*Oedura lesueurii*) preferred to nest in nest sites with hatched eggshells over empty nest sites (Pike et al., 2010; Brown & Shine, 2005), while skinks of the same species preferred to nest with conspecific eggs over conspecific eggshells (Elphick et al., 2013). In contrast, delicate skink (*Lampropholis delicata*) mothers showed no preference between conspecific eggs and eggshells (Paull, 2010; Paull & Doody, unpubl. data). Finally, in a longer-term field experiment with delicate skinks, four artificial nest sites (grinded crevices and drilled holes) were added to each of 40 original nest sites (sandstone crevices and holes) in the field to increase the number of nest sites. In the first few years, most mothers chose original sites, supporting conspecific attraction. In subsequent years, mothers began to use artificial nest sites; however, eggs were rarely spread among the four sites, but rather they were communally oviposited into only one of the four.

The weight of the evidence suggests that communal egg-laying in many reptiles is often underpinned by conspecific attraction to eggs or eggshells. Although communal egg-laying is likely to be (or have been) adaptive, we cannot be sure (see costs and benefits sections below). Regardless, it is now reasonably certain that the habitat saturation hypothesis cannot fully explain communal egg-laying behavior in reptiles. An important caveat is that habitat saturation and conspecific attraction are not mutually exclusive (P. T. Gregory et al., 1987; Doody, Freedberg & Keogh, 2009). A gravid mother searching for an uncommon nest site could save time or energy, or reduce predation risk, by identifying and being attracted to conspecific eggs, eggshells, or mothers (EEM) (Doody, Freedberg & Keogh, 2009). But, a scarcity of suitable nest sites over many generations would favor the evolution of conspecific attraction to EEM (Doody, Freedberg & Keogh, 2009). It is possible that the scarcity of suitable nest sites was a prerequisite for the evolution of conspecific attraction (to EEM) in many reptile species (and thus a key precursor to social behavior in reptiles generally). This idea generates the prediction that conspecific attraction to eggs would be particularly strong in species in which suitable nest sites are limiting or rare.

What are the proximate cues and mechanisms that attract conspecific mothers to communal egg-laying sites? An expectant mother could follow another mother to such a site; once a site was found, the eggs, eggshells, or laying mothers could serve as obvious visual cues. Many reptiles have keen nasal olfaction and vomerolfaction, however, allowing the prediction that communal egg-laying sites could attract mothers from a distance via chemical cues (Elphick et al., 2013). Black flies (Simuliidae) oviposit communally, and the behavior is mediated by a pheromone that is released from the eggs immediately after oviposition, and which attracts further gravid females to oviposit on the same substrate (McCall, 1995). Green iguana females exploring and then laying eggs on the communal nest site on Slothia in Panama engaged in both substrate and aerial tongue flicking, often associated with locomotion, suggesting that chemical cues are used (Burghardt et al., 1986). Whether the behavior was associated with caution and stress, or social or nest site recognition, is not known, however.

6.7.2. Costs of Communal Egg-Laying

Although conspecific attraction is likely an adaptive mechanism for communal egg-laying, we reiterate that demonstrating the former does not confirm an adaptive explanation for the latter. Revealing a (potentially) adaptive explanation would likely require cost-benefit analysis of solitary versus communal egg-laying, in situ. Conceptually, an evolutionary stable strategy (ESS) maternal benefits model demonstrated how alternative strategies (nest finders vs. nest freeloaders) would persist in a given population, despite any advantages of communal egg-laying (Doody, Freedberg & Keogh, 2009). In short, as the probability of finding a conspecific nest decreases, the likelihood of net benefits for freeloaders decreases, supporting the alternative strategy of finding one's own nest. Unfortunately, there have been no quantitative cost-benefit studies of communal versus solitary nesting in a reptile population (Doody, Freedberg & Keogh, 2009). The seemingly ubiquitous occurrence of solitary clutches in communally nesting reptile populations (e.g., A. S. Rand & Dugan, 1983), however, is consistent with predictions of the ESS model (Doody, Freedberg & Keogh, 2009).

Iguana females do compete over nest holes and often the fights escalate and can be prolonged. One of the first models of such competition, based on observational field data, in which energetic costs figure greatly, were the pioneering dispute settlement analyses of Rand and Rand (1976, 1978).

There is evidence that communal egg-laying can incur both costs and benefits in different species; we will cover costs first. High densities of nesting mothers associated with communal egg-laying can lead to inadvertent destruction of nests or eggs by conspecific nesting mothers in lizards, turtles, the tuatara, and crocodylians. Eggs from communal nests were three times as likely to be pushed from a sandstone crevice by conspecific mothers and desiccate than were eggs from solitary nests in the delicate skink (Cheetham et al., 2011). Communally nesting turtles utilizing ocean and freshwater beaches also excavate the eggs of conspecific mothers (Vogt, 2008; see review in Bernardo & Plotkin, 2007). For example, 21% of leatherback sea turtle nests were destroyed by nesting mothers at a major nesting beach in Guiana (Girondot et al., 2002). Nearly 20% of American crocodile nests were destroyed by conspecific mothers over an 11-year period in Cuba (Alonso et al., 2002). Similarly, about 25% of tuatara nests were excavated by conspecific mothers in New Zealand (Refsnider et al., 2009). Communal egg-laying in several species of American iguanas results in similar fates, and in some cases can even lead to the mother's death from burrow collapse (reviewed in Iverson et al., 2004; J. B. Moss et al., 2020). In at least some of these species and in the tuatara, nest defense appears to have evolved in response to conspecific nest destruction, even though this mortality appears to be accidental (see review in Iverson et al., 2004; Refsnider et al., 2009; J. B. Moss et al., 2020). In turn, the intense female aggression associated with nest defense against conspecific mothers in the marine iguana (*Amblyrhynchus cristatus*) may have led to the evolution of male-like coloration in females (Rauch, 1988), generating the idea that communal egg-laying can lead to the evolution of other social behaviors and associated coloration via a cascade of evolutionary events.

Aside from conspecific nest destruction, a cost of group living is increased disease transmission (C. R. Brown & Brown 1996). There is evidence that very high nest densities lead to elevated microbial activity in olive ridleys (Honarvar et al., 2008), causing very high nest mortality; hatching success was 2% ($N = 37$ nests) on one beach in Costa Rica (Valverde et al., 2010). However, rather than microbial attack, the mechanism underpinning the mortality appears to be low oxygen levels as a result of microbial decomposition (Honarvar et al., 2008; Bezy & Cole, 2014). Solitary nests likely avoid these sources of mortality, but more data are needed for comparison.

Parasitoid beetles seem to exploit communal nests of black rat snakes more readily than solitary nests (see above account, Blouin-Demers et al., 2004). The probability of fungal infection may also be higher in communal egg masses, but evidence is lacking (but see Patino-Martinez et al., 2012).

In birds, although there is support for the idea that communal egg-laying evolved to reduce nest predation, there is also support for the opposite idea—that predation can *increase* with increasing nest density due to communal nests being more conspicuous to predators than solitary nests because of visual, acoustic, or olfactory cues (Lack, 1968b; C. R. Brown & Brown, 2001; Varela, 2007). Reptile nests almost never have vocalizing parents nearby, and most reptile communal nests are hidden, but olfactory cues could make communal nests more vulnerable to predation. We found two published reptile examples: (1) raccoon predation was higher on clumps of simulated painted turtle (*Chrysemys picta*) nests (with quail eggs) than on scattered nests (Marchand et al., 2002; Marchand & Litvaitis, 2004); (2) in nature, communal nests of Johnston's crocodiles were more likely to be preyed upon by yellow-spotted monitors than were solitary nests (G. J. W. Webb et al., 1983; G. J. W. Webb & Smith, 1984; A. M. A. Smith, 1987). Although evidence that reptile communal nests incur higher predation than solitary nests is scarce, theoretical models predict that low numbers of nests might not be worth a predator's time and effort (Tinbergen et al., 1967; Lack, 1968b; Bernardo & Plotkin, 2007). Consistent with this idea, predation on softshell turtle (*Apalone mutica* and *A. spinifera*) nests was very low (<10%) at a site in which an abundance of large nesting beaches resulted in <5 nests per beach (Doody, 1995; Godwin, 2017).

Groups of reptile eggs could experience increased competition for soil moisture, as hypothesized by Marco et al. (2004) after finding lower and more variable water absorption in aggregated eggs of Iberian emerald lizard (*Lacerta schreiberi*) in the laboratory. The hydric environment can influence both egg survival and hatchling phenotypes (Deeming, 2004). However, aggregated eastern three-lined skink eggs absorbed less water in the laboratory (Radder & Shine, 2007).

Finally, by concentrating groups of hatchlings, communal egg-laying could incur a cost by increasing intraspecific competition. There is no doubt that for at least short periods of time groups of hatchlings can be found at communal egg-laying sites (e.g., Burghardt, 1977b; Burghardt et al., 1977). The latency to disperse away from the communal site, the feeding biology (are hatchlings feeding or simply utilizing internalized yolk?), and the abundance of food would mediate any effects of communal egg-laying on competition. Evidence for these in reptiles is lacking.

Animals often make critical decisions by copying the choices of conspecifics (Pruett-Jones, 1992; Danchin & Wagner, 1997), and social acquisition of information has been well documented in a diversity of animals (Giraldeau, 1997; Galef & Giraldeau, 2001). These nonindependent choices are believed to benefit "copiers" by offering a savings in time, energy expenditure, or survival, relative to independent choices (Giraldeau et al., 2002). Provided that the benefits outweigh costs, we might expect to find copying behavior in populations or species where it is free to evolve. A mother faced with the task of finding a laying site could realize such benefits by copying the egg-laying site choice of conspecifics (Doody, Freedberg & Keogh, 2009). Benefits could accrue in eggs, mothers, or hatchlings.

6.7.3. Adaptive Explanations—Benefits to Eggs

Various benefits of communal egg-laying have been proposed, of which the most popular is that high nest density reduces nest predation (reviewed in Varela et al., 2007). Some components of this idea may not be relevant to reptiles (e.g., group vigilance, communal defense). For reptiles, this hypothesis falls under the concept of attack abatement and has two components: encounter probability and dilution effect (G. F. Turner & Pitcher, 1986; Wrona & Dixon, 1991). The encounter probability predicts that a predator is less likely to find a single group of many eggs than many scattered small groups of eggs. The dilution effect, or "safety in numbers," predicts that as the number of eggs in a nest increases, the probability of each being preyed upon decreases. This benefit has been suggested to occur in reptilian mothers, eggs, and neonates (Fitch, 1954; Burghardt, 1977b; Burghardt et al., 1977; Shine, 1979; Wiewandt, 1982; Congdon et al., 1983; Burger & Zappalorti, 1992). In particular, the arribadas of communally nesting sea turtles have long been considered to satiate predators, including humans, with massive numbers of eggs (Carr, 1967; Pritchard, 1969; D. A. Hughes & Richards, 1974; Bustard, 1979; Hirth, 1980; see theoretical treatment in Bernardo & Plotkin, 2007).

Despite its popularity, evidence for this hypothesis is underwhelming—relatively few studies have tested the influence of spatial and temporal distributions on the probability of nest predation in reptiles (Burke et al., 1998; Doody, Freedberg & Keogh, 2009). Eckrich and Owens (1995) found higher predation rate on solitary olive ridley nests (51%) than on arribada nests (8%); predators were mainly coyotes (*Canis latrans*), raccoons (*Procyon lotor*), and coatis (*Nasua nasua*). Although the consequences are unknown, solitary nesters oviposited on two-week cycles compared to four-week cycles in arribada nesters (Pritchard, 1969; Kalb & Owens, 1994; Kalb, 1999), and solitary nesters used multiple beaches (Kalb, 1999) while arribada nesters produced larger clutches (Bernardo & Plotkin, 2007). Conversely, an increase in the number of nests constructed per day decreased nest predation in the common snapping turtle (*Chelydra serpentina*) (Robinson & Bider, 1988), and two studies on snapping turtles and diamondback terrapins (*Malaclemys terrapin*) found that nests within 1 m of one another suffered higher predation than more dispersed nests (Burger, 1977; Robinson & Bider, 1988).

In contrast to the above studies, Brown and Duffy (1992) found no differences in hatching success between communal and solitary nests of the mourning gecko (*Lepidodactylus lugubris*), and there was no evidence for an effect of nest density on predation rate in four species of freshwater turtles

(*Carettochelys insculpta*, *Kinosternon subrubrum*, *Pseudemys concinna*, and *Trachemys scripta*) (Burke et al., 1998; Doody, Sims & Georges, 2003) or two species of sea turtles (*Chelonia mydas* and *Eretmochelys imbricata*) (L. E. Fowler, 1979; Leighton et al., 2008). Nest density also did not influence predation rates by dingos (*Canis lupus*) on the Johnston's crocodile (Somaweera et al., 2011).

Mothers may use specific fitness components as cues for communal egg-laying (Danchin & Wagner, 1997). After finding that common keelback snakes nested preferentially with conspecific eggshells over empty nest sites, G. P. Brown and Shine (2005) proposed that mothers were assessing reproductive success by examining hatched eggshells (Magnusson & Lima, 1984) and nesting appropriately. However, proper controls were not used to make this claim—conspecific attraction to eggshells alone was adequate to explain their findings, irrespective of reproductive success. The reproductive success-based hypothesis was tested using the delicate skink, a species that nests communally in sandstone crevices in Sydney, Australia. Laboratory experiments showed that skinks preferentially oviposited in nest sites (crevices in bricks) with conspecific eggs or hatched eggshells over empty sites, but they did not prefer nest sites with eggs over those with eggshells (Paull, 2010; Paull & Doody, unpubl. data), rejecting the reproductive success-based hypothesis. Similarly, eastern three-lined skinks preferred to lay with conspecific eggs over conspecific eggshells in the laboratory (Elphick et al., 2013). In summary, there remains to be firm evidence for the reproductive success-based hypothesis for communal egg-laying in reptiles.

Communal egg-laying may confer a thermal advantage to ectotherms, and this had been used to explain the evolution of the behavior in some temperate pond-breeding frogs in North America and Europe, where egg masses at the center of a communal mass were 1°C to 7°C warmer than solitary masses or those at the periphery of the communal mass (Guyétant, 1966; Howard, 1980; Waldman, 1982; Waldman & Ryan, 1983; Håkansson & Lohman, 2004). Such warming is thought to be adaptive in preventing freezing and reducing the developmental times (Seale, 1982; Waldman,

1982; Waldman & Ryan, 1983; Håkansson & Lohman, 2004). In reptiles, several species of snakes nest communally in cold climates, as does the tuatara (Blouin-Demers et al., 2004; reviewed in Löwenborg et al., 2012; Kovar et al., 2016; Meek, 2017). Blouin-Demers et al. (2004) suggested that communal nesting conferred a thermal advantage to embryos in a population of snakes <20 km from the cold climate range margin. Communal nests were nearly 5°C warmer than solitary nests, and the authors used incubation experiments to suggest that the warmer temperatures would result in shorter incubation times, and longer and faster (but less aggressive) neonates. In contrast, there was no difference in nest temperatures between communal and solitary nests in the much smaller eggs of eastern three-lined skinks (Radder & Shine, 2007). The effect of the number of eggs in a group on metabolic heating may be negligible in small reptile eggs.

A key biogeographical pattern underlying this hypothesis is that these species breed either in cold climates or during the winter in milder climates (Caldwell, 1986; Doody & Young, 1995; Doody, Freedberg & Keogh, 2009). Contrary to the predictions of this hypothesis, most reptile breeding occurs in the warmer months, and communal egg-laying occurs in many tropical species (reviewed in Doody, Freedberg & Keogh, 2009; de Sousa & Freire, 2010; Sreekar et al., 2010; Karunarathna & Amarasinghe, 2011; Montgomery et al., 2011; Doody, James, Ellis, et al., 2014; Robinson et al., 2014; Doody, James, Colyvas, et al., 2015). Although this number is lower than that for temperate species, such a pattern would be expected due to the "boreal bias" (the fact that northern species are usually better studied; Platnick, 1991). An analysis is needed of the breakdown of the incidence of communal egg-layers across different climates, using only species for which the eggs in nature are known (Doody, Freedberg & Keogh, 2009).

A related hypothesis that may be more relevant to amphibians involves communal eggs offering improved resistance to desiccation rather than to extreme temperatures (Ryan, 1985; Doody, Freedberg & Keogh, 2009). Aggregated eastern three-lined skink eggs absorbed less water in the laboratory and produced hatchlings that were larger and ran faster (Radder & Shine, 2007; but see Marco et al., 2004).

6.7.4. Adaptive Explanations—Benefits to Mothers

Communal egg-laying may provide maternal benefits, such as a savings in energy and time or increasing the likelihood of survival. First, by adding their eggs to those of conspecifics, mothers could reduce energetic costs associated with searching for an egg-laying site, assessing potential egg-laying site characteristics, and, in some cases, carrying out nest excavation. Experimental evidence for such benefits is limited, but communally laying insects saved time and energy by ovipositing into host fruits with conspecific bore holes, despite the potential costs to offspring in increased competition and, consequently, survival (Papaj et al., 1992; Lalonde & Mangel, 1994). In extreme cases, the time and energy involved in solitary nest excavation can be prohibitive, as in some bee species (Danforth, 1991). Nesting migrations can be substantial in reptiles such as sea turtles or some iguanas (Bjorndal, 1982; reviewed in W. K. Hayes et al., 2004), thereby placing considerable energetic strains on those animals that may ultimately lead to selection for minimizing energy expenditure once the nesting areas are reached (J. J. Bull & Shine, 1979; Bjorndal, 1982; D. I. Werner, 1983; but see Congdon et al., 1989). At least some species of iguanas, monitor lizards, snakes, and the tuatara take several days to complete nesting (A. S. Rand 1968; Burger & Zappalorti, 1992; W. K. Hayes et al., 2004; Doody, James, Ellis, et al., 2014; Doody, James, Colvyas, et al., 2015; N. Nelson, personal communication). Soils are extremely hard or compact at some nesting sites, and, as a result, nesting burrow excavation requires considerable time and energy. Mothers of these species are known to deposit their eggs in the chambers of conspecifics (references above) and may save considerable energy and time, or ensure their survival, by doing so (Wiewandt, 1982; A. S. Rand & Dugan, 1983; Bock & Rand, 1989; Mora, 1989; Burger & Zappalorti, 1991; Doody, James, Colvyas, et al., 2015; J. B. Moss et al., 2020). The commonness of mothers excavating previous nesting burrows to save on excavation costs at communal nest sites is manifested in vigorous nest defense against conspecifics in some iguana species and in tuataras (Rauch, 1988; Refsnider et al., 2007; J. B. Moss et al., 2020).

Second, in addition to energetic costs, these tasks could involve allocating considerable time that could otherwise be spent foraging or performing other important activities such as feeding or thermoregulation. Third, communal egg-laying may be the result of predator avoidance if the landscape of predation is patchy. In at least some animals, mothers are especially vulnerable to predation during egg-laying. Dragonflies, for instance, have been known to reduce their vulnerability to predation by ovipositing communally in frog-free areas of ponds (McMillan, 2000). Evidence is lacking in reptiles, but communal egg-laying in complex burrow systems may allow lizard mothers to reduce their accessibility to predators during nesting (A. S. Rand & Dugan, 1983).

Communal egg-laying could increase opportunities for multiple mating if individuals aggregate prior to nesting and are sexually receptive. Multiple mating can provide numerous adaptive benefits (references in Bernardo & Plotkin, 2007). Although mating and laying are separated by weeks to months in reptiles, in the case of sea turtles, the timescale of migrations would increase the frequency of encounters between receptive individuals as the turtles concentrated in water near nesting areas; oceanic matings probably occur, but the probability of multiple mating encounters may be lower (Bernardo & Plotkin, 2007).

A few other adaptive hypotheses for communal egg-laying have been offered, but these are restricted to animals exhibiting parental care, and thus have limited applicability to our review (<2% of the species in our review are known to brood their eggs). The aggressive usurpation hypothesis involves later-nesting mothers displacing earlier-nesting mothers, thus allowing them to secure a nest site and possibly eat the conspecific's eggs (Vehrencamp, 1978; Koford et al., 1990). In green iguanas in the communal nest site on Slothia, the eggs of previous nesting females are excavated just like pebbles. They are not eaten, but vultures lurking nearby quickly fly down and eat them, often squabbling among themselves for the bounty (A. S. Rand, 1968). This is a particularly crowded site and nest excavations have shown that a single burrow may have several chambers so in this sense there could be space sharing as well as competition (A. S. Rand & Dugan, 1983).

The intraspecific brood parasitism hypothesis involves a mother laying her eggs with those of a conspecific and then deserting them, leaving the indiscriminate conspecific mother to brood (reviewed in Zink, 2000, 2001, 2005). The multiple defenders hypothesis posits that communal egg-laying may have evolved as a result of increased nest defense from predators in species that brood their eggs (K. J. McGowan & Woolfenden, 1989; Pilastro, 1992). As yet there is no evidence for the behaviors associated with these hypotheses in reptiles, except perhaps in crocodylians and various extinct taxa.

6.7.5. Other By-Product Hypotheses

Aside from habitat saturation, communal egg-laying could be an artifact of other behaviors. Social interactions among mothers could result in communal egg-laying. As we have seen in previous chapters, social behavior is widespread in reptiles, and any number of these behaviors could occur among females at or near communal nesting sites. Thus, in some cases communal nesting could be the byproduct of the close proximity of mothers to a potential nest site. In particular, social aggregations are relatively common in geckos (B. Greenberg, 1943; Cooper et al., 1985; Kearney et al., 2001; López-Ortiz and Lewis, 2002; Shah et al., 2003; Lancaster et al., 2006; Barry et al., 2014). In fact, geckos possess the highest incidence of communal egg-laying among lizards (see Fig. 6.1). In another example, gravid iguanas exhibit territorial displays in nesting areas but not in nearby non-nesting aggregations (C. C. Carpenter, 1966; A. S. Rand, 1968; Rand & Rand, 1976, 1978).

The kin selection hypothesis suggests that relatives contribute to communal clutches (Tallamy, 1985). Specifically, mothers from the same maternal lineage are expected to gravitate to the same egg-laying area, causing an aggregation of related females. The resulting communal egg-laying is not an adaptation in itself, because mothers or eggs may not directly benefit from laying with conspecifics. Graves and Duvall (1995) suggested that communal egg-laying in squamate reptiles may result from perennial laying due to natal homing (A. S. Rand & Dugan, 1983; Mora, 1989; Burger & Zappalorti, 1992). No study has addressed this in reptiles, but there is evidence for such relatedness in birds (Koenig & Stacey, 1990). A single velvet gecko returned to its birth site to nest (J. K. Webb et al., 2008), and there was some genetic structuring in a communally nesting population of green iguanas that was consistent with natal homing (Bock & McCracken, 1988). Molecular data suggest that natal homing produces aggregates of closely related nesting females in several turtle species (Meylan et al., 1990; Allard et al., 1994; Freedberg et al., 2005; Lohmann et al., 2013).

6.7.6. Adaptive Explanations—Benefits to Hatchlings

Hatchlings emerging en masse from a communal egg-laying site could satiate predators. Indeed, synchronous hatching among and within turtle clutches has been suggested to evolve to confuse predators or dilute predation risk (also known as predator swamping) (Pritchard, 1969; Ims, 1990; Eckrich & Owens, 1995; Spencer et al., 2001; Elphick et al., 2013). In particular, sea turtle hatchlings can be subject to high predation risk in their short journey to the ocean (D. W. Ehrenfeld, 1979; Stancyk, 1982; L. Brown & Macdonald, 1995), and so could benefit from mass emergence of hundreds to millions of hatchlings. Terrestrial predators are apparently quickly sated by this sudden availability of surplus prey, allowing the survival of a large proportion of hatchlings (Cornelius, 1986; Cornelius et al., 1991; Bernardo & Plotkin, 2007). See Chapter 8 for a more thorough discussion on synchronous hatching and emergence.

6.8. Behaviors Related to Communal Nesting

6.8.1. Traditional Site Communal Egg-Laying

In a review of communal egg-laying in reptiles and amphibians, Doody, Freedberg, and Keogh (2009) found that >33% of papers on communal nesting in reptiles reported eggshells from previous clutches within the year or from a previous year, or years, in the nest sites (see also newer references: e.g., Sreekar et al., 2010; Karunarathna & Amarasinghe, 2011; Luiselli et al., 2011). Traditional communal egg-laying is particularly evident in sites in which eggshells are preserved (e.g., in caves or rock crevices) or conspicuous (e.g., sea turtle mothers in arribadas). In soil-nesting reptiles, eggshells tend to disintegrate within a few years (Burger & Zappalorti, 1992; JSD, pers. obs.), perhaps leaving no detectable olfactory cues from previous years. Nesting in traditional sites likely reflects longitudinal communal egg-laying within individual mothers, and it could evolve for many of the same reasons as communal egg-laying. As stated above, kin selection and natal homing could be mechanisms underlying traditional nesting. Field experiments manipulating eggshells and olfactory cues associated with nesting mothers would be useful in determining the ultimate reason for traditional nesting, and perhaps could disentangle those reasons from the ones involved in communal nesting, if they exist.

6.8.2. Interspecific Communal Egg-Laying

In some cases, more than one reptile species lay their eggs together (Brodie et al., 1969; Vitt et al., 1997; Shea & Sadlier, 2000; Krysko et al., 2003; Doody, Castellano, Rhind & Green, 2013; Doody, Zavala, et al., 2018; Doody, 2019). For example, Vitt et al. (1997) found the eggs of South American clawed gecko *Gonatodes humeralis* with those of four other lizard species (*Norops trachyderma*, *Arthrosaura reticulata*, *Gonatodes hasemani*, and *Thecadactylus rapicauda*), and Doody (2019) found the eggs of three lizard species in one nest (*Intellagama lesueurii*, *Acritoscincus duperreyi*, and *Lampropholis guichenoti*). As in conspecific mothers, copying behavior by interspecific mothers could be beneficial, especially since suitable incubation conditions tend to be similar for most reptile eggs (e.g., 26°C–34°C, moist soil, or high humidity). Although interspecific communal egg-laying may be incidental and uncommon, it may reflect pheromonal cues. In the laboratory, common keelback snakes preferentially oviposited in nest sites with eggshells from a slaty-grey snake (*Stegonotus cucullatus*) over empty nest sites (G. P. Brown & Shine, 2005). The aggregation pheromones produced by black flies are identical in several closely related species (McCall et al., 1997). Perhaps communally egg-laying reptiles also possess aggregation pheromones (Elphick et al., 2013), and perhaps these are phylogenetically conserved enough to attract multiple species to a communal egg-laying site. This is probably not the case, however, in Panama on the islet Slothia, where the same small communal site had several dozen nesting green iguanas and also nests of an American crocodile and slider turtles (Burghardt et al., 1977).

6.9. Evaluation of the Evidence

In an otherwise authoritative treatment of the biology of lizards, Pianka and Vitt (2003) scarcely mention communal egg-laying and generally dismiss the behavior as an artifact of a shortage of nest sites. Since that time, laboratory manipulations have unambiguously revealed that conspecific attraction to eggs does occur in lizard and snake mothers (reviewed in Doody, Freedberg & Keogh, 2009). Nevertheless, our knowledge of the nests and eggs of reptiles pales in comparison to that of other terrestrial vertebrates. For example, only 6% of eggs and nests were known for Australian lizards (of the three families that commonly lay communally) in 2009 ($N = 411$ oviparous species), a figure that is probably similar to our knowledge of the same for other continents (indeed the estimated figure is 7% for all lizards, see Table 6.1). This is in stark contrast to some other tetrapods, such as birds, for which complete field guides to the eggs and nests are available for several continents (e.g., C. Harrison, 1975; Beruldsen, 1980; Baicich & Harrison, 1997). Comprehensive cost-benefit analysis of communal egg-layers is lacking, but researchers have confirmed benefits and costs in a number of species, along with intriguing speculation on potential costs and benefits. Adaptive explanations are always difficult to test, but we would like to highlight in particular the lack of evidence for predator dilution as the selective agent for the evolution and maintenance of communal nesting (Doody, Sims & Georges, 2003; Tucker et al., 2008). The notion of predator dilution certainly is reasonable for sea turtles in which dozens to thousands of hatchlings emerge simultaneously. Yet, predator dilution continues to be forwarded as a likely reason for communal nesting in species other than sea turtles (Tucker, 1997; Spencer et al., 2001; Elphick et al., 2013), species in which there is little evidence for simultaneous emergence (Doody, Sims & Georges, 2003; Tucker et al., 2008).

We hope the provocative and recent studies outlined here not only challenge conventional thinking, even among herpetologists, but, more importantly, stimulate further research. The systems used by many communally laying reptiles can facilitate advances in our understanding of egg aggregations by bringing nest site choice and its associated costs and benefits into sharper focus (Doody, Freedberg & Keogh, 2009). In this way, the diversity among reptiles provides a means for narrowing the gap between evolutionary theory and the supporting empirical data. This can occur by offering model systems for understanding both fundamental adaptive advantages for communal egg aggregations and their attendant behavioral mechanisms, as well as contributing to a better understanding of conspecific attraction and social behavior evolution in animals.

6.10. Future Directions

The egg-laying system of lizards is ideal for testing adaptive hypotheses for communal laying in both the laboratory and in the field. First, in addition to the predominant lack of parental care and associated behaviors, we note that, in reptiles, mating is generally uncoupled from the oviposition site. Second, the cue for communal egg-laying is likely to be the presence of eggs or eggshells, and these are typically easily manipulated. Maintaining a breeding colony of small lizards, such as skinks or geckos, is also straightforward, compared to birds for example.

How should future research into communal egg-laying in reptiles proceed? Although we argued above that the saturated habitat hypothesis cannot, in many cases, explain communal egg-laying, it should be investigated prior to exploring adaptive benefits (Doody, Freedberg & Keogh, 2009). There are surprisingly few studies that have compared attributes (e.g., temperature, moisture) between used and available nest sites to determine if the conditions encompassing the eggs do not exist immediately beyond the eggs, or in nearby areas accessible by nesting mothers. Focal observations of the movements associated with nesting, in conjunction with the above microhabitat analysis, might offer a useful approach for determining the sites available to mothers and their suitability for embryonic development (Doody, Freedberg & Keogh, 2009).

The habitat saturation hypothesis is also readily addressed by testing for conspecific attraction, although the two are not mutually exclusive. Conspecific attraction to eggs or eggshells has been demonstrated in every species tested, and this approach offers a good starting point for revealing potential adaptive benefits. Of particular interest is the possibility that aggregation pheromones exist in communally laying reptiles, as they do in insects (McCall et al., 1997). If so, could those pheromones be a way of surveying for communal nesting in secretive species for which nests are not known?

An appropriate framework for investigating the adaptive value of communal egg-laying would involve quantifying the associated costs and benefits to both mothers and offspring (Doody, Freedberg & Keogh, 2009). Parent–offspring conflict (sensu Hamilton, 1964a, 1964b; Trivers, 1974) could even occur in populations of communally laying species when the payoff for mothers (savings of time and energy or increased likelihood of survival) offsets or exceeds the costs incurred by the offspring (e.g., competition, survival) (Takasu & Hirose, 1993; Lalonde & Mangel, 1994). Using this framework, we would predict communal egg-laying to occur in species or populations with relatively high costs to ovipositing mothers, compared to those incurred by their offspring (Doody, Freedberg & Keogh, 2009).

Cost-benefit analysis of communal egg-laying would ideally include the energetic costs of nesting, which are generally not known for reptiles and amphibians (but see Rand & Rand 1976, 1978; D. I. Werner, 1983; Congdon & Gatten, 1989). What are the energetic and time costs associated with migration to an egg-laying site (D. I. Werner, 1983; Congdon & Gatten, 1989; Burghardt, 2004), assessment of egg-laying sites, and nest excavation (Wiewandt, 1982; Doody, James, Colyvas, et al., 2015; Doody, McHenry, Brown, et al., 2018; Doody, McHenry, Durkin, et al., 2018)? How do these costs differ between communal and solitary layers? Studies of the relative proportion of communal versus solitary layers would be particularly useful in species in which eggs are more easily located, especially in a comparative context. For instance, comparison of survival and other costs between two or more populations with differing proportions of communal layers might elucidate ultimate factors underpinning those proportions.

Trade-offs also occur within populations. Earlier-nesting Cuban iguanas (*Cyclura nubila*) dug deeper nests with higher hatching success than later shallower nesters, but the latter may reflect a reduced excavation cost for mothers (J. B. Moss et al., 2020). In the same study, larger females nested earlier in "priority oviposition sites" and defended them, while smaller females nested later but forewent extended nest defense. Tracking individuals and determining their decision-making under various contexts would thus be fruitful in species where this is possible or practical.

Could communally laying reptile mothers be copying the choices of conspecifics? Individual animals can copy the choices of other individual animals rather than make their own choices, and this has become a popular and fast-growing area of social behavior research. For example, in "mate choice copying," some females mate with males based on the male's observable mating success with other females, rather than based on more traditional traits exhibited by the male (Pruett-Jones, 1992), while in "social learning," individuals learn to perform a task by imitating conspecifics (Galef & Laland, 2005). There are three important and closely related components of such copying. First, individuals must retain the ability to perform a biological task without copying (there needs to be an original performing a behavior for others to copy). Second, individuals must trust the judgment or choice of another individual (anyone can be the original). Third, copying must be advantageous, or at least not detrimental, for the copier's evolutionary fitness. This copying framework can be applied to mothers in search of an egg-laying site. Egg-bound mothers can thus copy the choices of three types of individuals: conspecifics both within (communal laying) and among years (traditional laying), their own mothers (natal homing), or themselves (site philopatry). We must be diligent, however, in demonstrating that there is conspecific attraction, and thus copying, in each species (or perhaps population), as the most parsimonious explanation according to some is that communal egg-laying in reptiles is the sometimes imminent result of a shortage of suitable nest sites (reviewed in Graves & Duvall, 1995; Doody, Freedberg & Keogh, 2009).

Future endeavors in the field or laboratory could elucidate (1) the influence of the availability of egg-laying sites on the propensity to lay communally; (2) the relative cues offered by eggs versus eggshells (i.e., do eggshells signal nest sites of superior quality because eggs hatched successfully there?); (3) the existence of a dilution effect; (4) the relative amount of time and energy expended by communal layers versus solitary layers; (5) the extent to which mothers would trade-off nest site quality to achieve communal laying, given the apparent benefits; (6) the role of experience in the propensity to lay communally; (7) decision-making of individuals within a communally laying population; (8) the influence of the energetic state of mothers on their propensity to lay communally; (9) the occurrence of natal homing; and (10) behavior associated with communal egg-laying (e.g., movements and social interactions). To this, we add the need to reveal patterns in communal nesting in reptiles to better interpret the behavior in birds and dinosaurs (Burghardt, 1977b).

7 |

Parental Care

Parental care is probably the most mysterious part of reptile biology. Its distribution between and within taxa is difficult to explain, its origins are extremely uncertain (Trumbo, 2012, p. 85), and its extent is still underappreciated (see Chapter 1). Parental care and family interactions are, of course, major aspects of social behavior, and some claim they are also prerequisites for complex social organization (see, for example, R. D. Alexander, 1974).

The earliest evidence of parental care in amniotes comes from varanopids, primitive amniotes somewhat similar in general body shape to modern monitor lizards. Varanopids are usually considered to be synapsids, although a recent study (D. P. Ford & Benson, 2019) found them to be basal diapsids. An adult varanopid *Heleosaurus scholtzi* from the middle Permian (~272–260 million years ago) was found surrounded by four half-grown juveniles; the patterns of the fossil preservation strongly suggest parental care (Botha-Brink & Modesto, 2007). Another varanopid, *Dendromaia unamakiensis*, was found with a juvenile inside a hollow in a petrified tree trunk from the Late Carboniferous, ~310 million years ago, close to the origin of synapsids and diapsids, showing that parental care was already present in early amniotes (Maddin et al., 2020).

All of today's synapsids are called "mammals" (from a Latin word for "breast") because all of them have parental care involving females feeding offspring with milk; the length of lactation ranges from four days in hooded seals (*Cystophora cristata*), which abandon their offspring upon weaning (Stewart, 2014), to seven years in Sumatran orangutans, *Pongo abelii* (Williamson et al., 2013). Males are known to lactate in two species of bats (see Kunz & Hosken, 2009). The total length of parental assistance (including financial support and help with raising children) can exceed 50 years in some humans (*Homo sapiens*).

As for diapsids, a study by J. R. Moore and Varricchio (2016) found indirect evidence that the ancestral condition for this group was absence of parental care; crocodylians apparently evolved parental care independently from dinosaurs, and biparental care evolved from maternal care in the former but from paternal care in the latter; the earliest birds apparently inherited biparental care from their non-avian dinosaur ancestors. Tullberg et al. (2002) also arrived at the conclusion that biparental care was the ancestral condition in birds but found evidence that parental care is ancestral to archosaurs. However, all these suggestions are based on phylogenetic bracketing and very limited fossil data, and thus should not be taken as firm conclusions at this point. (Note that parental care in squamates has almost certainly evolved independently.) All modern reptiles are obligate egg-layers except for squamates, where many lineages contain both oviparous and viviparous forms. The early claims that live-bearing reptiles were ovoviviparous, merely retaining eggs within the female's body for thermoregulatory reasons, have been discarded for many taxa, as true placental connections are present that transmit nutrients and so forth (Blackburn, 2015). Yet this view persisted even well into the twentieth century, as shown by a letter written to *Science* by Chapman Grant, the founding editor of the journal *Herpetologica*, discrediting the rationale for a paper by GMB showing that food preferences were not transmitted to embryos via maternal experience (C. Grant, 1971).

It is generally believed that the first reptiles laid eggs. There is no direct evidence to support this claim, and the earliest reptiles for which the mode of reproduction is known were viviparous (Piñeiro et al., 2012), but it is a logical assumption given that (a) most extant amphibians lay eggs, and reptiles evolved from amphibian-like ancestors, and (b) switches from oviparity to viviparity have happened at least 140 times in vertebrates, while the opposite is relatively rare and postulated only for squamates (Pyron & Burbrink, 2014). Squamates have evolved viviparity somewhere between 35 (M. S. Y. Lee & Shine, 1998) and 141 (Blackburn, 1985; Pyron & Burbrink, 2014) times. Most extinct marine reptiles were likely viviparous (Blackburn & Sidor, 2015); many have evolved viviparity before even becoming marine (Motani et al., 2014), some as early as the Permian (Blackburn & Sidor, 2015). Mammals have evolved it either once or twice (it is unknown if marsupials and placentals evolved viviparity independently or inherited it from a common ancestor). Until recently, there were only two tetrapod genera (bushmasters, *Lachesis*, and sand boas, *Eryx*) for which a change from viviparity to oviparity was strongly suspected, although not proven (Fenwick et al., 2011). Recent phylogenetic analysis suggested in squamates that reversals in either direction happened with almost equal frequency (Pyron & Burbrink, 2014), but this conclusion is contradicted by numerous lines of other evidence (Blackburn, 2015). It has been suggested that oviparity has the benefit of blocking mother-to-offspring parasite transmission, but recently it was discovered that at least some parasites are capable of invading developing eggs (Feiner et al., 2020).

Currently, about 15% of reptiles are viviparous; that number reaches 100% in some particularly cold areas (Tinkle & Gibbons, 1977). Some viviparous lizards and snakes are known to have placentae; these include numerous skinks (Scincidae), Yarrow's spiny lizard (*Sceloporus jarrovii*), night lizards (*Xantusia*), De Vis's banded snake (*Denisonia devisi*), the ornamental snake (*D. maculata*), the rough earth snake (*Haldea striatula*), North American brown snakes (*Storeria*), and gartersnakes (*Thamnophis*) (Weekes, 1935; Rahn, 1939; Clark, et al., 1955; Guillette et al., 1981). Some skinks even developed complex placentae and mammal-like hormonal regulation (Brandley et al., 2012). But, mysteriously, no known member of archosaurian lineage (including turtles, which are currently believed to be a sister group of archosaurs) has ever evolved viviparity. As far as we know, all dinosaurs, pterosaurs, and extinct crocodylomorphs laid eggs, and so do all extant birds (about 10,000 species), crocodylians, and turtles, despite the fact that some of them suffer horrendous nest losses due to predation (see, for example, Snow, 2004). There must be some aspects of archosaurian reproductive biology that, once evolved, have made the switch to viviparity impossible, but despite many theories (some invoking differences in egg anatomy as possible explanation) and much discussion (such as in Tinkle & Gibbons, 1977), what these aspects are remains unknown.

Nevertheless, being unable to become viviparous did not prevent archosaurs from evolving extremely advanced and complex parental care. Indeed, some forms of archosaurian parental behavior are among the most spectacular in the animal world.

7.1. Archosaurs: Nests of Mystery

Among extinct archosaurs, reproductive behavior has been most studied in non-avian dinosaurs. Only very limited fossil evidence of such behavior has been found for other lineages. It appears that some Triassic archosaurs (possibly aetosaurs) already dug large open nests at particular sites and lined them with plant material (Avanzini et al., 2007). It also appears that at least some pterosaurs nested in large colonies (X. Wang et al., 2014, 2017), as did some birds as early as the Cretaceous (Naish, 2014), although, unlike modern birds, they placed their eggs in nests vertically, which means they did not turn them periodically like most modern birds do (Fernández et al., 2013). Such early evidence, and the presence of parental care in all extant lineages (see below), suggests that having parental care has always been widespread among archosaurs and likely was the ancestral condition for the group.

It has been suggested that flaplings (pterosaur hatchlings) required not just prehatching, but also posthatching parental care (X. Wang et al., 2017). However, this has been disputed: according to Unwin and Deeming (2019), flaplings were probably capable of flight and so posthatchling parental care, although possible, is yet to be confirmed.

Breeding behavior of dinosaurs has been the subject of much interest in recent years, and some fascinating new discoveries have been made, although we still know frustratingly little. The following account is mostly based on the overview in Brusatte (2012). Dinosaurs nested in extremely diverse locations, ranging from lakeshores to hydrothermal areas. Their nests were shallow open pits in the ground, rounded or elliptical in shape, sometimes with a rim or located atop a mound. The eggs could be haphazardly dropped into the nest or neatly arranged in a highly symmetrical fashion. Some species partially or completely buried their eggs or covered them with plant matter. A recent study by Tanaka et al. (2018) found that hadrosaurs and some sauropods built organic-rich mound nests that relied on heat generated by microbial decay for incubation. Other sauropods dug simple hole nests in sand that relied on solar or potentially geothermal heat for incubation; the ability of such nests to provide sufficient warmth could be a limiting factor in the dinosaurs' latitudinal distribution. The giant sauropod dinosaurs apparently divided their eggs among a number of small clutches, most likely to minimize predation losses (Ruxton et al., 2014).

Many species nested in colonies, sometimes with thousands of regularly spaced nests, and these colonies may have persisted at the same locations for thousands of years. So far, only derived bird-like theropods, such as *Troodon*, *Oviraptor*, and *Citipati*, are known to have brooded their eggs, sitting on top of them in bird-like fashion. This was not an easy job, considering that egg incubation times in these species likely exceeded 100 days (S. A. Lee, 2019). Parental care probably allowed for high egg survival, estimated at 60% in one particularly well-preserved theropod nesting colony (Tanaka et al., 2019). There is evidence that males brooded the eggs, which is interesting, because phylogenetic analysis suggests that paternal care is ancestral to birds, and the most basal extant birds—ostriches (*Struthio* spp.)—have paternal care (Varricchio et al., 2008). One species, *Byronosaurus jaffei*, is suspected of having been a nest parasite because its hatchlings have been found in the nest (among the eggs) of a larger dinosaur, *Citipati osmolskae* (Norell et al., 1994).

Dinosaur hatchlings were generally small (eggs of the largest sauropods are about the same size as those of some flightless birds), but evidence of posthatching parental care is limited (this is hardly surprising since preservation of such evidence requires very special circumstances). Adults of a few species have been found close to juveniles, but it is unknown if they were taking care of them or their bones simply happened to be deposited at the same place. More or less solid evidence has been found for just four species. An adult *Psittacosaurus lujiatunensis* has been found in association with 32 juveniles, all apparently buried in life-like postures while oriented in the same direction (Meng et al., 2004). Another such group, instantly buried by a lahar flow, contained six juveniles of two age classes, suggesting extended parental care or complex juvenile sociality (Qi et al., 2007). Bones of one adult and two half-grown juveniles of a small burrowing dinosaur, *Oryctodromeus cubicularis*, were found in the same burrow, strongly suggesting extended parental care (Varricchio et al., 2007). Juveniles of a larger, highly gregarious dinosaur, *Maiasaura*, hatched with underdeveloped legs and poorly ossified bones, and remained in their nests for at least a year while being fed by adults. Interestingly, the juveniles had shortened faces and large eyes; these "cute" features are typical for tetrapod babies dependent on parental care and serve to elicit parental behavior from adults (Prieto-Marquez & Guenther, 2018), although they are also found in many taxa with no parental care. Hatchlings of a prosauropod, *Massospondylus carinatus*, lacked teeth and also had small limbs and huge heads, so they were probably dependent on feeding by adults (Reisz et al., 2005). Track sizes at their nesting site suggest that these hatchlings remained in the immediate vicinity of the nest for a long time (Reisz et al., 2012). As for sauropods, the largest of all dinosaurs, their breeding behavior was apparently diverse, with some species covering their eggs with (possibly rotting) vegetation and others leaving them open (possibly for brooding); clutches were relatively small, suggesting multiple clutches per year and a possibly lack of posthatching parental care (Sander et al., 2008). However, juvenile sauropods had growth rates similar to those of juvenile whales; such rapid growth in plant eaters is difficult to explain unless some analogue of milk was provided by adults (Erickson et al., 2001).

Almost all modern birds provide extensive parental care, although given rapid maturation this is often limited to several weeks, whereas in crocodylians parental care can last for well over a year (see below); it can be speculated to have been much longer in large dinosaurs. A useful review by

J. D. Ligon (1999) discusses the evidence on the dinosaur–avian transition in brood care and summarizes what is known about parental care in birds. A few avian lineages have evolved facultative or obligatory brood parasitism, most advanced in some cuckoos (Cuculidae), and thus have secondarily lost parental ability, though their offspring are still dependent upon it. Megapodes (Megapodiidae) have apparently reverted to reproducing in the manner similar to that of many non-avian dinosaurs: males build large mounds where heat for egg incubation is provided by rotting vegetation or geothermal sources, and a constant temperature in the interior is maintained by adding or removing vegetation (males use their bills as thermometers), while females visit the mounds periodically to drop eggs in them. Enormous mounds on New Caledonia—some 50 m across and 4–5 m high—have been interpreted as nest mounds built by the recently extinct megapode-like bird *Sylviornis* (Mourer-Chauviré & Popin, 1985). Posthatching parental care is absent in megapodes and also reduced in a few genera belonging to other basal lineages; for example, hatchling mergansers (*Mergus*) can often survive on their own. Some nests, such as those of plovers (*Charadrius*), are nothing more than a flat patch of sand, and penguins of the genus *Aptenodytes* do not build nests at all, instead brooding their eggs in a special skin fold. But the general trend in bird evolution seems to be toward extensive biparental care (sometimes provided not just by the parents, but also by older siblings or other relatives), altricial young, and complex nest-building behavior. Nests of some passerine birds are marvels of engineering and look like exquisite works of art, while those of others are incredibly labor-intensive. Although some birds such as swifts (Apodidae) and albatrosses (Diomedeidae) receive no postfledging care, most continue to be fed by their parents after leaving the nest, sometimes for almost a year like in cranes (Gruidae). Pigeons (Columbidae), flamingos (Phoenicopteridae), and some penguins (Spheniscidae) secret a substance called "crop milk" to feed their offspring (Eraud et al., 2008).

Crocodylians have also evolved extensive parental care, but in a manner very different from that of birds. Females build nests that may be shallow burrows in sand or large mounds of dead vegetation. Some species may build both types in different places (Grigg & Kirshner, 2015). Crocodiles of genus *Crocodylus* sometimes nest in loose colonies. Schneider's dwarf caimans (*Paleosuchus trigonatus*) living in shady rain forests, where direct sunshine as the source of heat is not available, often build their nests next to, or even on top of, termite mounds, using the heat produced by decomposing organic matter inside the mound for egg incubation (Magnusson et al., 1985).

Females usually guard their nests (Plate 7.1a), but not always. Inexplicably, there is extreme individual variation in parental devotion: some females immediately abandon the nest, while others guard it with their lives (Platt et al., 2020). In at least eight species, males can participate in nest guarding (Charruau & Hénaut, 2012). Conversely, there is no evidence of nest attendance or defense in Johnston's crocodile (*C. johnstoni*), a relatively well-studied species (G. J. W. Webb et al., 1983), and no evidence of nest defense in slender-snouted crocodiles (*Mecistops*) or in some populations of the larger, more powerful and aggressive American crocodile (*C. acutus*), despite high predation pressure (Abercrombie, 1978; Murray et al., 2016, 2020). In other populations of the American crocodile, some females do defend nests (Charruau & Hénaut, 2012). Nest defense in Panama by an American crocodile included predation on nesting female iguanas close to her eggs, which were buried in a non-mound nest (filmed by GMB and included in the film *Dragon of the Trees* [Burghardt & Rand, 1980]). The earliest fossil evidence of nest guarding in crocodylians is from the middle Eocene (Hastings & Hellmund, 2015), but the behavior is likely much older.

Crocodylians have temperature-dependent sex determination (see Grigg & Kirshner, 2015 for an overview). Eggs are often laid in layers, and it has been suggested that by doing so the female ensures producing both male and female offspring, but this has never been proven. There are also anecdotal claims of the female's ability to maintain constant temperature in the nest by adding or removing plant matter (in the same way as male megapodes do it), but this has never been documented and probably is not true. On hot days the female will sometimes position herself briefly over the nest in such a way that the water from her body drips on the nest, possibly cooling and/or moisturizing it (McIlhenny, 1935); urinating on the nest could also be used but apparently has never been observed.

Baby crocodylians of all extant species begin to vocalize before hatching (Murray et al., 2020). If the mother is still around, she usually digs up the nest, helps the hatchlings out of their eggs, and carries them to the water (Plate 7.1b–c), sometimes up to 10 at a time (McIlhenny, 1935), although gharials apparently do not do this (Lang, 2015). This is not an easy task. As Herodotus (2008) was the first to note, ca. 420 BC, crocodylians are unique among extant tetrapods in "growing the largest after being the smallest," i.e., they have an extreme size difference between hatchlings and adults. A female estuarine crocodile (*C. porosus*) can weigh 10,000 times more than a hatchling. Still, she is capable of gently opening the egg, taking the hatchling into her massive jaws, and carrying it for a few meters, never harming the baby in any way (G. J. W. Webb & Manolis, 1989).

For the most part, the extent of posthatching parental care in crocodylians has been discovered only very recently, and for some species virtually nothing is known. In fact, though well-described in alligators almost 90 years ago (McIlhenny, 1935), it was denied as occurring by noted crocodylian specialists for decades afterwards (e.g., Neill, 1971). Females of at least some species put their young in multibrood crèches and take turns guarding them (Pinheiro, 1996); other species exhibit biparental care and can feed their young (Plate 7.1d) (Brueggen, 2001; N. Whitaker, 2007). Herzog (pers. comm. to GMB), in his observations of alligators in Florida in the early 1970s, described watching a mother hold a decomposing pig's head in her jaws while the babies tore off small pieces and ate them. As an example of the resistance of herpetologists to reports of such behavior, McIlhenny (1935) reported similar behavior in his pathbreaking work on alligator life history and this was arrogantly dismissed by a leading crocodylian biologist in his magisterial review (Neill, 1971; see also Burghardt, 2020). In the Indian gharial (*Gavialis gangeticus*), the dominant male guards a huge communal crèche (sometimes more than a thousand juveniles), coming from 5–25 broods (Lang, 2015; Lang & Kumar, 2016). Vocal communication between juveniles and adults is apparently very complex and involves numerous distinct signals (see Mathevon et al., 2013, for bibliography). Some individuals of both sexes are extremely protective, not only of their offspring but of any juvenile, and they would respond to a distress call of an unrelated hatchling (not necessarily conspecific) even at a great risk to their own life (crocodile hunters sometimes imitate those distress calls to attract adults) (Neil, 1971; Myrna Watanabe, pers. comm.). Female American alligators are sometimes surrounded by juveniles from two broods, hatched a year apart (VD, pers. obs.; see Fig. 11.3). Male Siamese crocodiles (*Crocodylus siamensis*) guard adopted hatchlings (John Brueggen, unpubl. data). Cuvier's dwarf caimans (*P. palpebrosus*) estivate with their young in burrows during the dry season (Campos et al., 2012).

However, some crocodylians are just bad parents. One female American alligator was observed to leave her brood unattended for hours at a time to join "dances" (group courtship) ~2 km away; this resulted in all of her offspring being lost, probably to heron predation (Dinets, 2010). American crocodiles apparently do not have posthatching care (Murray et al., 2016; VD, pers. obs.). The Sunda gharial (*Tomistoma schlegeli*) is believed to exhibit no parental care after laying the eggs, but its nesting has only been observed in captivity (Bezuijen et al., 1998); juveniles of this species produce distress calls (Bonke et al., 2015), indicating that some parental care might exist in the wild. Schneider's dwarf caimans guard their broods for only a week or two after hatching (Grigg & Kirshner, 2015).

Parental care can make a big difference in survival rates of crocodylian offspring (Joanen, 1969), so it should be strongly selected for. Why there is still so much variation in this important trait remains unknown and contradicts most evolutionary models. In some cases, the nests are far from the water and thus more difficult to defend (JSD, pers. obs.), but not all differences can be explained that way. Older, larger females are more likely to be good nest guardians (C. J. Baker et al., 2019), but that does not seem to explain all variation, either. Human hunting can lead to loss of nest guarding behavior (McIlhenny, 1935); however, in Bolivia, where the black caiman (*Melanosuchus niger*) was once hunted almost to extinction, while the yacare caiman (*Caiman yacare*) has not suffered so much hunting pressure, black caiman females guard their nests as vigorously as yacare caiman females (Alfonso Querejazu, pers. comm.). Perhaps selection against nest guarding was more intense where people frequently raided nests and killed guarding females, rather than just hunted adults; this might explain the lack of nest guarding in Johnston's crocodile (see above). Juvenile American crocodiles give distress calls more often in areas of high human disturbance (Boucher et al., 2020), possibly because females become less responsive there.

7.2. Other Reptiles: From Subtle Self-Sacrifice to Formidable Defense

Until recently, it was believed that most reptiles, except for crocodylians and some viviparous squamates, simply bury their eggs in soil and forget about them, resulting in no parental care after egg deposition (see Shine, 1988). The same was claimed for viviparous species that give birth and then leave their offspring to their separate fates. This is, indeed, what most reptile species do. Some tortoises do not even bother to bury their eggs if the ground is not soft enough: Galápagos tortoises (*Chelonoidis nigra*) would simply leave them between rocks with no protection whatsoever (Darwin, 1839). But considerable parental solicitude can still be involved, including well before birth or hatching. We are now beginning to realize that even reptiles that do not care for their offspring after the eggs are laid can still be devoted parents; moreover, the number of species known to exhibit more extended parental care is steadily growing. Some tortoises, such as gopher tortoises (*Gopherus*), defend their nests against predators, including venomous Gila monsters (*Heloderma suspectum*) (Somma, 2003; Gienger & Tracy, 2008; Grosse et al., 2012; reviewed in Agha et al., 2013; Radzio et al., 2017). Female painted turtles (*Chrysemys picta*) select the warmest nesting sites, significantly improving the survival of eggs and hatchlings (T. S. Mitchell et al., 2015); such ability might be very widespread. Females of this and sometimes other turtle species of the southeastern United States often lay eggs in nests of American alligators, where rotting vegetation provides extra warmth and the female alligator provides protection from nest predators; this behavior is so common that in some areas almost half of alligator nests contain turtle eggs (Deitz & Jackson, 1979;

Kushlan & Kushlan, 1980). Is this turtle egg and hatchling protection part of the parental care tactics of the mother?

Limited research has shown that, at least in some reptiles, gravid/pregnant females take the needs of the developing eggs or embryos into account and alter their thermoregulatory behavior accordingly. Vipers of both viviparous and oviparous species spend more time basking if their developing offspring need more warmth, even though it means increased risk of predation (H. W. Greene et al., 2002). The ability of gravid females to warm up the developing eggs by basking drastically improves the health and general viability of hatchlings (Shine & Harlow, 1993). Could the female warming up her embryos by her thermoregulatory behavior explain the higher incidence of viviparity among reptiles living in cold temperate regions? It is also possible that by basking the female reduces her physical burden through this speeding up the development of the embryos (Lorioux et al., 2013)? Of note, mother gartersnakes kept at cool temperatures give birth to fewer and more deformed young than those at warmer temperatures (Burghardt & Layne, 1995). Recently it was discovered (Tattersall et al., 2016) that tegus (*Salvator merianae*), large Neotropical lizards, become endothermic every night during the breeding season, with body temperatures rising above the ambient by as much as 10°C; the main purpose of this adaptation is likely to aid in egg production and incubation. Facultative endothermy, sometimes used only in cold weather and sustained by spasmodic muscle contractions, is also known in some species of large pythons (Pythonidae); it appears that the pythons evolved it multiple times (Hutchison et al., 1966; G. J. Alexander, 2018). Interestingly, egg-laying mammals also raise their body temperature during egg incubation (Nicol & Andersen, 2006).

Some turtles remain with their eggs for a few days after laying. Female yellow mud turtles (*Kinosternon flavescens*) remain in their underground nests for up to 36 days (longer in drought years), possibly to protect the eggs from predation and desiccation (Iverson, 1990). Female Inagua sliders (*Trachemys stejnegeri malonei*) can dig and loosen the soil around their nests just prior to hatchling emergence (Hodsdon & Pearson, 1943). In the giant Amazon river turtle (*Podocnemis expansa*), juveniles and adults exchange vocal signals from prior to until months after hatching and migrate together (Ferrara et al., 2013). Most turtles are not yet known to care for their offspring after burying the eggs. However, for many aquatic species, climbing ashore to dig a nest and lay eggs is extremely dangerous, and some sea turtles migrate for hundreds and even thousands of miles to reach the nesting beaches (Sale & Luschi, 2009). Small ridley sea turtles (*Lepidochelys*) may minimize predation of mothers and eggs by nesting synchronously, sometimes in the hundreds of thousands (Plotkin, 2007; see also Chapter 6, section 6.6.1).

The earliest records of parental care in lepidosauromorphs are for small aquatic reptiles called choristoderes. An adult female choristodere *Philydrosaurus* from the Early Cretaceous was found surrounded by six juveniles, all of the same size and too large to be hatchlings (Lü et al., 2014). An adult female of a related species, *Hyphalosaurus*, was also found associated with a juvenile (Ji et al., 2006); this species is known to be viviparous (Ji et al., 2010). It is worth noting that the taxonomic position of choristoderes is still uncertain; some authors consider them to be basal diapsids, leading Lü et al. (2014) to suggest that postnatal parental care might be the ancestral condition for the entire Diapsida clade (but see J. R. Moore & Varricchio, 2016). But as this chapter illustrates, presence of postnatal care is a trait prone to reversals, so the current fragmentary data on extinct lineages is inadequate for making sweeping statements about the age and origins of this behavior.

Early experiments (Noble & Mason, 1933; Noble & Kumpf, 1936) showed that five-lined (*Plestiodon fasciatus*) and broadhead (*P. laticeps*) skinks brood their eggs, retrieve them, and test them with tongues upon returning to the nest. Vomerolfaction was more likely than nasal olfaction (review in Burghardt, 1970). Great Plains skinks (*P. obsoletus*) helped hatchlings leave eggs and repeatedly licked their cloacal vents up to 10 days post hatching (L. T. Evans, 1959). The majority of lizards and snakes leave their offspring immediately after burying their eggs, laying them in some sheltered place (such as inside a termite mound or under a log), or, in case of viviparous species, after giving birth. Females can choose nesting sites with optimal moisture and temperature, thus enhancing the survival of the offspring and the environmental flexibility of the species (Pruett et al., 2020; Tiatragul et al., 2020). More advanced parental care occurs in many species (see Somma, 2003, and bibliography therein), but its distribution among lineages is mysterious. It looks like large, powerful species guard their eggs and/or neonates more often; snakes do so more often than lizards; species living in cold places tend to be "better parents" than tropical ones. However, all of these patterns might be artifacts of research bias toward charismatic and dangerous species and those living in nontropical countries (Platnick, 1991). Often two closely related species with similar ecology, two populations of the same species, or even individual females from the same population show remarkably different levels of parental care—the same unexplained pattern as described above for crocodylians. Such discrepancies are found not just in squamates but also in the tuatara (*Sphenodon punctatus*): females guard their nests in some populations but not in others (Gillingham et al., 1995), and there is much variation within populations (Refsnider et al., 2009). Long-tailed skinks (*Eutropis longicaudata*) from a population inhabiting a small island off Taiwan exhibit extended parental care, while females from other known populations leave nests immediately after laying (W.-S. Huang et al., 2012). In an overview of the nesting behavior of iguanid lizards, Iverson et al. (2004) found similarly large differences in the extent of nest guarding between closely related species, conspecific populations, and individuals from the same population and suggested that nests are guarded only in response to particular stimuli, such as close proximity of other nesting females that might accidentally damage the nest while excavating their own nest. Indeed, nest guarding by female tuataras is more likely and lasts longer in the presence of other nesting females, apparently for that reason (Refsnider et al., 2009). Gerber (1997) noted that in the Cayman Islands ground iguana (*Cyclura nubila caymanensis*), females nesting in rocky areas defend their nests, while those nesting in sandy areas do not; a possible explanation is that in sandy areas the females can build deeper nests, so the eggs are less exposed. However, female green iguanas guard their own nest from other females only until it is sealed; in high-density nesting colonies, this frequently leads to other females destroying nests by digging right through them. The females apparently find this a worthwhile trade-off between a higher probability of clutch loss and higher energy expenditure at a time when energy reserves are already severely depleted by egg production and nest digging (W. M. Rand & Rand, 1976).

A study of elongated tree iguanas (*Liolaemus elongatus*), small ovoviviparous Patagonian lizards, found that females differed substantially in their parental behavior. Some females lay eggs in burrows and then stay in and around them with the hatchlings, but others lay eggs in the open; when threatened by a predator, some females cover the burrow entrance while others try to move away, probably attempting to lead the predator away from the burrow; some females cover hatchlings with their body, while others do not (Halloy et al., 2007).

In one genus, *Varanus*, some species, such as the rock monitor (*V. albigularis*), leave the nest immediately after laying their eggs and never return (Paul Reitz, pers. comm.); others, such as the desert monitor (*V. griseus*) and the Komodo dragon (*V. komodoensis*), guard their nests from predators (including conspecifics) for up to three months after laying the eggs, but then leave (Pianka & King, 2004; Purwandana et al., 2020); yet others, such as the lace monitor (*V. varius*), return to the nest at the time of hatching to assist the young in getting to the surface (De Lisle, 1996; Carter, 1999). Yellow-spotted monitors (*V. panoptes*) lay eggs in very deep, complex communal burrows (Doody, James, Ellis, et al., 2014; Doody, James, Colyvas, et al., 2015), described in more detail in Chapter 6.6.4.

According to an overview by Somma (2003), taking care of eggs and/or neonates has been documented in ~130 lizard species from at least 14 families. Some, such as the Oudri's fan-footed gecko (*Ptyodactylus oudrii*), care for the eggs communally, while a few others, such as the tokay gecko (*Gekko gecko*), exhibit biparental care. Parental duties may include nest-cleansing, hydroregulation, thermoregulation, and defense. Removing dead eggs, assisting the hatching process, and grooming the neonates has been recorded in a few species, and some lizards have been observed leaving food for their young or nudging them toward food. So far, only one species, the armadillo girdled lizard (*Cordylus cataphractus*), has been observed actively feeding its offspring (Somma, 2003). *Egernia* skinks can remain with their offspring for many years (Chapple, 2003), forming stable families (C. M. Bull, 1994; Whiting & While, 2017; see chapter 3, section 3.3).

An overview by H. W. Greene et al. (2006) shows that in one family, Anguidae, over half of the species for which information is available exhibit parental care, and the pattern of its phylogenetic distribution suggests that it might be the ancestral condition for the family. Various members of this family protect their clutches and hatchlings, assist the hatching process, protect the clutch from rot by removing spoiled eggs and empty eggshells, and/or provide some degree of thermoregulation. High incidence of viviparity in this family confirms the theoretical prediction that parental care should increase the likelihood of evolving viviparity (Shine, 1985; Shine & Lee, 1999).

We know that by being aggressive toward conspecifics female black rock (*Egernia saxatilis*) and White's (*Liopholis whitii*) skinks improve the survivorship of their offspring at almost no cost to themselves, as long as the offspring remain in their mothers' territories (Langkilde et al., 2007; Sinn et al., 2008). This raises the question of why parental care is not more widespread in lizards. Perhaps it actually is, but it remains to be discovered in many taxa. Some data suggest that differences in parental care among species, populations, and even individuals can be explained by differential predation pressure acting as an evolutionary factor, as well as individual experience (W.-S. Huang & Wang, 2009; W.-S. Huang et al., 2012), but this probably cannot explain all observed patterns. One theory suggests that parental care allows for having a smaller number of larger offspring, but it is unclear if such a relationship exists in reptiles (see review by Shine, 1978). One notable example is the Solomon Islands prehensile-tailed skink (*Corucia zebrata*), in which females give birth to just one or two very large offspring and protect them for some time after hatching (K. M. Wright, 1993). However, small litter sizes are also known in many island species without parental care (Cody, 1966), and all anoles lay just one egg at time (Losos, 2009).

Hoare and Nelson (2006), who studied New Zealand common geckos (*Woodworthia maculata*), made a fascinating observation. These nocturnal lizards form large aggregations in daytime shelters. At dusk they disperse, climbing surrounding vegetation and moving from tree to tree, often in small groups. In a few instances, adult geckos were observed to grab twigs belonging to different trees, bridging the gap between them; neonate geckos then used these "bridges" to move from tree to tree without the necessity of walking on the ground where predation risk is higher. Unfortunately, it is unknown if the adults making the bridges were related to the neonates using them.

It would be interesting to know how widespread parental care is in fossorial snakes and lizards, since they are extremely cryptic and possibly enjoy low predation pressure. However, little is known about their behavior in the wild. There are isolated observations of (a) female mottled worm lizards (*Amphisbaena heterozonata*) guarding their eggs (Berg, 1898); (b) the closely related Darwin's worm lizards (*A. darwini*) forming adult-juvenile aggregations (Borteiro et al., 2013); (c) juvenile checkerboard worm lizards (*Trogonophis wiegmanni*) apparently remaining with their mothers for extended periods of time (J. Martín et al., 2011); (d) Texas blind snakes (*Rina dulcis*) caring for their eggs in communal nests (Hibbard, 1964); and (e) white-nosed blind snakes (*Liotyphlops albirostris*) laying their eggs in colonies of fungus-growing ants, *Apterostigma* cf. *goniodes*, where worker ants care for the eggs (Bruner et al., 2012; Vaz-Ferreira et al., 1973).

Parental behavior in more advanced snakes is summarized by M. J. Greene and Mason (2000; see also H. W. Greene et al., 2002) and Stahlschmidt and DeNardo (2011). In most snake species, the only known form of parental care is digging a nest or placing the eggs in a well-selected location, such as inside an American alligator nest in the case of some mud snakes (*Farancia abacura*). Increased basking has been documented in only a few species (see above), although it is probably very common in cold areas. A few snakes, such the carpet python (*Morelia spilota*) and the king cobra (*Ophiophagus hannah*), build nest mounds. Postlaying care of eggs or young (Fig. 7.1) is known in over 100 snake species belonging to five families, and for most of them this knowledge is based on just one or very few observations, indicating that the actual number of snakes providing postlaying care is much larger. True cobras (*Naja*) and some pitvipers are sometimes said to exhibit biparental nest guarding, but this is disputed (see, for example, Hill et al., 2006); there are observations of males visiting female rattlesnakes with litters, but it is unknown if the males are related to either the females or offspring (Schuett et al., 2016). A recent observation of a male timber rattlesnake (*Crotalus horridus*) displaying to an intruder in

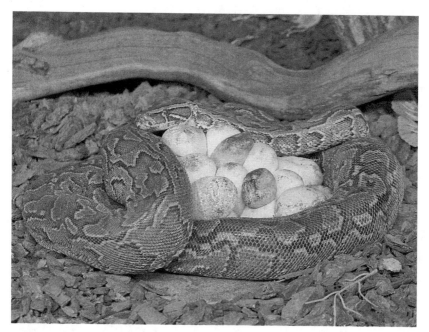

Fig. 7.1. Female African rock python (*Python sebae*) guarding her clutch. Photograph by J. Lanki (Creative Commons Attribution-Share Alike License).

front of a female with 17 babies (Plate 7.2) illustrates how little we know about even well-studied serpents (Hewlett & Schuett, 2019). Indeed, the male not only followed the observers as they retreated but then went over to another female and coiled beside her. She was also post parturition but had no babies next to her. Given that a typical litter size is 7–10, the authors surmise that the first female was guarding both litters. Thus, the possible communal behavior of snakes becomes even more mysterious as we get glimpses of their secret societies.

A Naga folk story from northeastern India (Petrovsky, 1964, p. 88) explains the origins of parental care in king cobras and pythons:

Once upon a time, jungle snakes decided to elect a king. There was much argument and even some fighting, but eventually all agreed that the great cobra should be the king. Only the python was unhappy: "Why should the cobra be the king if I am so much bigger and stronger?"

As soon as the cobra became the king, he decided that his children shouldn't be born in mud and sand like those of other snakes. He told his wife to build a palace for them, made of twigs and dry leaves.

The python said that his children were no worse than the cobra's princes, and told his wife to build a palace, too. But his wife was lazy. Instead of building a palace, she just coiled around the eggs, trying to look like a pile of leaves and twigs.

And that's how they've been taking care of their children ever since.

Two groups with particularly developed parental care are pythons (Pythonidae) and pitvipers (Crotalinae). All species of pythons coil tightly around their eggs until hatching (see Fig. 7.1), fasting throughout the incubation; this can result in body weight loss of ~40% (G. J. Alexander, 2018). Python eggs need high temperature (30°C–33°C) and humidity for development, and studies have shown that in cold weather some pythons can raise the temperature of the clutch by up to 7°C by becoming endothermic (Hutchison et al., 1966; G. J. Alexander, 2018). Female pythons constantly adjust their posture in a complex way in an attempt to maintain an optimal environment for the clutch. They also tend to become darker at that time, apparently to facilitate absorption of solar heat. Female African rock pythons (*Python natalensis*) do not become endothermic, but after their eggs hatch they stay with the brood for about two weeks and repeatedly transport heat from the outside into the nesting burrow by briefly warming up on the surface, crawling inside, and coiling around the hatchlings (G. J. Alexander, 2018). Pitvipers can coil around their eggs or cover them with their bodies, apparently providing camouflage as well as protection from the elements. The Malayan pitviper (*Calloselasma rhodostoma*) coils around eggs more often if the ambient air humidity is low, and it is also very defensive during this time (York & Burghardt, 1988). Hill et al. (2006) have also found that, despite their relatively small size, female Malayan pitvipers keep their eggs on average 1.5°C warmer than the environment, although this was not found by York and Burghardt. All viviparous nontropical pitvipers remain with their young until their first shedding of skin (Plate 7.3), leaving only to drink; the same extension of parental care is known in the Malaysian pitviper and the African rock python. Juvenile prairie (*Cro-*

talus viridis) and timber rattlesnakes follow their mothers into hibernation dens and stay with them all winter (H. W. Greene et al., 2002). Female western diamondback rattlesnakes (*C. atrox*) can maintain episodic contact with their sisters throughout their lifetimes, and they will sometimes share a den when one of them is giving birth, probably for extra protection (Schuett et al., 2016).

The fact that pythons and pitvipers exhibit extensive parental care more often than other snakes could be explained by their ability to effectively defend their offspring. However, very similar boas (Boidae) and true vipers (Viperinae) care for their broods to a much lesser extent, although there are some exceptions: a female rhombic night adder (*Causus rhombeatus*) coils around her eggs and a female horned adder (*Bitis caudalis*) stays with her offspring for a few weeks. A female rainbow boa (*Epicrates cenchria*) examined her newborns by gently pushing them with her snout, consumed a stillborn, and remained with live ones until they emerged from fetal membranes up to 36 hours later (Groves, 1981). One feature that pythons and pitvipers have, but boas and true vipers do not, is specialized heat-sensing organs. It has been suggested (A. L. Campbell et al., 2002) that these organs allow them to better regulate egg temperature, but this theory does not explain why viviparous pitvipers tend to care for their broods longer than viviparous true vipers. It could be that the heat-sensing organs facilitate brood defense by improving a mother's ability to detect mammalian predators.

Just like in lizards, high inter- and intraspecific variability of parental behavior in snakes remains unexplained. One recent study (Stahlschmidt et al., 2012) found that the extent of parental investment is a trade-off between higher offspring survival and more frequent breeding; such an equilibrium can be highly sensitive to microhabitat conditions and minor details of evolutionary and individual history.

In conclusion, parental care in reptiles seems to be generally more flexible and variable than in birds and mammals, but for now there is no generally accepted explanation of this difference. Fragmentary knowledge of parental care of reptile species and populations is hampering our understanding of its evolution, but the recent growth of information we do have suggests fascinating future comparative information.

8 |

Hatching and Emergence

A Perspective from the Underworld

Neonate animals with caring parents, such as mammals and birds, behave differently than those with little or no parental care, like most reptiles. In the former, the dependent young usually cannot survive even a few days without parental care, leaving the parent(s) with a vital role in feeding the altricial young, keeping them warm, protecting them, and allowing them to gain or develop skills needed for feeding, running, flying, climbing, hiding, and making sounds or other communicatory signals. In contrast, the young of precocial reptiles are self-sufficient at hatching or birth; most never meet their mothers or fathers (Burghardt, 1984, 1988). Remarkably, however, the sibling bond may be strong and profound, but in a secret place—embryos and hatchlings can communicate with one another beginning in the egg. Embryos can use mechanisms such as vibrations or vocalizations, and possibly even heartbeats, to synchronize, accelerate, or expedite hatching, and this provides natural selection the opportunity to improve survival during hatching or birth, and during emergence from the nest or nesting or birthing area. Here we consider increasing evidence for social interactions among sibling embryos and hatchlings within a nest, and the apparently rarer social interactions of emerging hatchlings or neonates from different nests.

8.1. Sibling Love: Social Behavior in Embryos

A recent but growing body of work has revealed that some embryos can choose their own birth date. Environmentally cued hatching (ECH), whereby hatching-competent embryos can alter their timing of hatching, may be widespread in animals (reviewed in Warkentin & Caldwell, 2009). For example, in several precocial bird species, embryos begin moving and vocalizing to one another, thereby synchronizing the hatching process (Brua, 2002; Reed & Clark, 2011). In aquatic amphibian larvae, the presence of predators, drying ponds, or other threats can both speed up hatching as well as alter the morphology and size of the animals (Wilbur & Collins, 1973; Sih & Moore, 1993). Asynchronous hatching can occur when incubation starts before the completion of egg-laying; however, in many waterfowl species incubation is initiated prior to the completion of egg-laying, yet hatching is more or less synchronous (reviewed in Reed & Clark, 2011). Vocalizations are a mechanism by which the embryos stimulate higher metabolic rates in sibling embryos (Vleck et al., 1979; Brua, 2002). Vocalizations, however, also play a role in communication between embryos and parents, which occurs in birds (Gottlieb, 1991; Bolhuis & Van Kampen, 1992), as well as in humans (Hepper et al., 1993). Some precocial birds may learn to discriminate their own mother's vocalizations from those of other parents while still in the egg, which may facilitate imprinting to the parent after hatching. Vocalizations near the end of incubation can stimulate parents to maintain adequate incubation temperatures (reviewed in Abraham & Evans, 1999; Bugden & Evans, 1999).

Reptiles do not currently feature prominently in our understanding of ECH (Doody, 2011). However, the nests, eggs, and embryos of reptiles are poorly known relative to other vertebrates, largely due to the secretive egg-laying habits of reptiles (Shine, 1991; Doody, Freedberg & Keogh, 2009). Reptilian ECH, triggered by embryonic vocalizations, was first suspected in crocodylians (Reese, 1915), but it took many years to provide experimental evidence for its function (Herzog & Burghardt, 1977; Vergne & Mathevon, 2008). Bustard (1972) was the first to identify vibration-induced ECH in reptiles, noting that in nests of the green sea turtle (*Chelonia mydas*) eggs in the middle of the (large) clutch should develop faster than peripheral eggs due to metabolic heating, yet hatchling emergence was more or less synchronous. He hypothesized that embryo–embryo communication caused retarded peripheral eggs to "catch up" with the more advanced eggs in the center of the clutch. Using experiments, Bustard (1972) then stimulated hatching in *C. mydas* by vibrating eggs, demonstrating that vibrations of embryos could cause embryos to catch up with their clutchmates, a finding that was repeated later (Spencer et al., 2001). Meanwhile, herpetologists were noting the apparent coincidence of lizard eggs hatching in their hands. In fact, it was eventually determined that the handling of the eggs induced hatching; lizard embryos were interpreting handling as imminent predation risk (Doody, 2011). Since then, one quantitative study and a spate of anecdotes have documented early hatching in lizards, including in skinks, geckos, and anoles (Doody & Paull, 2013; Doody & Schembri, 2014a, 2014b; Doody, Ellis & Rhind, 2015; Hernandez et al., 2017; Doody, Coleman, et al., 2019).

ECH can be roughly categorized into three types. In some species embryos delay hatching until environmental conditions are (more) suitable for neonates, while others hatch early in response to the immediate threat of a predator or pathogen (Warkentin, 1995, 2000; K. Martin, 1999). A third type of ECH involves siblings hatching synchronously (Doody, 2011). The relative costs-benefits of dynamic stage-specific threats can determine when an embryo should remain within the egg versus hatch into a new environment, and the embryo will hatch accordingly, provided that they can perceive changing risks by sampling environmental variation (cues) from within the egg (Warkentin & Caldwell, 2009).

Which of these three types of ECH involve social interactions? There is evidence for social interactions in all three, although to what extent is not known. Here we review the evidence for the three types of ECH in reptiles. In each section we discuss the role (or possible role) of embryo–embryo communication and avenues for future research.

8.1.1. Hatching Early in Response to an Immediate Risk

Embryos can advance hatching date in response to an increase in risk (or perceived risk) such as that induced by predators, pathogens, or flooding (reviewed in Warkentin & Caldwell, 2009; Warkentin, 2011b). Early hatching generally involves an external cue (e.g., vibrations) that will cause embryos to hatch earlier than their "spontaneous hatching period" (Gomez-Mestre et al., 2008). Early hatching can increase survival, suggesting an adaptive function (Warkentin & Caldwell, 2009). For example, some frog embryos can distinguish between vibrations that are critical (i.e., those made by egg predators such as snakes or wasps) versus benign (i.e., those made by the wind or rain falling) and hatch early (or not) accordingly (Warkentin, 1995, 2000). There are costs to early hatching in these frogs, with early hatchers more vulnerable to aquatic predators (reviewed in Touchon et al., 2013). Early hatching has been demonstrated in other animals (reviewed by Doody, 2011; Warkentin, 2011a; Doody & Paull, 2013), but the ability of those embryos to distinguish between important and benign cues is not known.

In a review of ECH in reptiles, Doody (2011) found early hatching to be the most common type, having been reported in 27 species; since then, it has been found in a few more reptile species (Doody & Schembri, 2014a, 2014b; Doody, Ellis & Rhind, 2015; Hernandez et al., 2017; Doody, Coleman, et al., 2019). Although early ECH does not explicitly involve embryo-embryo communication, it seems likely that embryo-embryo communication does exist, and it occurs in at least one species. The pig-nosed turtle (*Carettochelys insculpta*) delays its hatching until the wet season, and then hatches explosively in response to (hypoxic) flood conditions (G. J. W. Webb et al., 1986; Doody, Georges, Young, Pauza, et al., 2001). However, vibrations associated with even slight disturbance of the eggs can also stimulate hatching, and experiments demonstrated that groups of eggs can hatch and emerge faster than solitary eggs when immersed in water or perfused with gaseous nitrogen (Doody, Stewart, et al., 2012). A model demonstrated that this effect is likely to occur in nature, and the cue appears to be vibrations (Doody, Stewart, et al., 2012), although vocalizations cannot be ruled out.

Although we might be tempted to categorize the pig-nosed turtle as exhibiting delayed hatching or synchronous hatching (see sections 8.1.2 and 8.1.3, below), siblings are likely communicating to expedite hatching. In nature hatching and emergence in the pig-nosed turtle was often asynchronous; hatchlings emerged singly (Doody, Georges, Young, Pauza, et al., 2001). In theory, early hatching in response to increased risk could not only increase survival in the individual(s) perceiving the risk, but the risk could also be communicated to sibling embryos via embryo–embryo communication.

Assuming early ECH does involve embryo–embryo communication, where should we look for it? In a review, Doody (2011) suggested that, although the life cycle of reptiles, lacking metamorphosis, is simple relative to amphibians and insects, early hatching should still evolve (barring constraints) when embryos can perceive an imminent direct attack, even though many predators of reptile eggs would also readily consume hatchlings (Doody, Freedberg & Keogh, 2009; Doody, 2011). This assumes that (1) the embryos have reached a hatching-competent stage; (2) predation risk is typically high during the hatching competence period; and (3) the hatching embryos are able to escape the predator (Doody, 2011). For example, early hatching delicate skinks (*Lampropholis delicata*) burst from the egg running (Doody & Paull, 2013). In contrast, sluggish, slow-hatching embryos that are small relative to an agile predator would likely not escape (Doody, 2011).

8.1.2. Delayed Hatching: Hiding during a Time of Increased Risk

In delayed hatching, embryos defer hatching spontaneously, waiting to hatch later during a more favorable period. For example, the marbled salamander (*Ambystoma opacum*) lays its eggs in a depression that will later fill with water; the eggs wait out the dry period after becoming hatching-competent, but hatching must be triggered by hypoxia associated with inundation by rising pond levels (Petranka & Petranka, 1980). In a review on reptiles and birds, Ewert (1991) distinguished between delayed hatching and embryonic estivation: "delayed hatching tends to last only a very brief period, a few days in birds to three weeks in reptiles," while "aestivation occurs when the embryo remains in the egg for very long periods." In both cases, metabolism either plateaus or decreases, although in estivation metabolism may be lower (Ewert, 1991). Because delayed hatching and embryonic estivation essentially differ only in the magnitude of the delay and the extent of metabolic depression, Doody (2011) considered embryonic estivation to be a subset of delayed hatching, rather than a parallel category. In reptiles, pig-nosed turtle embryos remain in suspended animation within the egg during the dry season, delaying hatching until the more favorable conditions of the approaching wet season (G. J. W. Webb et al., 1986; Doody, Georges, Young, Pauza, et al., 2001). In the laboratory, hypoxia induces eggs to hatch explosively within minutes (G. J. W. Webb et al., 1986; Doody, Georges, Young, Pauza, et al., 2001); in most other turtle species, the hatching process spans hours to days (Bustard, 1972; Ewert, 1985). The precise adaptive benefit to synchronizing hatching with the wet season is not known but is probably an increase in the availability of food and/or a reduced risk of predation associated with flooded conditions; hatching at a time when available water is clear and shallow during the dry season may be risky for small turtles (Doody, Georges, Young, Pauza, et al., 2001). Temporal buffering is probably the ultimate reason for embryonic estivation in this species. Timing of nesting can vary by up to five weeks and is related to the magnitude of the previous wet season(s) (Doody, Georges & Young, 2004). After small wet seasons, late nesting causes considerable nest failure due to flooding associated with the onset of the next wet season; after big wet seasons, early nesting causes embryos to reach the onset of hatching competence during the dry season (Doody, Georges, Young, Pauza, et al., 2001; Doody, Georges & Young, 2004). Estivation thus decouples the highly variable seasonal timing of nesting from the timing of hatching, thereby allowing turtles to both nest early and hatch later under more favorable conditions.

There are a number of reptilian species that appear to time their hatching with the onset of the wet season; however, the strength of the evidence for delayed hatching in these species varies (Doody, 2011). For example, in the Fijian iguana (*Brachylophus vitiensis*), two clutches with dates of oviposition 63 days apart hatched within two days of one another (Morrison et al., 2009). Although an increase in developmental rate associated with a seasonal increase in air temperatures can reduce the difference in duration of incubation between early and late clutches (Doody, 1995; Pezaro et al., 2016), this thermally based catch-up is not sufficient to offset a 63-day difference in incubation. Rather, embryos in the early clutch almost certainly reached hatching competence weeks prior to embryos in the later clutch but delayed hatching until the environmental trigger of heavy rainfall occurred (Morrison et al., 2009). Some reptile species appear to synchronize hatching with the onset of the wet season and possess long and variable incubation times, suggesting delayed hatching (Doody, 2011). These include African spurred (*Geochelone sulcata*), leopard (*G. pardalis*), and radiated (*Astrochelys radiata*) tortoises (V. Wilson, 1968; Cloudsley-Thompson, 1970) and Gray's monitor (*Varanus olivaceus*) (Auffenberg, 1988). The pattern of nesting during the dry season and hatching with the onset of wet season flooding (e.g., in the pig-nosed turtle) occurs in the Indian flapshell turtle (*Lissemys punctata*), another riverine species, and in all six species of hole-nesting crocodylians (Vijaya, 1983b; Thorbjarnarson & Hernández, 1993b). Hatching in response to flooding or hypoxia may occur but has not been confirmed in these species (in this scenario, crocodylians would possess both the ability to hatch in response to flooding and maternal assistance in hatching). Hatching in response to flooding/hypoxia does occur in Indian (*Nilssonia gangeticus*) and Indian peacock (*N. hurum*) softshell turtles, however. Although these two riverine species do not nest exclusively during the dry season, hatching is associated with rainfall and flooding, and eggs immersed in water in the laboratory hatched almost immediately (Andrews & Whitaker, 1993).

Reptilian embryos showing a peak of oxygen consumption at about 90% of development followed by a modest decline in oxygen consumption could either possess delayed hatching or be preadapted to evolve it (Birchard et al., 1984; Black et al., 1984; Thompson, 1989; Ewert, 1991). Researchers have considered this pattern to underpin delayed hatching within clutches, thereby facilitating synchronous hatching (Thompson,

1989; Whitehead & Seymour, 1990). Indian black (*Melanochelys trijuga*) and loggerhead musk (*Sternotherus minor*) turtles, Johnston's (*Crocodylus johnstoni*) and estuarine (*C. porosus*) crocodiles, and the American alligator (*Alligator mississippiensis*) may delay hatching based on this pattern of embryonic metabolism (Whitehead & Seymour, 1990; D. Booth & Thompson, 1991).

As with early hatching, the paucity of evidence for embryo–embryo communication in the above examples may reflect a lack of scientific attention rather than the absence of embryo–embryo communication. Assuming delayed ECH does involve embryo–embryo communication, where should we look for it? Although predicting which species might respond in these ways is challenging, the occurrence of one type—ECH in response to hypoxia via flooding of eggs—is somewhat more predictable.

Eggs of terrestrial reptiles that are not exposed to flooding conditions may not have evolved flooding-induced ECH, whereas riparian-nesting species, such as all freshwater turtles and crocodylians and some species of lizards and snakes, could experience mortality due to nest flooding and might evolve delayed ECH (Doody, 2011). Flooding is the leading cause of mortality in many freshwater turtle species and in some crocodylians (Magnusson, 1982; D. Moll & Moll, 2004). If the timing of flooding is somewhat predictable, some of these species may delay hatching during a dry period and hatch during nest flooding associated with the onset of the wet season. Thus, we should search for flooding-induced ECH in species in which hatching coincides with the onset of the wet season, and in which nests are vulnerable to flooding (Doody, 2011). It is also possible that species that do not synchronize hatching with the onset of the wet season, but nevertheless experience a reasonable probability of nest flooding, could evolve flooding-induced ECH. Experiments comparing embryos' responses to hypoxia between species hatching during the wet versus the dry season or between those typically exposed to flooding versus those not typically exposed would be a starting point.

Aside from dealing with flooding, many reptiles (e.g., monitor lizards, green iguanas, crocodiles, and some turtles) living in regions with wet and dry seasons clearly synchronize their hatching with the onset of the wet season (e.g., A. S. Rand & Greene, 1982), and hatching-competent embryos of at least some of these species probably delay hatching during the dry season (Doody, 2011). Thus, future investigations should include species with long and variable incubation periods and those in which the appearance of hatchlings seems to be synchronized with the onset of the wet season (e.g., Auffenberg, 1988).

8.1.3. Synchronized Hatching and Emergence: The Norm or Adaptation?

Although asynchronous oviposition can produce developmental asynchrony in birds (Brua, 2002), in reptiles the eggs are laid in one brief bout. Thus, all eggs in a clutch should hatch at the same time, assuming negligible individual variation in developmental rate (null hypothesis). In small reptiles with small clutches of small eggs, this may be the norm. However, thermal gradients occur in shallow-nesting reptiles with large three-dimensional clutches (Whitehead, 1987; Thompson, 1988), and developmental asynchrony occurs because embryonic development is temperature-dependent (Ewert, 1979). Shallower eggs within these nests can experience higher and more variable temperatures, both of which reduce developmental time relative to deeper eggs (Georges et al., 1994). Yet, hatching and emergence can be more or less synchronous in clutches with thermal gradients, indicating the evolution of some additional phenomenon that offsets developmental asynchrony (Thompson, 1989; Spencer et al., 2001; Glen et al., 2005; P. Colbert et al., 2010; Spencer & Janzen, 2011).

Hatching and emergence can be synchronized within or among clutches; here we consider within-clutch hatching and emergence first. Within-clutch synchronized hatching and emergence, especially in reptiles with large three-dimensional clutches possessing thermal gradients, is proposed to be adaptive by swamping or diluting predators (Fitch, 1954; Spencer et al., 2001; Tucker et al., 2008; Spencer & Janzen, 2011; Aubret, Blanvillain, et al., 2016; Santos et al., 2016). However, evidence is conspicuously lacking or absent (see below, "Evidence for Synchronous Hatching and Emergence and Its Application to Adaptive Hypotheses"). Indeed, references used in support of a dilution effect of synchronous hatching or emergence in reptile embryos have nothing to do with reptile eggs (e. g., S. J. Arnold & Wassersug, 1978, and Dehn, 1990, cited in the introduction in both Spencer et al., 2001, and Aubret, Blanvillain, et al., 2016).

We contend that robustly addressing the validity of the predator swamping hypothesis requires (1) defining and distinguishing synchronous hatching and synchronous emergence; (2) revealing the evidence for synchronous hatching and emergence for reptiles in nature and its application to adaptive hypotheses; (3) deciding if synchronous hatching and emergence might be adaptive by revealing which species possess large enough clutches to exhibit thermal gradients within the nest (small clutches of small eggs in shallow nests would occupy the same thermal niche such that synchronous hatching and emergence would be expected due to the physiology of development, precluding the need for an adaptive explanation); and (4) understanding the social mechanisms involved.

Here we break down these goals, noting two things: first, the vast majority of hatching and emergence data are from turtles and to a lesser extent crocodylians (our knowledge of hatching and emergence in squamates is severely limited, especially in snakes); and second, it is difficult to record hatching in the field in subterranean nests, so emergence patterns are assumed to reflect hatching patterns in our discussions below. We distinguish between the two whenever possible, noting that several days separate hatching and emergence in many species (Hendrickson, 1958; but see Doody, Georges, Young, Pauza, et al., 2001, for an exception).

(i) Defining Synchronous Hatching and Emergence

Determining the degree of synchronicity of hatching and emergence is critical for interpreting the evolutionary significance of synchronized hatching and emergence. Aubret, Blanvillain, et al. (2016), citing Spencer et al. (2001), stated that "despite (thermal) gradients in nests and the fact that development is strongly linked to temperature, all eggs hatch in synchrony." Yet, asynchronous hatching and emergence is actually quite common (see below). What constitutes synchronous hatching or emergence within a clutch? By synchronous do we mean siblings hatching or emerging within days, hours, minutes, or seconds? This distinction is important. We propose that predator swamping would require synchronous emergence in which siblings emerge within seconds or a very few minutes of one another. Emergence intervals ranging from several minutes to days would likely not confuse, satiate, or swamp predators.

Most researchers have taken care to distinguish between hatching and emergence; for example, clutches that emerge synchronously may have hatched asynchronously—earlier hatchers may wait on their later-hatching siblings and emerge together. However, many authorities have not carefully defined "synchronous," including those that argue for an adaptive dilution function (e.g., Tucker, 1997; Spencer et al., 2001; Tucker et al., 2008; P. Colbert et al., 2010; Santos et al., 2016). Others have defined within-clutch synchronous hatching or emergence as occurring within one day, or 24 hours (e.g., Doody, Georges, Young, Pauza, et al., 2001). In a review of synchronous hatching in turtles, Spencer and Janzen (2011) defined synchronous hatching as "any response of an embryo to the developmental stage or hatching behavior of clutchmates, which may not necessarily result in spontaneous hatching, but does result in the adjustment of incubation times." While appropriate for their review, that definition provides little utility for addressing the predator swamping hypothesis. Frustratingly, the lead-in for most of the abovementioned papers invariably sells predator swamping as the interesting biological phenomenon that would promote synchronous hatching and emergence. Importantly for this chapter, the reviews of Spencer and Janzen (2011) and Doody (2011) also implicate embryo-embryo communication. In areas of hundreds of nests, emergence from nests singly or at intervals could still provide a swamping effect. Still, individual animals emerging from the same nest nearly simultaneously, as in green iguanas, would be more convincing evidence.

(ii) Evidence for Synchronous Hatching and Emergence and
Its Application to Adaptive Hypotheses

The classic adaptive hypothesis for why synchronous hatching and emergence evolves is predator swamping or dilution (Fitch, 1954; Spencer et al., 2001; P. Colbert et al., 2010). The dilution effect, or "safety in numbers," predicts that as emergence, and thus hatching, synchronicity increases, the probability of each neonate being preyed upon decreases. The exact dilution mechanism can involve predator satiation, distraction, or handling time, all of which can allow siblings to escape. For example, researchers have contended that synchronous hatching in turtles is common and widespread, and thus, likely to be an ancestral trait (Spencer et al., 2001; P. Colbert et al., 2010; Santos et al., 2016). However, direct evidence of a dilution effect of synchronous hatching or emergence in reptiles remains elusive. One study found that the risk of being depredated by crabs was significantly greater in smaller groups of emerging green sea turtles (Santos et al., 2016). Indirect evidence is also scarce; predation rates were higher for solitary nests than for communal nests in the olive ridley sea turtle (*Lepidochelys olivacea*) (Eckrich & Owens, 1995), but not in the mourning gecko (*Lepidodactylus lugubris*) (S. G. Brown & Duffy, 1992). Moreover, in several turtle species, emergence occurs one at a time and/or spans more than one day (Congdon et al., 1983; Butler & Graham, 1995; Doody, Georges, Young, Pauza, et al., 2001; but see Tucker, 1997). Finally, field experiments revealed no effect of dilution on survival of red-eared slider (*Trachemys scripta*) hatchlings, although the expectation of finding such an effect experimentally is ambitious (Tucker et al., 2008).

An alternate adaptive hypothesis for synchronous hatching and emergence is that asynchronous hatching releases smells in the nest that would attract predators that would then consume unhatched eggs (Lack, 1968b; Congdon et al., 1983; Vitt, 1991). This hypothesis is supported by the fact that most predation in turtle nests occurs during two distinct periods: soon after laying and during hatching (Ernst & Lovich, 2009; reviewed in Riley & Litzgus, 2014). During these two periods, nests smell like egg-laying fluids and hatching fluids, respectively.

A third hypothesis is that in some deeper-nesting species, such as sea turtles, the ability to escape the nest may be improved by the joint efforts of many hatchlings over the separate efforts of individuals (Carr & Hirth, 1961). Joint efforts may also be required in the deep-nesting yellow-spotted (*Varanus panoptes*) and Gould's (*V. gouldii*) monitor lizards; in most nests, which average 2.4–3.0 m deep, there is one burrow through which hatchlings emerge, and it does not follow the mother's nesting burrow (Doody, James, Ellis, et al., 2014; Doody, James, Colyvas, et al., 2015; Doody, McHenry, Brown, et al., 2018).

Synchronous hatching and emergence may also be adaptive because staying in the nest may be detrimental due to depleting energy stores (Hays et al., 1992) or the greater likelihood of exposure to dipteran larvae (A. McGowan, Broderick, et al., 2001; A. McGowan, Rowe, et al., 2001; Koch et al., 2008).

Most data addressing synchronous hatching and emergence in reptiles are from turtles, presumably because their nests are easier to find, disturb, and monitor than those of lizards, snakes, and crocodiles. Spencer et al. (2001) stated that "juvenile turtles usually emerge together, despite the high probability of different rates of development within single nests." We need to quantify "together" for reasons that will become clear below. Gleaning the literature revealed 19 studies on 16 turtle species with data sufficient to address synchronous hatching or emergence (Table 8.1). The best data stem from field studies that used cameras to determine exact individual emergence times to the nearest minute (Doody, Georges, Young, Pauza, et al., 2001; Greg Gellar, unpubl. data), followed by studies that checked for hatching or emergence several times a day in the field, followed by laboratory data.

Three patterns emerge. First, within-nest hatching or emergence invariably ranged over several days (Table 8.1). Second, the proportion of nests in which all siblings emerged within 24 hours was usually <50%. Third, when intersibling emergence intervals were available, they reflected a "trickle forth" pattern of nonsimultaneous emergence. Even in sea turtles there can be asynchronous emergence: 81% of flatback nests exhibited protracted emergence (>3 hours) (Koch et al., 2008). Within this study, 27.5% and 31.3% of green turtle hatchlings in 1997 and 1998, respectively, and 19.9% and 36.9% of loggerhead hatchlings in 1997 and 1999, respectively, emerged in groups of fewer than 10 hatchlings; these hatchlings accounted for 71%–80% of all emergence groups in both species (Glen et al., 2005). Asynchronous emergence may be beneficial if there are different development rates within the clutch because this asynchronous emergence

Table 8.1. Review of studies quantifying the degree of synchronous hatching and/or emergence within nests of various turtle species

Species	% Within-nest emergence within 24 hrs (# of nests)	Within-nest emergence range (days)	Within-nest mean sibling emergence interval (range; number of nests)	Reference
Apalone mutica	92 (26)[a]	1-2	NA[b]	Plummer (2007)
Caretta caretta	? (26)	1-12	NA?	Houghton & Hays (2001)
Caretta caretta	26-71 (48)	1-7	NA	Glen et al. (2005)
Carettochelys insculpta	79 (62)	1-4	12 m (0.7-47 m; 14)[c]	Doody, Georges, Young, Pauza et al. (2001)
Chelodina longicollis	asynchronous (16)[a]	NA	NA	Spencer (2012)
Chelonia mydas	22-31 (38)	1-7	NA	Glen et al. (2005)
Chelydra serpentina	65 (20)	2-3	NA	Congdon et al. (1987)
Chelydra serpentina	25 (4)	1-20	NA	P. Baker et al. (2013)
Chrysemys picta	100 (5)	NA	NA	Christens & Bider (1987)
Chrysemys picta	43 (7)	1-9	NA	Lindeman (1991)
Chrysemys picta	30-32 (260)	1-77	NA	P. Baker et al. (2013)
Emydoidea blandingii	46 (14)	1-8	NA	Congdon et al. (1983)
Emydoidea blandingii	mostly asynchronous	2-11	NA	Standing et al. (1997)
Emydura macquarii	synchronous (10)[d]	NA	NA	Spencer et al. (2001)
Gopherus polyphemus	36 (11)	1-20	NA	Pike & Seigel (2006)
Graptemys geographica	44 (9)	1-31	NA	Nagle et al. (2004)
Graptemys geographica	15 (20)	1-36	NA	P. Baker et al. (2013)
Graptemys ouachitensis	78 (18)	1-2	218 m (0-4,058 m; 18)	G. Geller, unpubl. data
Macroclemys temminckii	20 (5)	1-12	NA	Holcomb & Carr (2011)
Malaclemys terrapin	some	1-11	NA	Burger (1976)
Malaclemys terrapin	NA	1-11	NA	Burger (1977)
Natator depressus	67 (79)	1-8	NA	Koch et al. (2008)
Natator depressus	14 (21)[a]	1-9	NA	Koch et al. (2008)
Trachemys scripta	22 (27)	1-64	NA	P. Baker et al. (2013)

[a] Data from artificial nests.
[b] Data not available.
[c] Data only from nests with all emergence within 24 hrs.
[d] Data from hatchlings buried in artificial nests.

reduces unnecessary energy consumption from waiting in the nest for siblings to emerge (Hays et al., 1992). More protracted emergences of several hours or days are not uncommon and have consistently occurred on some nesting beaches for loggerhead sea turtles (*Caretta caretta*) (Hays et al., 1992; Houghton & Hays, 2001; Glen et al., 2005), green sea turtles (Balazs & Ross, 1974; Glen et al., 2005), and hawksbill sea turtles (*Eretmochelys imbricata*) (Diamond, 1976). Whether this is more common than originally believed is unknown because few studies have recorded the emergence of each hatchling from a clutch (Koch et al., 2008). Although a continuum likely exists with asynchronous and synchronous hatching at opposite ends, we might predict an adaptive function for any mechanism that counteracts asynchrony due to thermal gradients.

Thus, within-clutch hatching and emergence do not appear to be synchronous to the level required to distract, swamp, or satiate predators, except in some turtles and, perhaps, green iguanas (Plate 8.1) (Burghardt, 1977b; Burghardt et al., 1977). We argue that, within nests, distracting, swamping, or satiating predators of freshwater turtle and lizard nests would require synchronous hatching and emergence within seconds to minutes; this would be especially true in trying to satiate a predator with relatively small clutch sizes (e.g., most turtles and lizards lay clutches of ~5-20 eggs). In field experiments releasing hatchling red-eared sliders, Tucker et al. (2008) found no evidence for predator swamping. An important assumption here is that there are sufficient numbers of hatchlings *among* nests to satiate, distract, or occupy predators. For example, in some populations of sea turtles or iguanas tens to thousands of hatchlings can emerge from many nests in a short period of time (Carr, 1967; Burghardt, 1977b; Christian & Tracy, 1981); in these cases, any within-nest emergence (and thus hatching) synchrony could be adaptive due to among-nest dilution.

Why might there be so much interspecific variation in the degree of synchronicity of hatching or emergence? The degree of synchronicity of within-nest hatching and emergence (Table 8.1) may simply reflect low within-nest variation in developmental time and temperature. From this idea we would expect larger nests (via thermal gradients) or shallower nests (because of greater temperature fluctuations) to exhibit less synchronous hatching and emergence, and this receives some empirical support. As the thermal variance decreased, synchrony increased in loggerhead sea turtles (Houghton & Hays, 2001). Hatchlings in the shallower nests of flatback sea turtles (*Natator depressus*) exhibited greater emergence asynchrony (Koch et al., 2008). In the gopher tortoise (*Gopherus polyphemus*), asynchronous emergence patterns were more common in nests located close to the burrow mouth and buried shallower than nests with synchronous emergence patterns (Pike & Seigel, 2006; Koch et al., 2008). Consistent with these results, daytime emergence increased emergence duration and reduced synchrony in loggerhead and green sea turtles (Glen et al., 2005). For the former, nests with longer emergence durations were found to have a higher number of dead hatchlings; however, a layer of dead hatchlings in the subsurface layers of the sand may have acted as a barrier, inhibiting the emergence of siblings from underneath (Glen et al., 2005). Because excavation of a nest emergence hole may be the result of a joint effort of the hatchlings (Legler, 1960), Congdon et al. (1987) hypothesized that nests with poor incubation conditions might produce smaller or weaker hatchlings that may be more likely to exit the nest asynchronously. Synchronous nests, in comparison, might have fewer undeveloped eggs and dead hatchlings remaining in the nest. Three of seven (48%) asynchronous nests and 8 of 13 (62%) synchronous nests had partial hatchling emergence. Asynchronous nests had five detectable failures (4%) out of 138 eggs, whereas synchronous nests had 20 failures (5%) out of 397 eggs. Synchronous and asynchronous nests produced an average of 25 and 19 hatchlings per nest, respectively. In addition, hatchlings emerging from asynchronous and synchronous nests were not significantly different in body mass (Congdon et al., 1987).

The degree of the synchronicity of within-clutch hatching and emergence synthesized from Table 8.1 is more consistent with the "predator preclusion" hypothesis than the predator swamping hypothesis. The predator preclusion hypothesis suggests that if within-clutch hatching and emergence are too asynchronous, the first hatchings or emergence(s) would release smells that would attract olfactory-driven predators to the nest that could consume unhatched eggs or hatchlings (Congdon et al., 1983; Vitt, 1991; Doody, 2011). This hypothesis was first proposed by Lack

(1968b) for synchronous hatching in birds. The hypothesis predicts synchronous hatching and emergence, but it does not require hatching and emergence within seconds or minutes, as does the predator swamping hypothesis. The predator preclusion hypothesis is supported by the fact that most predation in turtle nests occurs when nests smell like egg-laying fluids and hatching fluids (Riley & Litzgus, 2014). Indeed, one of us (JSD) has often found apparent evidence of predation on turtle and lizard nests in the form of scattered eggshells on the surface; closer inspection, however, revealed that eggs successfully hatched (as evidenced by pipping slits in the eggshells) prior to being excavated by the predator. In at least two of these populations, the hatchlings were known to have emerged from the nest successfully (Doody, 1995; JSD, unpubl. obs.). Thus, hatching and emergence released smells that attracted would-be predators to nests, but the predators were too late. Our analysis leads us to conclude that synchronous hatching and emergence in turtles, and perhaps other reptiles, is maintained by higher hatching and emergence success due to the reduction of the window of opportunity for olfactory-driven nest predators, rather than by higher hatching success due to predator swamping. The former would operate on the timescale of minutes to hours to days, while the latter would be important on a scale of seconds to minutes.

Despite the better fit of the predator preclusion hypothesis than the predator swamping hypothesis to the data in Table 8.1, why would some hatchlings emerge two to four days later than their siblings, assuming that predation risk increases with time since first emergence? One distinct possibility is that hatchlings require some time to internalize their yolks and are less vigorous and quite inferior in locomotion prior to this internalization; later-hatching siblings may have to trade off the increased predation risk in the nest with reduced predation risk out of the nest due to superior locomotor ability.

Another adaptive hypothesis for synchronous hatching and emergence may be relevant to only sea turtles. Social facilitation has been used to describe the combined digging effort of sea turtle hatchlings during nest escape (Rusli et al., 2016), and the "social facilitation hypothesis" (our name) posits that synchronous hatching and emergence can minimize energy expenditure for individual hatchlings (Carr & Hirth, 1961; Koch et al., 2008; Rusli et al., 2016). Sea turtles lay up to 150 eggs in a nest, and despite thermal gradients in the nest, eggs hatch relatively synchronously (Hendrickson, 1958; Carr & Hirth, 1961). Synchronous hatching within a clutch results in simultaneous digging activity of individual hatchlings. Social facilitation assists nest escape because the synchronous effort of many individuals might be needed to dig successfully through the column of sand above the nest chamber (Carr & Hirth, 1961). The spontaneous digging activity of one individual triggers the individuals around it to also start digging, resulting in cohort-wide synchronous digging (Carr & Hirth, 1961; Bustard, 1967). Group digging is hypothesized to result in hatchlings emerging from the nest in large numbers, which can lead to a predator swamping phenomenon that may decrease the overall predation rate of hatchlings as they leave the beach and swim out to sea (Bustard, 1972; Rusli et al., 2016).

Experimental support for the social facilitation hypothesis is compelling; Rusli et al. (2016) found an increase in group size from 10 to 60 hatchlings caused a ~50% decrease in both the time taken to escape the nest and mean metabolic rate during this time, resulting in reduced energy expenditure during nest escape. The authors concluded that because a finite amount of energy is available to hatchlings upon hatching, the energy saved by synchronous digging could be allocated to other activities such as the frenzied off-shore swim. Most reptiles excavate shallower nests than sea turtles (Doody, James, Ellis, et al., 2014), but a few deep nesters may receive similar benefits from hatching synchronously. For example, in Gould's and yellow-spotted monitors that nest 1–4 m deep, small groups of hatchlings ($N = 4$–8) excavate their own escape burrows (Doody, McHenry, Brown, et al., 2018; Doody, McHenry, Durkin, et al., 2018).

In an example that does not fit the above synchronous hatching strategy, late-term embryos of the eastern three-lined skink (*Acritoscincus duperreyi*) enter temporary torpor during cold periods (e.g., at night in a cool climate). This torpor apparently allows embryos in shallow nests to synchronize hatching with aboveground conditions that facilitate successful emergence from the nest (Radder & Shine, 2006). The cue for this "diel-sensitive" hatching is temperature, which varies considerably across a

24-hour period in the nests of most reptiles (e.g., Whitehead, 1987; Thompson, 1988). It is worth noting that synchronizing hatching with environmental conditions is different from, but not mutually exclusive of, synchronizing hatching with sibling embryos.

Costs of synchronous hatching are virtually unknown. However, P. Colbert et al. (2010) found that sibling turtles that hatched earlier exhibited longer righting times compared to later-hatching sibs and concluded that this difference in neuromuscular development (motility) reflected a cost to early hatching. Some precocial birds similarly shorten incubation times and subsequently exhibit reduced motor skills (Vince & Chinn, 1971; Cannon et al., 1986).

(iii) Thermal Gradients in Reptile Nests and a Nonadaptive Explanation for Synchronous Hatching and Emergence

How has the synchrony of hatching and emergence evolved? P. Colbert et al. (2010), noting that shortening incubation could incur developmental (fitness) costs, stated that "it is curious that hatching synchrony occurs at all." Conversely, we contend that synchronous hatching and emergence would be the most parsimonious scenario in oviparous reptiles, in general. A null model for synchronous hatching and emergence in reptiles is that sibling embryos of small species with small clutch sizes, with their very similar genetics and experiencing identical or very similar developmental temperatures within shallow nests, would hatch and emerge synchronously, without any need to evoke adaptation. Although data are lacking, more reptile species (than not) produce small clutches of small eggs that would exhibit negligible thermal gradients in natural nests, because their nests occupy such a small vertical space in the soil (or under bark, rocks, or logs, in shallow rock crevices, etc.). How do we know this? There are more species of small lizards and snakes with small clutches of eggs than there are species of turtles, crocodylians, and large lizards and snakes with large clutches of eggs (Uetz et al., 2020). In other words, only eggs in nests with thermal gradients would be challenged to achieve hatching synchrony—in nests without such non-negligible gradients, synchronous hatching (and thus, emergence) would be the norm, or the null model.

Thus, any discussion of adaptation of hatching and emergence synchrony should be preceded by a demonstration of thermal gradients within the nest of the species in question. We hypothesize that the nests of most reptiles will not exhibit thermal gradients due to their composition (small eggs) and position (shallow) and thus would be expected to hatch and emerge synchronously, without the need to invoke adaptation.

(iv) Social and Physiological Mechanisms of Synchronous Hatching and Emergence

By what mechanism might siblings within a nest time their hatching together? Two experiments on turtle species with thermal gradients in their nests indicated that less-advanced sibling embryos in the presence of more developed siblings pipped or hatched earlier than expected, suggesting synchronous hatching in nature (Spencer et al., 2001; P. Colbert et al., 2010). In the Macquarie river turtle (*Emydura macquarii*), embryos placed with more advanced siblings hatched about five days earlier than control embryos, although they still lagged behind their more advanced siblings by about five to six days (Spencer et al., 2001). Later experiments corroborated these results (McGlashan et al., 2011; see also McGlashan et al., 2015, for a similar story in the eastern longneck turtle *Chelodina longicollis*). Similarly, embryos of the painted turtle (*Chrysemys picta*) placed with more advanced siblings pipped about two days earlier than did control eggs, although pipping in these eggs still lagged behind their more advanced siblings by about four days (P. Colbert et al., 2010). It thus appears that a threshold of developmental asynchrony exists, beyond which embryos cannot fully catch up to their more developmentally advanced clutchmates (P. Colbert et al., 2010). Asynchrony in the above experiments was seven days for Macquarie river turtle and 11 days for painted turtle (Spencer et al., 2001; P. Colbert et al., 2010). Also in the laboratory, eggs of the viperine water snake (*Natrix maura*) hatched earlier than control eggs in response to being incubated in physical contact with more advanced eggs (Aubret, Blanvillain, et al., 2016). Collectively, these experiments revealed what has been termed the "catch-up hypothesis" (Spencer et al., 2001). The opposite mechanism may also occur; after finding synchronous hatching with no evidence of metabolic compensation in embryos of the painted turtle, McGlashan et al. (2018) suggested that less-advanced embryos hatched at earlier embryonic stages. Finally, Spencer (2012), after establishing developmental asynchrony, found no adjustment of developmental time in the eastern longneck turtle (*Chelodina longicollis*).

By what mechanism do siblings communicate with one another? Thus far, studies have suggested or implicated vocalizations, vibrations, and heartbeats. The first mechanism revealed for reptiles was embryonic vocalizations in crocodiles, which can be heard a few days before hatching (Deraniyagala, 1939; Cott, 1961; Pooley, 1962; D. Lee, 1968; Magnusson, 1980; Britton, 2001). Given that hatchlings also vocalize and thus elicit the mothers to open the nest (reviewed in Mathevon et al., 2013), could embryo vocalizations simply reflect a "warming up" phase for the emission of hatchling calls? Embryonic and hatchling calls differ in duration, frequency, and intensity, and playback experiments demonstrated that embryos respond (call) to the calls of other embryos, suggesting they function in embryo–embryo communication to synchronize hatching and thus emergence (Vergne & Mathevon, 2008). In support, synchronous hatching is common in birds (Lack, 1954), and avian embryos vocalize to synchronize hatching using "peeps" and "clicks" (Oppenheim, 1972).

The recent finding of vocalizations in turtle embryos suggests that other reptiles produce prehatching calls that could also fine-tune hatching and emergence synchrony. Embryos of the giant Amazon river turtle (*Podocnemis expansa*) began to vocalize up to 36 hours prior to hatching, emitting a sound every three minutes, on average (Ferrara et al., 2013). Similar calls were recorded in embryos of leatherback sea turtles (*Dermochelys coriacea*) and in the embryos or hatchlings of several other turtle species (Ferrara, Vogt, Giles & Kuchling, 2014; Ferrara, Vogt, Harfush, et al., 2014). McKenna et al. (2019) found similar vocalizations in embryos of Kemp's ridley sea turtles (*Lepidochelys kempii*), citing a unique "pulse" sound. Somewhat illogically, the researchers concluded that the sounds were not involved in synchronous hatching or emergence because hatchlings continued to vocalize after emergence from the nest. The recent discovery of vocalizations in the adults of several other turtle species (Ferrara, Vogt, Giles &

Kuchling, 2014; Ferrara et al., 2017) predicts that chelonian embryonic calls may be ubiquitous or at least common and widespread.

The second type of embryo–embryo communication implicated in synchronous hatching is vibrations. As mentioned above, Bustard (1972) stimulated hatching in *Chelonia mydas* by vibrating eggs and proposed that the vibrations of embryos could cause embryos to catch up with their clutchmates. Eggs of the pig-nosed turtle hatched early in response to transport in a boat, handling, a lab shaker, and thunder, and several eggs of the related smooth (*Apalone mutica*) and spiny (*A. spinifera*) softshell turtles hatched in response to vibrations during transport in a car (Doody, 2011; JSD, pers. obs.). Eggs of the Indian flapshell turtle hatched in response to vibrations caused by thunder (Vijaya, 1983a, 1983b), and hatching in Nile crocodile (*Crocodylus niloticus*) eggs was induced by vibrations of a light airplane (Myburgh & Warner, 2011). Whether vibration-induced hatching in each case served to synchronize hatching versus expedite hatching—see section 8.1.1, above—is not known. Similarly, the vibrations that caused early hatching in several lizard species (Doody, 2011) may or may not be utilized to promote synchronous hatching.

Intriguingly, in a third type of embryo–embryo communication, embryos in contact with one another increased heart rates relative to solitary eggs in viperine water snake (Aubret, Blanvillain, et al., 2016), and embryos of the Macquarie river turtle increased heart rates and metabolic rates (measured as carbon dioxide production) to catch up with more advanced siblings (McGlashan et al., 2011). However, heart rates did not differ between "synchronous" and "asynchronous" embryos in two other studies (McGlashan et al., 2015, 2018). Also apparently related to heart rates, solitarily incubated eggs produced less-sociable young snakes than their siblings that were incubated in a cluster: the former were more active, less aggregated, and physically contacted each other less often than the latter (Aubret, Bignon, et al., 2016).

In some species, we have very little data on hatching synchrony and instead must rely on observations or quantification of emergence synchrony, and it is reasonable to predict that in some species there are social interactions that facilitate emergence and perhaps reduce predation risk.

While sea turtles emerging at night may not be able to see conspecific hatchlings, day-emerging species can. In a remarkable example, hatchling green iguanas (*Iguana iguana*) emerging in small groups establish visual contact with other pre-emergent groups from nearby nests (Plate 8.2) (Burghardt, 1977b; Burghardt et al., 1977; H. W. Greene et al., 1978; Drummond & Burghardt, 1983). When one hatchling began to emerge, others did also and moved in groups. If one was startled and quickly ran from the clearing (or later from other areas when in groups), others did also, typically in the same direction (H. W. Greene et al., 1978). Collectively, visual contact among hatchling iguanas can be considered a fourth type of communication, albeit among hatchlings rather than embryos. Moreover, groups of iguana hatchlings would often gather together in the reeds at the end of an islet, waiting for hours prior to making a journey to the main island in a sudden swarm, both by swimming and running bipedally over the water (Burghardt et al., 1977; Drummond & Burghardt, 1982) (Plate 8.3). However, iguanas also left nests singly and sometimes emerged in greater numbers during the night than during the day, perhaps to reduce predation (Drummond & Burghardt, 1982); at night they sometimes performed an unusual hopping behavior seen in no other context (illustrated in Burghardt, 2004).

8.2. Among-Clutch Synchronized Hatching and Emergence: Social or Incidental?

In theory, synchronous hatching and emergence could evolve to satiate predators with mass emergence among clutches, provided that the mass emergence is sufficiently synchronous to do so. Female nesting synchrony could be favored by natural selection if predators are swamped (H. W. Greene et al., 1978; Ims, 1990; Tucker et al., 2008; Santos et al., 2016). Given sufficient numbers, any level of synchronicity could swamp or confuse predators. The most obvious candidate for this is mass emergence in sea turtles. There are certainly sufficient numbers of simultaneously emerging hatchlings as a result of arribadas (Carr, 1967; Pritchard, 1969), and this may hold for other species of reptiles at smaller scales, as with

Plate 3.1. The Australian sleepy lizard (*Tiliqua rugosa*) exhibits seasonal monogamy for up to (at least) 27 years. Photograph by Bill Bachman / Alamy Stock Photo.

Plate 3.2. A (presumed) family group of black rock skinks, *Egernia saxatilis*, basking in eastern Victoria, Australia. Photograph by Bjorn Svenssen / Alamy Stock Photo.

Plate 4.1. Male black caiman (*Melanosuchus niger*) roaring, Rancho Karanambu, Guyana. Photograph by Vladimir Dinets.

Plate 4.2. Dewlap displays: **a.** Fitch's anole (*Anolis fitchi*). **b**. Many-scaled anole (*A. polylepis*). **c.** Blemished anole (*A. mariarum*). **d.** Goldenscale anole (*A. chrysolepis*). **e.** Fan-throated lizard (*Sitana ponticeriana*). **f.** Green iguana (*Iguana iguana*). Photographs by: *a,* L. A. Coloma (Creative Commons Attribution License); *b,* Steven G. Johnson (Creative Commons Attribution–Share Alike License); *c,* Weimar Meneses; *d,* Bernard Dupont (Creative Commons Attribution–Share Alike License); *e,* Nikhilpaigude (Creative Commons Attribution–Share Alike License); *f,* Vladimir Dinets.

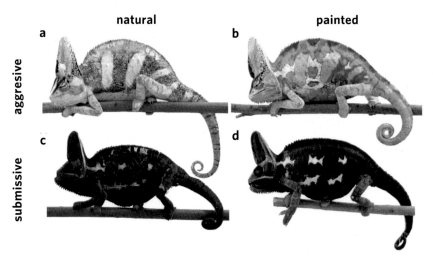

natural | painted

aggresive

submissive

Plate 4.3. Color signaling by male veiled chameleons (*Chamaeleo calyptratus*): aggressive and submissive coloration, as well as animals painted by researchers to test the meaning of each coloration. From R. A. Ligon & McGraw, 2016.

Plate 5.1. "Dance" of American alligators (*Alligator mississippiensis*) in Everglades National Park, Florida. Photograph by Vladimir Dinets.

Plate 5.2. Male collared lizard (*Crotaphytus collaris*) in an aggressive display over a female. Alabaster Caverns State Park, Oklahoma. Photograph by Vladimir Dinets.

Plate 5.3. Male Montpellier snake (*Malpolon monspessulanus*) guarding a female. Photograph by Erica Ostanek.

Plate 5.4. Mating aggregations in snakes: *top*, four males coiled around a much larger female anaconda (*Eunectes murinus*) in the Llanos of Venezuela; *bottom*, a larger female gartersnake (*Thamnophis sirtalis*) being courted by at least 10 males at the snake pits in Narcisse, Manitoba, Canada. Photographs by Tony Crocetta (*top*) and Christopher Friesen (*bottom*).

Plate 6.1. **a.** Arribada gathering of olive ridley sea turtles (*Lepidochelys olivacea*) near Ostional Beach, Costa Rica, in November 2016. More than 100 turtles can be counted in the photograph, which was taken from a height of 35 m. A photograph from a height of 100 m at the same location at the same time revealed more than 1,500 individual turtles. **b.** Diurnal arribada nesting of olive ridley sea turtles (*Lepidochelys olivacea*) at Ostional Beach, Costa Rica, on November 14, 2006. This arribada was estimated at 62,757 (CI 95% = 53,106–72,408). Photographs by: *a,* Vanessa Bezy; *b,* J. P. Baltodano Diaz.

Plate 6.2. Selection and packaging of olive ridley sea turtle (*Lepidochelys olivacea*) eggs during an arribada. Eggs are sold and distributed in the Costa Rican market. The increase in human harvesting of eggs is actually a conservation measure because it reduces the number of nests that are inadvertently excavated by later-nesting females; spoilage from excavated nests attracts microbes that are associated with very low hatching success. Photograph by Roldan Valverde.

Plate 6.3. Communal egg-laying in the giant Amazon river turtle (*Podocnemis expansa*), along the Purus River, Abufari Biological Reserve, Amazonas, Brazil. Photograph by Camila Ferrara.

Plate 6.4. Excavation of a communal egg-laying warren of the yellow-spotted monitor (*Varanus panoptes*), in the Pilbara region of northwestern Australia. A total of 130 nests were located in the deepest area (with two people shown), at depths of 2.1–3.6 m. Photograph by Angela Simms from Doody, McHenry, Durkin, et al. (2018).

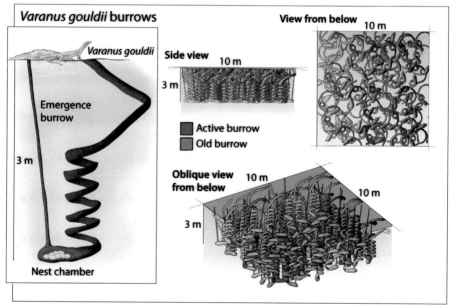

Plate 6.5. Communal nesting in Gould's monitor (*Varanus gouldii*), including a composite helical burrow with hatchling emergence burrow (*left*) and multiple views of the spatial arrangement of communal nests at one nesting warren. Figure drawn by Mesa Schumacher.

Plate 6.6. **a.** Communal nest of the Sri Lankan golden gecko (*Calodactylodes illingworthorum*), from a cave wall in Nilgala, Sri Lanka. **b.** Traditional communal egg-laying in this species results in egg scars that can number in the thousands over many years. Photographs by Steve Wilson.

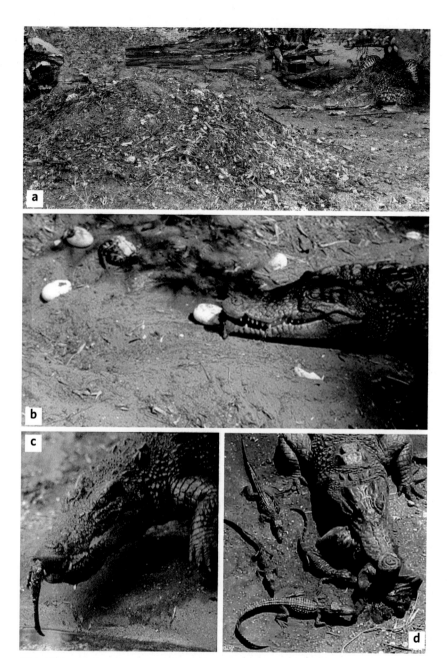

Plate 7.1. Parental care in crocodylians: **a.** female Cuban crocodile (*Crocodylus rhombifer*) guarding her nest (Zoo Miami, Florida). **b, c.** Female Siamese crocodile (*Crocodylus siamensis*) helping her hatchlings get out of eggs and then carrying them to the water. **d.** Male Siamese crocodile feeding his offspring. Photographs by: *a,* Vladimir Dinets; *b, c,* John Brueggen; *d,* David Kledzik.

Plate 7.2. Adult male timber rattlesnake (*Crotalus horridus*) in defense display to human observers (not visible). Note the postparturient female and 17–20 neonates in the background. Photograph by John B. Hewlett.

Plate 7.3. Female rattlesnakes with their babies: *top*, prairie rattlesnake (*Crotalus viridis*); *bottom*, timber rattlesnake (*C. horridus*). Photographs by Amanda Stronza (*top*) and Harry Greene (*bottom*).

Plate 8.1. Hatchling green iguanas (*Iguana iguana*) emerge synchronously from a nest after a long period of many hatchlings taking turns poking their heads out and looking around. Photograph by Gordon Burghardt.

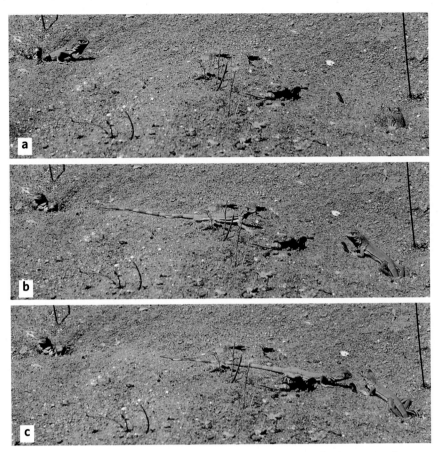

Plate 8.2. Sequence of green iguana (*Iguana iguana*) hatchlings dispersing together on the islet Slothia in Panama. **a.** Hatchlings begin to emerge from two separate nests. **b.** Hatchling from one nest heads toward hatchling from the second nest. **c.** Conspecific contact is made between hatchlings from different nests.
Photographs by Gordon Burghardt.

Plate 8.3. Green iguana (*Iguana iguana*) hatchlings gathered together in reeds over the water on the islet of Slothia in preparation for migration to the main island (Barro Colorado) in Panama.
Photograph by Gordon Burghardt.

Plate 9.1. Hatchling green iguanas (*Iguana iguana*) often forage and sleep in groups for months before leaving the nesting area. Here two iguanas sleep on top of one another on Barro Colorado Island, Panama. Note the eyespot on the eyelid of the lower animal compared to the alert lizard on top. The eyespot is gone within a year. Subsequent research revealed that the top animal is likely a male and the lower lizard his sister. Photograph by Gordon Burghardt.

Plate 10.1. Social feeding of American alligators (*Alligator mississippiensis*) in Everglades National Park, Florida. Photograph by Vladimir Dinets.

Plate 10.2. Huge numbers of Broadley's flat lizards (*Platysaurus broadleyi*) gather near certain rivers in South Africa every summer to feed on small flies (Augrabies Falls, South Africa). Photograph by Vladimir Dinets.

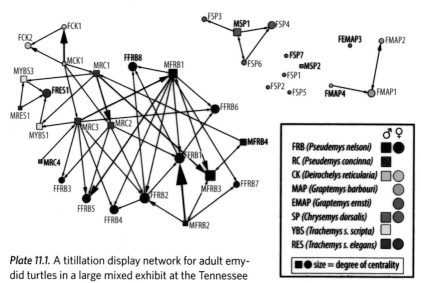

Plate 11.1. A titillation display network for adult emydid turtles in a large mixed exhibit at the Tennessee Aquarium in Chattanooga during the second year of a two-year study. The size of the nodes (circles) represents "indegree," which is a measure of centrality. Larger nodes indicate turtles that received more titillation displays and smaller circles indicate those that received fewer displays. Note that most titillation behavior was directed at conspecifics, even across subspecies, and that individual differences in rate and preferences were large. Animals listed in black were new additions the second year. Prepared by K. M. Davis (based on K. M. Davis, 2009).

green iguanas (Plate 8.2) (Burghardt, 1977b; Burghardt et al., 1977; H. W. Greene et al., 1978). In the communal nest site on Slothia in Panama, hatchlings from different clutches actively approach and physically touch and tongue flick each other and move out of the clearing en masse. They follow one another in virtually the same footprints and gather together in reeds and bushes on the edge of the islet before departing to the main island where they will live. After up to several hours of gathering at the edge, first one and then up to a dozen or more hatchlings will quickly enter the water and swim to the other shore, some even running over the water like basilisk lizards. Over several years, no predation by predatory birds or the lurking small crocodiles was observed during these excursions. Hatchling predation was only noted on hatchlings that were more solitary, suggesting that confusion, not just dilution, was operative. It was also observed that more predators were in the area as the weeks of nest emergence went on, as if the predators were learning that there was an ephemeral bounty at hand. Mass or prolonged emergence may thus attract more predators than smaller emergence in any species. Given the hundreds of hatchlings that could leave the small Slothia clearing in a season, and the fact that a female can lay 40 or more eggs, solitarily laid clutches could possibly avoid the downside of predator attraction while still having hatchlings emerge in sizable groups.

8.3. Conclusions and Future Directions

Social communication begins in the egg, and embryo–embryo communication may underpin the three types of environmentally cued hatching: early hatching in response to perceived predation risk, delayed hatching, and synchronous hatching, but these areas are either recently discovered (early hatching) or remain understudied in many reptiles (synchronous hatching). With respect to communication between hatchlings, we agree with P. Colbert et al. (2010) that synchronous hatching and emergence is likely to be an ancestral trait. We would add to this that the plesiomorphic status of these traits may be extended to all reptiles because of the tight relationship between temperature and development and the lack of thermal gradients in most shallow reptile nests. For exactly this reason, however, there is serious doubt for an adaptive function for synchronous hatching and emergence. Most reptiles have small bodies (e.g., skinks, geckos, anoles, many snakes, and turtles) and lay small clutches of small eggs; as such, nests generally occupy a single layer under cover or in a shallow nest, leading to similar if not identical thermal profiles among eggs. Thermal gradients and metabolic heating in these clutches are thus negligible, and given similar genetics and the tight relationship between temperature and developmental rate, we would expect those eggs to hatch synchronously. Thus, in most reptiles the most parsimonious explanation for synchronous hatching and emergence is a deterministic nonadaptive one. In larger species with large clutches of large eggs, eggs laid in deeper (subterranean) nests experience both thermal gradients and metabolic heating. In these nests we would predict asynchronous hatching and emergence; synchronous hatching and emergence in these nests strongly suggests adaptation via some mechanism (e.g., catch-up). However, within-nest emergence is often poorly synchronized, with hatchlings trickling forth across minutes, hours, or even days, resulting in a poor fit to predator swamping models. For example, in the same clearing on Slothia in Panama where the iguanas hatch in their splendid social manner, *Trachemys* turtles also hatch and emerge. But they do so rather sporadically, move slowly, and if several are out at the same time, seem to have no social interactions with each other (Burghardt et al., 1977). In species with mass emergences, however, among-nest synchrony could offset this trickling effect by swamping predators. More data on thermal gradients and metabolic heating are required to clarify the need to evoke adaptive hypotheses, and more in situ data from natural emergence in systems with some predation risk are needed to clarify the ultimate explanation for synchronous hatching and emergence in reptiles. Finally, most of what we know is from turtles and iguanas; data from other lizards, snakes, and crocodylians are needed to plug massive knowledge gaps in the hatching and emergence ecology of reptiles.

9 |

Behavioral Development in Reptiles

Too Little Known but Not Too Late

This book has covered a wide range of topics on social behavior across the diverse lineages of reptiles. We have amply documented that reptiles show much greater diversity in physiology, life history, reproduction, parental care, and many other traits than are found in mammals or birds. We have largely focused, up to now, on the behavior of adult animals, except for parents and offspring in Chapters 7 and 8. Here we delve into the topic of behavioral development. Development, or ontogeny, is one of the major aims of ethological analysis as formulated by Tinbergen (1963) and others. However, although the focus of considerable research with mammals and birds, development was not considered a major area in evolutionary biology until recently (e.g., West-Eberhard, 2003).

Here are some of the questions that researchers can ask about development. How do neonates and hatchlings change in behavior as they mature into juveniles and adults? How do these changes alter survival and selection in various settings? Is there major selection operating on different life stages that can channel or drive other behavioral systems in adulthood? How do hormones and environmental and social cues and experiences affect developmental trajectories? Behavior development is a major topic in human psychology, with often several departments in a university offering courses in child development, infant behavior, perceptual and motor development, abnormal development, toxicology, and so on. Behavior development, focused largely on mammals (rodents, carnivores, and non-human primates), but also birds (especially birdsong), is a major field called developmental psychobiology and has both a society and a journal. Development is also a major field of interest in invertebrates, as many insects, spiders, and other animals with indirect development go through numerous larval stages and metamorphosis. The transformation of caterpillars into butterflies is the iconic example of the latter, but many amphibians also go through major metamorphic morphological and behavioral changes, and comparisons with puberty in mammals have been drawn (Coppinger & Smith, 1989). In fact, neural and endocrine changes in adolescence are major areas of current research with rodents and humans. Much knowledge also has been accumulated on development of behavior in domestic mammals, who have lived in association with humans for centuries as livestock, agents of work, transportation, and sports (hunting, racing, vermin control), and as companions. Indeed, some of the first papers trying to separate learned from innate aspects of instinct in animal behavior were done with chickens and pigs (Spalding, 1872, 1875, reprinted in Burghardt, 1985).

Comparable studies are also needed with reptiles. In the field, observations of newly hatched birds in the nest and of changes over weeks have been recorded for centuries. This was abetted by the frequent visibility of avian nests and parental behavior, especially in many songbirds, waterfowl, and raptors. In addition, the often rapid growth and developmental change in birds, with fledging or independent life typically reached in weeks, allowed for accumulation of such developmental data with relative ease. Non-avian reptiles are a different story. Dens and nests are used for egg deposition and incubation, but most behavioral development occurs after dispersal from such areas. Nests are typically hard to locate, although many predators are more proficient than humans. Many reptiles, especially juveniles (even of large conspicuous species), are small and hard to find (Perry & Dmi'el, 1994), let alone to repeatedly relocate. This poses methodological problems, now being alleviated by radiotelemetry, GPS loggers, PIT tags, and other technologies that are continually improving.

There are additional challenges to following the social behavioral development of reptiles in wild populations. These include the typically prolonged development to adulthood (in most cases more than a year, often substantially more), small size not amenable to attachment of radio transmitters, secretive lifestyles (primarily to avoid predation), rapid dispersal from natal areas, and lack of ready field identification, from a distance, of known individuals. Additionally, many reptiles have large litters or clutches of eggs with correspondingly high mortality in the juvenile stages. Thus, marking hundreds of individuals may result in limited recaptures and few reaching maturity in field studies (Jayne & Bennett, 1990). The great financial and time commitments required to obtain limited information on animals often deemed of less value to study have hindered scientific and grant support.

There are also historical and conceptual factors that have negatively hindered appreciation of nonsexual social behavior as a promising area of research with reptiles. One factor is that parental care is often considered the key to adult complex social behavior, individual recognition, and cooperation. Such important nonsexual behavior as parental care, Kalmus wrote, "probably occurred in family groups and originated from parental behavior." He said that he doubted that "natural animal communities ever started in any other way" (Kalmus, 1965, p. 5). In a similar vein, MacLean asserted that positive emotions arose only with parental care, and thus reptiles were limited to fear, rage, and anger (MacLean, 1985), although when he made these claims the complexity and long duration of parental care in crocodylians was already well documented (Burghardt, 2020; also see Chapter 7). This shows the bias against reptiles having complex social behavior and shows that they are "often viewed as being born as miniature versions of the adult (e.g., Bellairs, 1970) that do almost all the same things, behaving in a simple 'instinctive' manner in which parental care is quite irrelevant even where faint sidesteps to it appear" (Burghardt, 1978, pp. 150–151). Although in some species developmental behavioral changes do appear minimal outside of the consequences of larger body size and sexual behavior (Rivas et al., 2016), this may actually not be true of many reptiles, especially those that live in social groups at some point in their lives or those that live as adults in habitats different from when they were juveniles (e.g., arboreal juvenile Komodo dragons). In addition, our understanding of the subtle and complex nature of development necessitates in-depth and careful comparative, descriptive, molecular-genetic, and experimental analysis.

In this chapter, we focus primarily on social aspects of behavioral development but will touch on other topics as well. This chapter should also be considered an update, with some relevant borrowing of previous reviews (Burghardt, 1977a, 1978, 1988; Burghardt & Layne, 1995; Burghardt & Layne-Colon, 2021) that also may be consulted for additional citations of primary sources, historical background, and extended discussions not appropriate here. In this book, the behavior of animals before hatching or birth and possible social influences are covered in Chapter 8 and parental behavior in Chapter 7. Here we focus on behavioral changes and underlying processes after birth or nest emergence.

It is interesting to compare more recent findings with what Burghardt wrote over 40 years ago in the first overview of behavior development generally in all groups of reptiles. He covered research on grouping, gregariousness, incipient territoriality, and displays, but then the author looked toward future research on ontogeny and social factors in reptiles in all of two pages, reprinted here in its entirety (Burghardt, 1978, pp. 168–169):

The processes of dispersion of hatchlings and juveniles need to be assessed. Simon and Middendorf (in press) have made a start with the lizard *Sceloporus jarrovi*. The related spiny lizard, *S. undulatus,* has been the subject of an experimental study on effects of density and crowding on juveniles (Tubbs & Ferguson, 1976). Courtship and copulation have not been described for most reptiles so it is not surprising that developmental studies are rare. Carpenter (1977) estimated that courtship had been described in only 3% of all species of snakes. Reptiles reared in social isolation do often appear to court and mate appropriately in captivity. But the situation in natural social groups may be different. While the sexual "drive" and behavior patterns may be endogenously organized and hormonally induced via external stimuli as temperature and day length, social factors, including experience, may play an important but subtle role (e.g., *Anolis carolinensis*, Crews, 1975). Evans (1961) reports that testosterone injections enhanced agonistic behavior and dominance in young turtles. Sexual dimorphism can be marked or absent, and more study, especially of its ontogenetic aspects, might illuminate many areas of reptilian social behavior.

As humans, we are biased toward visual distinctions between animals. It is also important to consider chemical, auditory, and other modalities. There is some evidence that the sexes can distinguish each other by odor (e.g., snakes) and visually (e.g., *Sceloporus* lizards, Noble & Bradley, 1933). Other physical differences also appear (see for example Fitch & Henderson, 1977).

While some aspects of agonistic behavior and territoriality occur at or shortly after birth, this is not always the case (as in iguanas). I am aware of no developmental information on combat rituals in snakes (Carpenter, 1977), whose elaborate "dances" seem restricted to males.

Parental behavior or, more generally, solicitous behavior of adults directed toward young needs to be studied. Turtles are considered the oldest living reptile group and seem to be without such behavior either before or after the eggs hatch. In all crocodilians, it probably occurs; Meyer (1977) has evidence for adult-young nurseries that last up to two years. In between, in some lizard and snakes the females brood and guard the eggs. Indeed, in some snakes the male also guards the eggs (e.g., cobras, M. A. Smith, 1943). These phenomena are often ignored in comparative treatments of parental care and investment (e.g., Maynard Smith, 1977).

One could list many other phenomena that we should know about before even trying to make generalizations and predictions about the evolutionary ontogenetic strategies and tactics employed and exploited by reptiles. Descriptive mathematical approaches (e.g., Singer & Spilerman, 1976) might be useful. The fact that young natricine snakes seem to be considerably less capable than adults, physiologically, of sustained activity (Pough, 1977, unpublished) may have far-reaching implications, pointing to the need for an area of reptilian developmental physiology (see also Gans, 1976, for crocodilian examples). Intra- and inter-individual variability may also be greater in newborns as compared to adult reptiles. Certainly, this appears to be the case in responses to prey chemical stimuli in some snakes (Burghardt, 1975; Gove & Burghardt, 1975). Object play and curiosity, often considered nonexistent in reptiles and highly important in mammals, may need to be reassessed as naturalistic observations of reptiles (e.g., Lazell & Spitzer, 1977) and appropriate sensory considerations (e.g., Chiszar, et al., 1976) accumulate. Thus, investigations of ontogenetic processes in reptiles will gradually move to detailed long-term studies of field populations, comparative studies of closely related species (e.g., following King, 1961), and sophisticated experimental studies of mechanisms, all eventually leading to the formulation of specific and testable functional evolutionary ideas to explain the diverse ways reptiles go from zygote to senescence. In addition, studies of carefully selected reptiles may allow testing the limits of theories about development which differentially stress processes such as parental care, socialization, food habits, encephalization, habitat, or developmental rate.

Today we have much more information on some of these topics, and we can extend, qualify, or correct the above. But too often we are limited to comparisons of hatchlings and neonates with adults (cross-sectional studies) and the limited, but now numerically increasing, studies tracking individuals throughout the juvenile to adult period (longitudinal studies). These can be either descriptive or experimental manipulations of the developmental processes themselves. In addition, unlike birds and mammals, squamate reptiles (constituting 95% of all reptile species) contain taxa that are either oviparous, ovoviviparous, or viviparous, opening up many avenues for research integrating development with environmental factors, such as temperature and climate, as well as phylogeny. Indeed, in some species, females may lay eggs and give birth to live young, and recently this has been reported in one individual lizard mother (Laird et al., 2019). What factors drive such variability in some taxa while others are seemingly locked into one mode or the other?

9.1. Some Principles and Concepts in Behavior Development

How does behavior change over an individual's lifetime, and what processes underlie such changes? This is the focus of several major areas of study in psychology, endocrinology, neuroscience, epigenetics, and the growing field termed evo-devo in evolutionary biology. West-Eberhard's book (2003) was a major impetus to the latter and integrated much information across many taxa, but social behavioral ontogeny in reptiles did not feature prominently in her volume.

Due to the lack of study, a comprehensive description and review of social developmental processes in reptiles is not warranted. However, some basic issues and terms can be introduced to set the stage for some examples and for fascinating opportunities for future work. In exploring this topic with reptiles, the work of those focused on mammalian developmental processes can provide useful guideposts. Smotherman and Robinson (1988) view the uterus as an ecological environment and Gottlieb (1991) broadens the approach to include eggs. Most relevant here, however, may be the perspective of J. R. Alberts and Cramer (1988). These authors point out that the embryo, neonate, and juvenile have their own evolved adaptations for their species' typical niche. For example, the newborn with the mother in the nest has all the behaviors associated with survival there, including nursing, competition, thermoregulation, and so on. However, the developing animal also has to prepare itself for the niche it will find itself occupying as an adult. Such a shift can include changes in perception, locomotion, foraging, food ingestion, predator avoidance, habitat use, and so on. While such shifts may be less dramatic in reptiles than in many birds, mammals, and amphibians, this does not lessen the need to pay attention to these often critical phenomena. Problems in successful introductions of head-started animals in conservation efforts may be due to ignoring developmental processes in animals that seem successfully raised in captivity (Hayward & Somers, 2009).

The traditional approach to behavioral origins has focused on the role of innate or congenital factors and the interaction with experience in channeling the way behavior is expressed throughout life. We do know that in reptiles much behavior is very precocial: hatchlings and neonates have rather advanced perceptual/sensory capacities and locomotor abilities at birth. In mammals and birds there is a continuum from quite precocial to highly altricial offspring. Examples of the former include birds such as chickens, ducks, and ostriches, where bipedal locomotion, vision, and behavior patterns such as pecking, preening, approach, avoidance, and vocalizations are operational shortly after hatching. Birds born in a very altricial state include many songbirds, parrots, and raptors, among others. Here, after hatching, birds may be blind, nest bound, nearly incapable of thermoregulation, and highly dependent upon parental care for survival. Some behaviors, such as gaping for food in order to stimulate feeding by parents, are highly developed and essential for survival. Almost all juvenile birds, however, are developmentally constrained in that they cannot fly. This raises the question of the role of experience and learning in flight behavior. Spalding (1875) showed through experiments that flying ability in larks is largely a maturational, not learned, behavior, and such studies were independently carried out in pigeons by Grohmann, a student of Lorenz (Lorenz, 1937, translated in Burghardt, 1985). These studies established that neuromuscular maturation was important in the development of flying in birds, and experience in terms of learning or practice in the nest was minimal for the onset of flying, although experience may be quite essential for perfected flight behavior, as in attacking prey or mobbing predators.

A similar situation occurs in mammals, where some species are quite precocial. Many ungulates and all cetaceans are able to locomote at birth, and their senses are quite functional. They are still often highly dependent on parents, principally the mother, for food and protection. Some other mammals are even more precocial. For example, guinea pigs (*Cavia porcellus*) have little need of maternal milk in that they can feed on solid food at two days of age and are able to reproduce when three months old; hooded seals (*Cystophora cristata*) are weaned and abandoned at four days of age. On the other hand, many mammals, including other rodents, canids, and bears, are born in a very altricial state: their eyes are not open and they are hairless, they are unable to thermoregulate, and they are otherwise in need of intense parental care. Rats and mice are weaned within several weeks, but black bear cubs do not even open their eyes until about seven weeks of age (Rogers et al., 2020). However, the degree of altriciality is not tied to the length of parental care. Primates, including humans, are altricial in terms of needing food provided by mothers and warmth, but they are born with functioning senses, vocal abilities, and so on. Still, they may not be weaned for many months or years and do not reach social reproductive maturity for more than a decade (see Nowak, 2018, for overviews of mammal life histories).

Besides the need for careful descriptions of the behavior of animals as they traverse life from birth or hatching, experiments are needed that focus on (1) separating maturation from the role of experience, (2) the limits and constraints of developmental plasticity, and (3) learning processes in behavioral ontogeny. Behavior at birth or hatching, even in highly precocial species, may involve, even necessitate, experiential processes to hone skills or adapt behavior to different contexts. As William James wrote in his still useful treatise *Principles of Psychology* (W. James, 1890, p. 390), in all animals capable of memory, every instinctive act *"must cease to be 'blind' after once being* repeated." The issue is more rightly framed as how flexible or plastic the behavior is after birth or hatching. This can vary greatly. In reptiles, even those with parental care, offspring are more precocial at birth or hatching than any bird or mammal. They are capable of doing much of what they need to do in order to survive on their own. This should not mislead us. Sociality among siblings or other juveniles can be important, perhaps even more so than parental care. This may explain the frequently great variation in parental care within those reptiles that have it. Perhaps the costs and benefits of parental care to parents may lead to discounting current offspring in favor of future offspring.

The typical contrast between the roles of maturation versus experience in improving or facilitating performance does not do justice to the complexity of the issues. There are many ways in which these two processes operate after birth or hatching. Specific experiences or their absence may have diverse effects. One contrast often made is between experience-dependent and experience-independent aspects of behavior, a distinction that is used as a replacement for the outmoded innate-learned, or nature-nurture, dichotomy (Burghardt & Bowers, 2017). Specific experiences, such as food, social interaction, and predator exposure, may delay, accelerate, eliminate, or qualitatively change behavior, including responses involving food, predators, habitat, and conspecifics. It is also the case that even the most basic, seemingly hard-wired, behavior patterns can be altered in both obvious and nonobvious ways. Temperature-dependent sex determination (rather than pure genetic sex determination) is a clear example of an alteration in sexual/reproductive behavior that no one considered a serious possibility 60 years ago but is now is found in turtles, crocodylians, tuatara, and many lizards with serious conservation consequences (J. J. Bull, 1980; Janzen, 1994). This is a nonreversible change in most instances.

Typically, developmental studies begin at birth or hatching, but fetal and embryonic behavior, not just environmental inputs such as temperature or moisture, can influence developmental outcomes. These are issues to be experimentally investigated as shown in Chapter 8. At least three hypotheses for embryonic behavior have been proposed (Smotherman & Robinson, 1988). One is the null hypothesis that embryonic behavior is an incidental by-product of structural or neuromuscular maturation, the *epiphenomenal* view. Another is the *preparatory* hypothesis that such behavior is needed for essential practice, learning, or experience for successful postembryonic life. Such was the proposal that pecking in baby chicks is derived from heart beat induced movements in the egg (Lehrman, 1953). This is closely related to the *functional* hypothesis that embryonic behavior is important in embryonic development itself. These are important questions and also involve, as shown in Chapter 8, social influences from clutchmates, as well as hatching synchrony. In viviparous species such as garter snakes, sex differences at birth may be mediated by whether a male or female is bordered in the womb by same or opposite sex littermates, as has been shown in mammals (Smotherman & Robinson, 1988).

There are many ways in which something nonobvious, at least initially, influences reptile behavior. For example, antipredator responses in snakes seem to be quite heritable (S. J. Arnold & Bennett, 1984; Burghardt & Schwartz, 1999; Waters et al., 2017). However, larger baby snakes are bolder (Mayer et al., 2016) and postnatal experiences can quickly change antipredator responses (Herzog, 1990; Placyk & Burghardt, 2011). Habituation is also important, especially with responses to stimuli in the environment (Herzog et al., 1989; Waters et al., 2017). In habituation, repeated encounters with stimuli lead to reducing or extinguishing responses. Sensitization is the opposite, where repeated presentations lead to a more pronounced or more rapid response. Both can also be important in social settings. More formal models of how behavior changes during development have been proposed, though not readily applied to reptile behavior at this point. Stamps and Frankenhuis (2016) provide a useful primer on recent tools that have been utilized in such studies.

These are just a few of the aspects of behavioral development involving sociality that have been extensively studied in mammals, birds, amphibians, and fishes but less so in non-avian reptiles. In the following sections, organized by major taxonomic groupings, we review some of the empirical findings involving behavioral ontogeny and sociality in reptiles. These tantalizing glimpses show the diversity of studies available.

9.2. Archosaurs

Extant birds have numerous social behaviors that are typical for particular ages, including imprinting (Lorenz, 1937, translated in Burghardt, 1985), vocal learning, parent following by precocial chicks, and begging in response to movement by altricial ones (see an overview by Starck & Ricklefs, 1998). In some cases, young adults help their parents' breeding rather than breed themselves (cooperative breeding reviewed by Koenig & Dickinson, 2004); in many species subadults are nomadic and gregarious, while adults live in territorial pairs (Heinrich, 1988). It can be safely assumed that many extinct archosaurs also had complex ontogenetic changes in social behavior, especially considering that some of them (e.g., pterosaurs) were dependent on parental care when young (see Chapter 7), and considering many large species had to change their diet repeatedly while growing up (Díaz et al., 2012). Fossil evidence suggests that some dinosaurs formed herds segregated by age (Varricchio et al., 2008; Myers & Fiorillo, 2009), while others were gregarious only while young (Q. Zhao et al., 2013); there is also a possibility (Currie, 1998) of division of labor between young and old individuals during group hunts, as sometimes observed in crocodylians today (see Chapter 10, section 10.1).

Extant crocodylians are remarkable in that parental care is common and well documented, from constructing elaborate nests to nest guarding, emergence facilitation, and guarding and leading clutches for many months, even years (Chapter 7, section 7.1). Different types of communication occur between parents and hatchlings and juveniles, but vocal signals have been most studied. Initially grunts or contact calls by hatchlings and distress calls, similar to those in chickens and other precocial birds, were noted, but early analyses did not seem to show sonographic differences (H. W. Campbell, 1973). However, acoustic differences were found between contact and distress calls in American alligators (*Alligator mississippiensis*) (Herzog & Burghardt, 1977), and various hypotheses for their role were outlined. Furthermore, distress or separation calls differed acoustically in different species, and these differences could be quite complex (Herzog & Burghardt, 1977; Bonke et al., 2015). In fact, in the Nile crocodile (*Crocodylus niloticus*), separate "screech" and moan components were found. More recently, Vergne et al. (2007), studying recordings of distress calls of black caimans (*Melanosuchus niger*), found limited evidence of individual signatures that could be used by mothers to identify their offspring. Though clearly difficult to carry out, behavior choice experiments are really needed before accepting this conclusion. The authors did find that the calls changed acoustically with age, and that this could provide information to parents even in the absence of individual recognition. Various hypotheses for distress calls have been invoked, including the idea that they can serve as warning calls for animals in the crèche and perhaps even morph into altruistic signals in adults (Mathevon et al., 2013), but it has also been argued that neither kin selection nor altruism need to be invoked (Staton, 1978). A fine, rather recent, review by Vergne et al. (2009) documents the current state of knowledge about crocodylian vocal signals, particularly in juveniles, notes the similarities with avian vocalizations, and highlights the key outstanding issues in understanding them in both mechanisms and functions.

While young crocodylians are gregarious and forage and move in groups, they do have temperament differences in agonistic behavior. An extensive set of observations documented the differences between two Australian species, with hatchlings of one, the estuarine crocodile (*Crocodylus porosus*), being particularly intolerant and aggressive with other crocodiles as well as their own species. The other, the Australian freshwater crocodile (*Crocodylus johnstoni*), is far more placid (Brien, Lang, et al., 2013; Brien, Webb, Lang & Christian, 2013; Brien, Webb, Lang, McGuinness & Christian, 2013). While there is much information, relatively, on parental behavior, there is far too little on the social ontogeny of any of these animals.

9.3. Turtles

While turtles live in many habitats, turtle diversity is far more limited than in squamates, though higher than in crocodylians. They are typically considered rather asocial, since they often seem to live solitary lives, and this view has been extended to neonates. All turtles lay eggs in nests and then abandon them shortly thereafter. When the hatchlings emerge, they are basically on their own. However, along with the recent findings on vocalizations in turtles (see Chapter 4), there are the findings of Ferrara, Vogt, and colleagues on maternal return to communal nest sites on rivers in Amazonia by giant Amazon river turtles (*Podocnemis expansa*) (Ferrara et al., 2013; Ferrara, Vogt, Sousa-Lima, et al., 2014). Not only do the mothers return from many kilometers downstream to retrieve hatchlings, using vocalizations, but they also apparently guide the hatchlings downriver to foraging sites. This would be remarkable for any species, but certainly was unexpected in an entire group of animals where no parental behavior had been previously recorded. Clearly, the hatchlings also have to fit into the ontogenetic niche of vocalizing even while still in the nest, respond to adult behavior, follow adult females, and travel with them in large social groups. These groups can include hundreds of females and thousands of hatchlings. At this point, we have no information, however, as to whether the hatchlings and/or mothers recognize their family members by vocalizations or any other cues. This is certainly a question worth pursuing. For years it was thought that in caves where thousands of Mexican free-tailed bat (*Tadarida brasiliensis*) mothers and their babies were roosting, mothers returning from foraging flights would feed offspring randomly, as there was no way they could find their own babies given the darkness, the cacophony, and the movements of the offspring. McCracken (1984) used genetic markers to show that mothers could recognize and locate their offspring in the caves. Subsequently, it was shown that pup calls showed stable individual signatures (Gelfand & McCracken, 1986).

We now know that turtle vocalizations are quite common in both adults and hatchlings (Ferrara, Mortimer & Vogt, 2014; Ferrara, Vogt, Giles & Kuchling, 2014; Ferrara, Vogt, Harfush, et al., 2014), and thus many more exciting findings on parental and social behavior of hatchlings and juveniles may be expected. Whether any ontogenetic changes occur in the calls, and if so how and why, has not been explored at all.

Young snapping turtles (*Chelydra serpentina*), at least in captivity, develop dominance hierarchies (Froese & Burghardt, 1974) that quickly result in some animals becoming many times larger than others, even clutch mates, yet all remain healthy due to relatively indeterminate growth, something which could not happen with birds and mammals. Western pond turtles (*Clemmys marmorata*) give open mouth threat signals, but juveniles respond to those of adults more strongly than those given by other juveniles (Bury & Wolfheim, 1973). Whether this is just due to the greater signal strength rather than the social standing of the adult is not established. Hatchling and juvenile gopher tortoises (*Gopherus polyphemus*) can be aggressive to one another at and inside burrows, which are a valuable resource (Radzio et al., 2016). The conclusion of the authors in this video-recorded field study, that "Social interactions appear to play a greater role in the ecology of hatchling and juvenile gopher tortoises than previously recognized" (p. 231), is but another reiteration of the theme of this chapter. Titillation or foreclaw displays in emydid turtles are frequent in courtship, especially by males, but can be performed by females as well (K. M. Davis, 2009). However, such behavior is also found in hatchling and juvenile turtles and has been considered a form of precocial play (Kramer & Burghardt, 1998).

There clearly needs to be more research on social behavior development in turtles. The paucity of information reflects, perhaps, the view that turtles are asocial (Wilkinson et al., 2010). Yet given all the breeding and rearing of turtles, including tortoises, for conservation and commercial purposes, the lack of studies is hard to fathom and many research opportunities must certainly exist.

9.4. Tuatara

The highly endangered and relict tuatara (*Sphenodon punctatus*) is all that exists of the rhynchocephalians, as noted earlier. The limited knowledge of its social, territorial, and reproductive behavior is reviewed elsewhere, but Cree's (2014) magisterial volume, *Tuatara: Biology and Conservation of a Venerable Survivor*, which reviews all aspects of tuatara life, ecology, and evolution, including the species' discovery and cultural significance, is probably the most detailed treatment of a single species of reptile ever written. In terms of behavioral development, we know little from the field and somewhat more from captive animals. Recent morphological studies show that the tuatara fits the criteria of being a "living fossil" due to its slow lineage evolution, originating in the Triassic and with major diversification in the Jurassic (Herrera-Flores et al., 2017). These animals are unusual in that they live at cool temperatures—temperatures at which many other reptiles could not thrive or even survive. There are both descriptive and experimental studies on adult social and reproductive behavior (Gans, Gillingham & Clark, 1984; Gillingham et al., 1995), but not much is known about juvenile behavior. These animals can live to 100 years and reproduction may not occur until at least 15–20 years of age. Indeed, 12-year-old animals are still considered juveniles (Cartland et al., 1994; J. T. Nelson et al., 2006). However, based solely on growth rings, maturity at 9–13 years has also been proposed (Castanet et al., 1988).

Juvenile locomotor behavior is affected by body size, incubation temperature, and other factors, with animals from natural populations performing better than captive reared animals (Cartland et al., 1994; J. T. Nelson et al., 2006). In the field, social network studies show that social connectivity, sex, and age affect parasite loads, but little social behavior was reported. It would appear that the long generation time and low rates of behavior in this cold-adapted, low metabolic rate species mitigates against detailed studies of behavioral development.

There is one study, a thesis (Wörner, 2009), that looked at social interactions in one-year-old captive hatched and reared tuataras paired with same sex, opposite sex, and different sized animals. As they have temperature-dependent sex determination, it was possible to rear and test animals in same and mixed sex groups. This is important because in the field sex identification may not be obvious until the animals are 15 years old. However, sex of all animals was confirmed in this study with laparoscopy. This was a complex study lasting about a week and looking at social encounters, food competition, and space use. Males were more aggressive to one another than to females, regardless of size, but larger animals dominated smaller ones, which gave submissive responses. Most dyads developed dominance relationships. Food competition was variable but size and sex were factors, with large males securing more food than smaller males or females, whereas female-female pairs were more likely to share equally. In space-use experiments with various shelters available, no territoriality was found; in fact, juvenile (one-year-old) tuataras seem to have "obligatory clustering" (p. 80). Interestingly, urinations and defecations, particularly by males, were very frequent during the first 24 hours after being introduced to the novel housing and perhaps could be scent-marking (Wörner, 2009). Clearly, these animals deserve more study, but the general picture of social interactions is comparable to that recorded in some turtles and lizards (Froese & Burghardt, 1974; Burghardt, 1977a;). Tuataras are territorial as adults, and the transition from a more social-dominance regulated group to a territorial system may take many years in this long-lived and slowly maturing species.

9.5. Lizards

We perhaps know more about social behavior and its development in juvenile lizards than in any other group of reptiles. But given the thousands of living lizard species, that is not saying a great deal. For example, virtually nothing is known about social behavior, let alone its ontogeny, in the hundreds of species of fossorial lizards. Our rudimentary understanding of the ontogeny of the greatly divergent social behavior and social organization of the many saurian taxa (never mind their morphological, dietary, ecological, sensory, and physiological adaptations) is almost sinful. Given this diversity, we will just discuss examples where social behavior and communication in newly born or hatched lizards and juveniles have been studied. Those examples dealing with the development of color dimorphism used in signaling and social cues, such as mating, dominance, and territoriality, will not be reviewed here, although clearly such changes and their hormonal and neural substrates are important.

Although the effects of incubation temperature on sex determination in lizards is well known, the effects on nonsexual behavior is little studied, and the implications might be great as the climate changes. In small but suggestive studies, groups of bearded dragons incubated at different temperatures within the normal range (27°C and 30°C) showed stark developmental differences when reared at 29°C for several months after hatching, although sex ratios were similar. As expected, eggs incubated at the warmer temperature hatched sooner, but the difference was remarkable—they hatched an average of 30 days earlier (58 versus 88 days). After hatching, the cooler incubated lizards grew faster, but the hotter incubated animals completed foraging tasks more quickly and ran faster (Siviter et al., 2019). In a comparable group of animals, the ones incubated at the cooler temperature both learned and completed a social learning task (see Chapter 11) more quickly (Siviter et al., 2017). Indeed, temperature effects during early development in both oviparous and viviparous populations of the same species of Australian skink (*Saiphos equalis*) showed significant, but mixed, results on learning and exploratory behavior (Beltrán et al., 2020). Light may also have developmental and cognitive effects (Y.-P. Zhang et al., 2020). Such studies have implications for climate change effects on populations of less vagile species.

There are now numerous studies documenting incubation temperature effects on behavior and learning in lizards. Many are reviewed in Siviter et al. (2019), but others on diverse species are those by T. S. Mitchell et al. (2018) and D. W. A. Noble et al. (2018). Dayananda and Webb (2017) are among those whose data explicitly implicates the deleterious effects of climate change on cognition. The profound effects of incubation temperature on sexual and agonistic behavior in leopard geckos (*Eublepharis macularius*) are especially indicative of how temperature can skew behavioral development in one of the most popular pet reptiles (Rhen & Crews, 1999;

Sakata & Crews, 2003). Also, some lizard eggs may overwinter before hatching and emerging, with incubation times about equal to human gestation of nine months. These findings in the Gila monster (*Heloderma suspectum*), certainly not a North Temperate species, also raise important and interesting developmental questions (DeNardo et al., 2018).

Many lizards of several families engage in visual headbob displays. Charles Carpenter pioneered the objective analysis of these displays and the variation seen across species, populations, sex, and other factors in many lizard families (see, for example, C. C. Carpenter & Ferguson, 1977; C. C. Carpenter, 1978, 1982; N. Greenberg & Jenssen, 1982). There are various types, including assertion displays, courtship displays, territorial displays, and so on. These can also be individually different in territorial male green iguanas within a population (Dugan, 1982a). Hatchling green iguanas (*Iguana iguana*) will perform simplified headbob displays within minutes of emerging from the nest (Burghardt et al., 1977). This has been observed in anoles (*Anolis*), among many other species (Cooper, 1971; Stamps, 1978; N. Greenberg & Hake, 1990). Hatchling eastern fence lizards (*Sceloporus undulatus*) perform simplified headbobs from hatching on, but these seem to be initially nonsocial in that they are not directed to any individual and no context has been identified (Roggenbuck & Jenssen, 1986). However, Australian frillneck lizards (*Chlamydosaurus kingii*) direct frill erection and hand-waving displays toward other hatchlings at one day of age (Christian et al., 1999). In the latter study, dominance among hatchlings was not related to androgen levels, but there was far less sex difference among the juveniles than in adults. Hormones have been implicated in the ontogeny of headbob displays in green iguanas, and some displays seem to be only found in juveniles (Phillips et al., 1993). In fact, such social displays have been considered play when performed before any serious role for performing them can be identified (Burghardt, 2005).

Research on the bronze anole (*Anolis aeneus*) by Stamps (1978) is an excellent study of social, spacing, agonistic, and display behavior comparing juveniles with adult males and females. One of the many findings is that juveniles react more aggressively to intruders than do adults. There are also changes in the display details, which nonetheless show little variability.

Further studies illuminated other topics. For example, due to predator avoidance, juvenile anoles often had overlapping territories, but this came at the cost of slower growth (Stamps, 1984). On the other hand, young green iguanas (which for the first year of life, at least, are not territorial) that stayed closer to others actually grew faster (Burghardt & Rand, 1985). Thus, here are two groups of iguanid lizards that differ greatly in juvenile sociality.

Juvenile lizards of different species may also engage in rubbing and marking behavior (presumably pheromonal) when very young, indeed belly rubbing on shrub branches was recorded, and filmed, in green iguanas shortly after leaving the nest site (GMB, unpubl. obs.). Detailed chemical and behavioral studies are now needed. Tongue sampling during social interactions is known in some species, and some experiments have been carried out. When veiled chameleon (*Chamaeleo calyptratus*) hatchlings were reared in social isolation and compared to those reared in groups, the isolates were more submissive socially and were slower to approach and capture prey (Ballen, Shine & Olsson, 2014). This may reflect acquired personality traits of social dominance and boldness, both of which have been extensively studied in lizards (Waters et al., 2017).

The advent of parental care and more social family lifestyles in lizards makes lizards even more central to understanding the development of sociality. For example, tree skinks (*Egernia striolata*) are born asynchronously in small litters, up to four, and wild groups can consist of parents and offspring, juveniles, or adult male-female pairs. Juveniles, then, can survive in the absence of postnatal parental care. An experiment examined if social rearing would affect the behavior of juveniles (Riley, Noble & Byrne, 2017). Juvenile skinks were reared from birth singly or in pairs and tested in a series of behavioral assays and growth measures several times up to 12 months of age. It was found that one lizard in a pair became dominant and so comparisons were made among isolate-reared, dominant, and submissive animals. The basic finding was that behavioral and personality traits were not consistently affected by rearing condition and also that behavioral repeatability was low. Some other findings were that isolate-reared animals were more social than subordinates in tests. The latter became more aggressive over time. Dominant animals became bolder over time, perhaps due to their experience in winning. The conclusion from these various studies is that the role of social experience during development can vary greatly and thus generalizations should not be drawn until much more work is done. Another study on this species predicted that isolate-reared juveniles would perform more poorly in a spatial learning task but found no difference from animals reared in pairs (Riley, Noble, Byrne & Whiting, 2017). The authors speculated that since in the wild juveniles often live alone, the species is buffered from the kinds of social isolation defects found in animals with obligate parental care.

Tree skinks are not the only family-living lizards. So far, extended family living has only been found in two families: almost all are skinks (Scincidae) and one a night lizard (*Xantusia vigilis*, Xantusiidae), but there is evidence that such behavior might be more widespread, according to a recent review of sociality in lizards (Whiting & While, 2017). While the details of the social arrangements and ecological factors have been explored, little has yet to be reported on the development of behavior. It was noted that lizards reared in family groups may disperse later and that this may have some adaptive consequences. There are other interesting findings with skinks, a huge family of lizards. For example, young and adults of a rather solitary species, the eastern water skink (*Eulamprus quoyii*), were tested in an instrumental task (learning to displace a lid to obtain food) and an association test (learning to choose a blue lid over a white lid). In both tasks some animals were able to observe a conspecific making the correct choice (D. W. A. Noble et al., 2014). It was found that age did not affect learning the instrumental task and neither did social observation. Most animals learned the association task within 24 trials. Neither age nor social observation affected the latency of response, but this was not true of the number of trials needed to learn the task. Here the older animals did not differ between the social and control groups (if anything, the social group took more trials to learn), whereas the young lizards in the social group learned the task in about 40% fewer trials than did the controls. There were additional differences favoring the young lizards. While an admirable and lonely experimental study, the fact that testing was not blind and no attempts were made to avoid observer bias (Burghardt et al., 2012) compromises the study, along with the fact that the stimuli were not balanced in that white as well as blue cues were not used. Interestingly, a study of social learning in the highly social tree skinks did not find any effects of social demonstration using both the instrumental and association task and then a blue to white reversal task (Riley et al., 2018). The latter study did implement blind scoring of video-recorded trials, but still only one color, blue, was used as the initial stimulus and reliability scores of independent observers were not reported.

Finally, the level of prosocial behavior and even altruism in reptiles needs to be assessed. For example, after dispersing onto the main island in Panama, juvenile green iguanas continue to forage and sleep together (Plate 9.1). Iguanas that sleep more closely to one another also grow faster, suggesting sociality enhances foraging and growth. Furthermore, juveniles often sleep in the same shrub, even the same branch or on top of one another. In trying to capture these sleeping animals at night, grabbing one would, through the slightest movement of a leaf or branch, awake the others, who could then escape. But there may be even more going on, as there is some evidence that siblings migrate together. Intriguing experimental data support the view that in sibling groups of juvenile green iguanas, a brother, when threatened by a predator, will jump on and cover his sister and put himself at risk to save her (Rivas & Levin, 2004). Why might he do this? The mating system of these animals may provide an answer (Dugan, 1982b; Rodda, 1992). In the harem-type system of green iguanas, only the largest and most dominant males establish territories during the mating season that many females, often more than 10, may visit. Females are reproductively active as early as three to four years old, whereas males reach full development of their sexually dimorphic traits, large size, and dominant status some years later. Furthermore, whereas virtually all females who reach reproductive maturity breed, not that many males do, even if they survive the additional years. Thus, a juvenile male's chance of breeding is far lower than his sister's, so it would appear a good strategy to favor her and increase the chances of half of his genes surviving (Rivas & Burghardt, in prep.). Also, Dugan (1982a, 1982b) reported that there are small female-mimicking males (Weldon & Burghardt, 1984) that try to accost and forcibly copulate, often unsuccessfully, with females when the large territorial male is patrolling elsewhere. The developmental processes involved in the two types of male strategies also need to be studied.

In short, there is a great need to explore the role of social development in many species and groups of lizards. They are excellent laboratory animals with a diversity greater than that found in either mammals or birds.

9.6. Snakes

Social behavior in snakes has been grossly underestimated, as discussed earlier in this book. For example, courtship and mating in snakes were considered the only social behavior they performed. It has long been known that aggregations of snakes are often found at den sites, and under cover objects (Gillingham, 1987, 1995). Many species do this. However, until 50 years ago, the prominent view was that snakes aggregate together due to a scarcity of places with the proper microclimate, such as humidity, temperature, and soil type; thus, aggregations of snakes were due only to physical resource availability and had no social component (Prater, 1933). In an authoritative review by Brattstrom (1974), 40 years after Prater's assertion, more examples of possible sociality, primarily anecdotal, were discussed, although the entire section on snakes was only about half a page long. Brattstrom did open the snake section on a prescient note: "Snakes may be the most asocial of all reptiles. This may be a function of their elongate bodies (with few structures of the body for utilization for display) and largely secretive nature, or it may be only a function of the difficulty of studying these forms" (pp. 44–45). Brattstrom also claimed that outside of aggregation behavior, "nobody has, in my opinion, designed the appropriate experiments to study social behavior in snakes properly" (p. 45).

As with lizards, snake eggs have been incubated at different temperatures and feeding and other behavioral differences noted. The extensive series of studies by Burger (1991a, 1991b, 1998) on threatened and declining pine snakes (*Pituophis melanoleucus*) are noteworthy; she brought early attention to behavioral effects that we now realize are related to climate change and unrelated to sex determination. In terms of effects on social behavior, Aubret, Bignon, et al. (2016) found that hatchlings from *Natrix maura* eggs incubated singly were more active and contacted siblings less often than those hatchlings incubated in clutches. There seem to be no comparable behavioral studies on viviparous snakes, although maternal body temperature had profound effects on survival and congenital defects in common gartersnakes (Burghardt & Layne, 1995).

An early experimental study of social aggregation behavior in snakes was carried out with adult brown snakes (*Storeria dekayi*) and eastern ribbon snakes (*Thamnophis sauritus*) (G. K. Noble & Clausen, 1936) and focused on the sensory systems involved. The stimulus for aggregation, shaking the cage, an earthquake simulation, was hardly "natural." The study did show that animals were more likely to group with conspecifics. Although many experiments were reported, the stimulus deprivation experiments were often severe and presumably stressful, and conclusions somewhat suspect (L. T. Evans, 1959; Burghardt, 1970a). The authors did report that grouped animals had lower oxygen consumption than isolated animals, and thus identified a possible functional benefit of the groupings.

A more naturalistic study involved adult prairie ringneck snakes (*Diadophis punctatus arnyi*) tested in a large outdoor arena with identical cover objects (Dundee & Miller, 1968). Not only did snakes begin to cluster under certain covers, but this attraction was observed outside the courtship season (see Plummer, 1981, for similar findings in gravid rough green snakes, *Opheodrys aestivus*). The authors also found that snakes in some way marked the soil under the cover objects and this served as an essential cue in experiments where soil was transferred among cover objects. The snakes preferred areas where other snakes had been, and Dundee and Miller called this phenomenon habitat conditioning. But what about neonate snakes? A laboratory study based on Dundee and Miller was carried out with 20 newly born common gartersnakes (*Thamnophis sirtalis*) and 20 newly born brown snakes (*Storeria dekayi*) released into the same arena with eight identical cover objects (Burghardt, 1983). The snakes not only developed clumping patterns over several days but they preferentially grouped with conspecifics and had preferred cover objects. Multispecies aggregations did occur, and many involved more than 10 (up to 26) individuals. Experiments on individually marked neonatal common gartersnakes that were tested for sociability and boldness, as well as in a Dundee and Miller inspired apparatus, showed that both measures affected aggregation measures and that preferential partnerships developed (Skinner & Miller, 2020). Thus, social aggregation is found from birth in this species, and such aggregation shows some ontogenetic development that seems affected by individual differences. It would now be useful to follow up on the development and alteration of such grouping preferences throughout life, including when reaching reproductive maturity.

That such changes can occur outside major hormonal or growth/age issues was shown by allowing juvenile plains gartersnakes (*Thamnophis radix*) to compete for a single prey item with one of two cagemates. Three snakes were housed together and their positions noted. The loser changed its social aggregation preference from the winner to the control snake (Yeager & Burghardt, 1991). Another study showed that Butler's gartersnake (*T. butleri*) juveniles preferentially aggregated with conspecifics that ate a diet different from theirs (Lyman-Henley & Burghardt, 1994). Although snakes are generally solitary hunters, they may return to areas with conspecifics, such as under the same boards. This behavior may be adaptive, as was shown in an experiment in which newly born banded watersnakes were, from birth, housed and fed individually, housed and fed in groups, or housed in groups but removed and fed singly (Burghardt, 1990). All snakes were allowed to eat small fish ad libitum for an hour weekly for 12 weeks. At this point, all snakes were reweighed; the snakes housed socially, but fed individually, had significantly higher body masses than the other two groups, which were almost identical. Thus social grouping, but foraging separately, may be an adaptive strategy for at least some snakes.

Trailing behavior is also found in many snakes, given their exquisite tongue-vomeronasal system (vomerolfaction) (N. B. Ford, 1986; R. T. Mason & Parker, 2010), and much work has been done on use of such cues in juvenile snakes locating dens used by adults (Gillingham, 1987, 1995). Kin recognition among litter and clutch mates has been demonstrated in a few species (Schuett et al., 2016), and kin recognize each other after two years of separation (R. W. Clark, 2004).

In a most thorough and stimulating review of snake sociality, Schuett et al. (2016) provide many examples and concepts of social ethology, along with a fine introduction to terminology and suggestions for where future work should be focused—ontogeny is one of these areas. Aspects of social behavior development in snakes other than kin recognition and aggregation are rare, however. Courtship, displays, and especially male-to-male combat have been described (Schuett, 1996), but as far as we know, the development of such behaviors has not been studied. Although most snakes are not territorial, exceptions do occur, and their development needs study (Schuett et al., 2016). Parental behavior, at least for short periods after birth, is known (H. W. Greene et al., 2002; Schuett et al., 2016). In rattlesnakes, mothers stay with offspring until their first shed. The young stay with her and she may defend them. But egg-laying species also may show parental care. In the southern African python (*Python natalensis*), mothers not only stay coiled around the eggs until they hatch but remain with the hatchings for up to two weeks, and the babies stay aggregated and return to the mother's coils at night (G. J. Alexander, 2018). This entails such a heavy cost on mothers (they do not eat or drink for months and lose up to 40% of their mass) that benefits to the offspring and their survival must be significant. Comparative studies of hatchling conspecifics without such maternal investment are now needed.

9.7. Final Thoughts

In the sections above, we have presented a mix of ontogenetic information based on longitudinal description of changes in behavior as animals mature, cross-sectional comparisons of animals of different ages or life stages, and experiments showing how experiences during life alter, or do not alter, behavior. Individual, sex, population, ecological, and genetic factors are possible sources of differences in ontogenetic trajectories, as well as maternal effects and epigenetics. A turtle that does not mature until five years old is going to have a different life path than a small lizard that matures in several months and is ready to breed the season after he or she was born. Thus, as Brattstrom (1974) noted years ago, we need more than just informal descriptive information—we need systematic observations and experiments.

The processes of development are complex, and as we discover more about them, the complexities compound. Except for the profound effects of incubation temperature, the basic questions have pretty much remained the same across 40 years (Brattstrom, 1974; Burghardt, 1977a, 1978), but the answers have not—especially simple and dichotomous ones. We also need to integrate work on ontogeny with the other of four main ethological aims of evolution, adaptive function, control mechanisms, and private experience (Tinbergen, 1963; Burghardt, 1997).

Most studies of development in reptiles focus on hatchlings, neonates, and juveniles over a period of days or weeks. These are primarily laboratory studies. Cross-sectional studies comparing age and sex classes are more typical of field studies. With more methods available to follow individuals in the field for extended periods, studies comparable to those in birds and mammals are beginning. The field studies on hatchling turtle navigation are innovative and uncovering the importance of early learning experience in field settings (review in Roth et al., 2019). The field studies on sleepy lizards (*Tiliqua rugosa*) by Whiting and his group (e.g., Whiting & While, 2017) are fascinating examples of the complexity of lizard social life. However, the ontogeny of even elemental social behavior, such as aggregation, has to be studied in a wide assortment of species. Investigation on any type of sociality seems to lead to finding more complexity than initially imagined, and such complexity also needs to be examined developmentally. The bottom line is that we may see the role of multiple factors in social development with more clarity in reptiles than we do in birds and mammals, and it is unclouded by parental care in many species (Burghardt, 1977a, 1978). As noted in this chapter, there are findings in all extant reptilian groups on social ontogeny that go against the stereotypical biases of many scientists, and even herpetologists, committed naturalists, and zoo curators. This is a frontier just beginning to be explored.

The Reach of Sociality

Feeding, Thermoregulation, Predator Avoidance, and Habitat Choice

Some aspects of reptile sociality have received very limited attention from researchers. As so little is known about them, we decided to merge overviews of these disparate aspects into one chapter.

10.1. Social Feeding

Social feeding is one area in which the difficulties of studying reptile biology are particularly extreme. Most reptiles are predators. Observing predation events in the wild is notoriously difficult, even in mammals, and reptiles usually eat 5–10 times less than mammals of similar size, so they hunt much less often; very large crocodiles and pythons can survive on just one good meal per year (Garnett, 2009). Moreover, in extant reptiles, social feeding usually represents only a small fraction of feeding episodes by the animal. It is hardly surprising, therefore, that many herpetologists have never seen social hunting despite spending decades in the field, and the total number of publications on social hunting in reptiles is minuscule. Much of what we know about social hunting is based on anecdotal accounts, nature documentaries, and amateur videos.

Fortunately, the emergence of (human) social networks and smart-phones has made it possible to solicit accounts of accidental observations from large audiences of researchers, breeders, filmmakers, wildlife photographers, tourist guides, park employees, and amateur naturalists worldwide. Their stories are sometimes unreliable and make little sense by themselves, but if multiple similar observations are independently reported, certain patterns begin to emerge.

Bailey et al. (2012) proposed the following three categories of social hunting, with increasing level of behavioral complexity: cooperation, coordination, and collaboration. *Cooperation* is simply hunting in the same place at the same time, *coordination* means that individual predators relate in space and time to each other, and *collaboration* refers to animals performing different complementary actions focused on the same prey. Distinguishing between these three categories during field observations can be difficult. All three have been reported for crocodylians and monitor lizards (see below), but only rudimentary cooperation has been observed in other reptiles.

Despite frequent claims to the contrary in popular media, virtually nothing is known about social versus solitary feeding in extinct reptiles. Certain death assemblages suggest pack hunting by dinosaurs, such as mid-size *Deinonychus* (Maxwell & Ostrom, 1995) and large *Albertosaurus sarcophagus* (Currie, 1998), and group scavenging by juvenile *Tyrannosaurus rex* (McCrea & Buckley, 2011). Multiple trackways are often interpreted as evidence of movement in herds by nonpredatory and sometimes predatory (McCrea et al., 2014) dinosaurs (although there is only circumstantial evidence that the tracks were made at the same time), so it is likely that they also fed socially.

Of course, countless extant bird species feed socially and in some cases cooperatively, with coordinated hunting reported in raptors (Bednarz, 1988; Leonardi, 1999), corvids (Bowman, 2003; Yosef & Yosef, 2010), pelicans (*Pelecanus*), and cormorants (*Phalacrocorax*)—with the latter two sometimes engaging in interspecific coordination during fishing (Crivelli, 1981). Many mammals also feed in large aggregations: migrating herds of reindeer/caribou (*Rangifer tarandus*), saiga antelopes (*Saiga tatarica*), springboks (*Antidorcas marsupialis*), and other grassland herbivores once reached hundreds of thousands, if not millions, while feeding aggregations of common dolphins (*Delphinus* spp.) and harp seals (*Pagophilus groenlandicus*) can number in the thousands (Nowak, 2018). Mammals known to exhibit coordination and collaboration during hunting include the gray wolf (*Canis lupus*), lion (*Panthera leo*), bottlenose dolphin (*Tursiops truncatus*), killer whale (*Orcinus orca* s. l.), chimpanzee (*Pan troglodytes*), and human (*Homo sapiens*) (Dinets, 2015). Irrawaddy dolphins (*Orcaella brevirostris*) and killer whales sometimes hunt in collaboration with humans (Nowak, 2018), while coyotes (*C. latrans*) often hunt in collaboration with American badgers (*Taxidea taxus*) (Minta et al., 1992). However, the vast majority of mammals, including all monotremes, virtually all marsupials, and most members of the two largest groups of placentals (rodents and bats), are solitary foragers, even in cases when, like most bats, they roost and breed communally (Nowak, 2018).

Among living reptiles, social hunting is best known in crocodylians. Two methods of social hunting are most frequently observed and have been described (F. W. King et al., 1998; Dinets, 2010, 2014).

One method is "chain-fishing," in which a group of animals forms a chain across a stream or around a stream outflow into a lake, facing the current or the direction of travel of migrating fish. Any fish carried by the stream or swimming toward its spawning grounds ends up being easily caught by one of the reptiles. If one predator leaves, another one takes its place, showing that the hunters are taking each other's positions into account and trying to optimize the efficiency of blocking the passage. This behavior can sometimes be observed at particular locations at certain times of year for decades and even centuries, suggesting cultural transmission. Well-known chain-fishing sites include the outflow of St. Johns River from Lake Dexter, Florida (American alligators), the Narrows of St. Lucia Lagoon, South Africa (Nile crocodiles, *Crocodylus niloticus*), Mzima Springs in Tsavo West National Park, Kenya (Nile crocodiles), and culverts under the Transpantaneira Highway, Brazil (Yacare caimans, *Caiman yacare*). But in other cases, the conditions for chain-fishing are created just once, and the reptiles clearly improvise to take advantage of the opportunity. Such serendipitous chain-fishing has been observed in Guyana, where three spectacled caimans (*Caiman crocodilus*) chain-fished a narrow outlet created during road construction, and in Brazil, where broad-snouted caimans (*Caiman latirostris*) chain-fished a breach in a pond dam.

Another method is fishing in drying-out ponds where large shoals of fish are trapped by the progression of the dry season. Such fishing events can attract impressive numbers of crocodylians (over a thousand in some cases) and result in the prey being fished out in a matter of hours or days. They have been observed in at least 10 species of crocodylians, including normally territorial and "asocial" ones, such as the black caiman (*Melanosuchus niger*) and the estuarine crocodile (*Crocodylus porosus*); the former would sometimes stun the fish by leaping out of the water and falling flat on the belly.

Here are descriptions of chain-fishing and fishing in drying-out ponds by Bartram, who recorded his observations of American alligators as early as in 1774:

It is scarcely credible what an immence number of Fish these monsters destroy, especially at these passes, the River being here[,] as I observed before[,] very Narow. The Trout [largemouth black bass] who pass here in their way to & from the numerous lakes & endless Lagoons & Marshes towards the head of this Vast River, where they go to spawn. The Alegator post themselves forming a line across whe[re] we see them opening their voracious Jaws into which the fish are intrap't. They heave their heads and upper part of their body upright[,] opening their throats to swallow them, & I have seen them with two or three great Trout in their mouth at a time[,] choping them up[,] the fishes tail hanging out. (Bartram, 1943)

In and about the Great Sink, are to be seen incredible numbers of crocodiles [alligators] . . . they are so abundant, that, if permitted by them, I could walk over any part of the bason and the river upon their heads, which slowly float and turn about like knotty chunks or logs of wood, except when they plunge or shoot forward to beat off their associates, pressing too close to each other, or taking up fish, which continually crowd in upon them from the river and creeks draining from the savanna, especially the great trout [largemouth black bass], mudfish, catfish, and various species of bream. (Van Doren, 1928)

The scene of a large-scale social fishing event (Plate 10.1) often looks like a chaotic feeding frenzy, and usually there seems to be no coordination between animals whatsoever, but this appearance might be deceiving. At least in some cases, crocodylians clearly use coordination and even collaboration. In lakes with deeper and shallower parts, they can divide themselves into "teams" of larger and smaller animals; the larger individuals repeatedly form a moving chain and drive the fish into the shallows, where they are either intercepted or forced to turn back toward their pursuers by the smaller reptiles. This behavior has been observed on a few occasions in the American alligator, once in the Johnston's crocodile (*Crocodylus johnstoni*), and once in the Nile crocodile. Cooperative chain-forming by American alligators has recently been recorded on video (S. G. Platt & Elsey, 2017). There is one observation of Nile crocodiles apparently cooperating in this way with a large flock of wading birds (Harry Roberts & Izanne Dalle Ave, pers. comm.). Mugger crocodiles (*Crocodylus palustris*) can use a technique reminiscent of the "bubble-fishing" used by some whales. They surround a school of fish and swim in a circle of gradually diminishing diameter, driving the fish toward the center of the circle, then take turns cutting across the center and snatching the fish one by one (Dinets, 2015). Nile crocodiles have also been observed using this technique (Harry Roberts & Izanne Dalle Ave, pers. comm.).

Two spectacled caimans have been reportedly observed to hunt frogs together: one caiman walked along the lakeshore flushing frogs, and the other swam parallel to the shore, catching the frogs as they jumped into the water (Dinets, 2015).

Nile crocodiles often gather to bring down large mammals or to dismember their carcasses (in which case, the crocodiles take turns performing a "death roll" while others are holding the carcass), and they have also been observed carrying a prey item together (Pooley & Gans, 1976). Cuban crocodiles show a tendency for "pack hunting" in captivity (Dieter, 2000). Estuarine crocodiles have been observed to conduct what seemed to be a well-coordinated hunt: one large individual used a threat display to drive a domestic pig toward two smaller reptiles concealed in apparent ambush. This case was particularly remarkable because the large crocodile could not see the smaller ones and probably just assumed that they would be in position to intercept the pig. Indeed, the smaller crocodiles were perfectly positioned and did catch the pig, which was shared by all three hunters (Dinets, 2015). Unfortunately, this was a singular observation so it is open to alternative interpretations. Captive Cuban and Nile crocodiles have been observed using a "pincer maneuver" to attack their handlers (Murphy et al., 2016).

In non-archosaurian reptiles, social hunting has been so far reported in monitor lizards, sea snakes, boas, vipers, and colubrid snakes. Characteristically, most of these reports are based on second-hand accounts, amateur reports, or footage filmed for TV documentaries, rather than scientific observations. There are reports of pairs of Nile monitors (*Varanus niloticus*) collaboratively hunting for eggs of Nile crocodiles, with one lizard distracting the female crocodile guarding the nest while the other steals the eggs (Pitman, 1931; Cott, 1961; Horn, 1999, based on film footage). Komodo dragons (*V. komodoensis*) are often said to hunt in packs, but there seems to be little, if any, supporting evidence for this claim, although group feeding on carcasses (with up to 17 lizards at a kill) is common and regularly observed (Auffenberg, 1981). There is a video allegedly showing a group of Komodo dragons feeding on a carcass of Timor sambar (*Cervus timorensis*) after mobbing and collectively killing it, but the hunt itself was not filmed (Tim Isles, pers. comm.). There are also second-hand reports of groups of yellow-spotted (*V. panoptes*) and lace (*V. varius*) monitors hunting European rabbits (*Oryctolagus cuniculus*), with some lizards entering the burrows and others catching rabbits on surface as they try to escape (M. James & Fox, 2007). A strikingly similar observation was reported to VD by a Turkmen shepherd who saw a desert monitor (*V. griseus*) enter a burrow system occupied by a colony of great gerbils (*Rhombomys opimus*) while another monitor remained on the surface and caught one of the escaping rodents at a burrow entrance.

In an extraordinary event that repeats every summer, Broadley's flat lizards (*Platysaurus broadleyi*) congregate in great numbers to feed on blackflies (Simuliidae) that swarm near certain rivers in South Africa (Plate 10.2); the lizards can be seen performing acrobatic leaps catching the flies on the wing.

Blunt-nosed vipers (*Macrovipera lebetina*), normally strongly terrestrial, are sometimes found in groups of up to five in isolated trees near water sources in arid mountains, where they ambush small birds that are migrating during spring and fall (Nedyalkov, 1967; VD, pers. obs.). Copperheads (*Agkistrodon contortrix*) have recently been discovered to congregate on summer nights at the sites of mass emergence of cicada larvae (Tompkins, 2016). An intriguing observation (Groen et al., 2020) of two European common vipers (*Vipera berus*) raiding a bird nest in rapid succession hints at the possibility that the vipers are capable of following each other's scent trail to a food source.

Yellow-bellied sea snakes (*Hydrophis platurus*) are known to form immense aggregations in the ocean; these aggregations do not appear to be breeding-related and might be feeding gatherings (Kropach, 1971). Another species of sea snake, the black-banded sea krait (*Laticauda semifasciata*), has been filmed in huge (hundreds of individuals) mixed packs hunting in cooperation with two species of predatory fish, yellow goatfish (*Mulloidichthys martinicus*) and bluefin trevally (*Caranx melampygus*) (BBC, 2009).

In 2016, a BBC filming crew documented (BBC, 2016) social hunting of hatchling marine iguanas (*Amblyrhynchus cristatus*) by Galápagos racers (*Philodryas biserialis*). There was no evidence of coordination or cooperation, and the snakes seemed to interfere with each other (Elizabeth White, pers. comm.). There is one report of common gartersnakes (*Thamnophis sirtalis*) hunting in apparent coordination: "About six or eight of them would swim together, some rift along the shore scaring frogs and small fish from the shoreline while others would be one to three feet off shore catching what tried to escape" (Thomas de Capo, pers. comm.).

At least two species of boas engage in social hunting of bats leaving or entering cave roosts: this behavior has been observed in Puerto Rican (*Chilabothrus inornatus*) and Cuban (*C. angulifer*) boas (Hardy, 1957; Rodriguez & Reagan, 1984). A recent study (Dinets, 2017) found that Cuban boas take into account each other's positions when choosing hunting positions in cave passages, so that boas hanging from the passage ceiling form a row across the passage and more effectively block the bats' flight path, thus significantly improving the efficiency of hunting.

Among herbivorous reptiles, social feeding has been reported in turtles and tortoises, particularly in Galápagos tortoises (*Chelonoidis nigra*), which can often be seen grazing in groups (Darwin, 1839). Day geckos (*Phelsuma*), although mostly insectivorous, would sometimes gather in small groups to feed on tree sap or rotting fruit (VD, pers. obs.). Feeding in iguanid lizards is a contagious behavior, and lizards can learn to eat novel food items from conspecifics and even from lizards of other species (N. Greenberg, 1976). Lilford's wall lizards (*Podarcis lilfordi*) prefer to feed on fruit in groups, and apparently use the presence of conspecifics as an indirect clue for finding food (Pérez-Cembranos & Pérez-Mellado, 2015).

Censky's ameivas (*Ameiva corax*) feed socially on large food items such as seabird eggs, and it has been suggested that individuals capable of procuring such food items occupy key roles in the social network (Eifler et al., 2016; Garrison et al., 2016). Feeding in some species of lizards can involve regular attempts to steal or wrestle food from each other (N. Greenberg, 1976); similar behavior in birds has been shown to facilitate more equal distribution of food and to correlate positively with brain size (Morand-Ferron et al., 2007). Similar kleptoparasitism is found in snakes, at least in captivity (Burghardt & Denny, 1983; Yeager & Burghardt, 1991).

Green iguanas in Panama are found in groups throughout the first year of life and may be seen eating leaves in the same bushes and sleeping in close proximity, even touching or on top of each other (see Plate 9.1; Burghardt, 1977b). Recapturing sleeping-marked hatchlings over several months and noting which ones were in proximity to others documented that those in such groups grew faster than more solitary individuals (Burghardt & Rand, 1985), suggesting social facilitation or social stimulus enhancement may be taking place.

10.2. Cannibalism

Cannibalism is seldom discussed in overviews of social behavior, but in many vertebrates it is an important part of the social system, and sometimes an essential part of the species' survival mechanism (Weygoldt, 1980; Gilmore, 1993; Pfennig et al., 1993). In mammals, cannibalism has been recorded in at least seven orders (most often in carnivores, rodents, and primates); infanticide is particularly common and often functions as part of a reproductive strategy (Polis et al., 1984).

The earliest, and rather dramatic, evidence of cannibalism in reptiles (X. Wang et al., 2005) dates to the Early Cretaceous: a semiaquatic choristodere, *Monjurosuchus splendens,* was found with seven skulls of conspecific juveniles in the abdominal cavity; it is possible that all seven were taken from the same nest.

Among archosaurs, evidence of cannibalism has been found in three species of large carnivorous theropod dinosaurs, although it might represent scavenging (Rogers et al., 2003; McLain et al., 2018; Drumheller et al., 2020). In birds it is also known mostly in carnivorous species, although occasionally in others (Paullin, 1987; Stanback & Koenig, 1992). The killing and consuming of the youngest chick by its older siblings has apparently evolved in three carnivorous lineages (accipitrid raptors, falcons, and owls) as a method of brood reduction at the times of food shortages (Mock et al., 1990; Anderson et al., 2015), and in some cases, it became obligatory even when food is abundant (Mock et al., 1990), although not all cases fall into these two categories (Bortolotti et al., 1991).

Bite marks apparently produced by conspecifics are sometimes found in fossil crocodyliforms, including marine ones (J. E. Martin, 2013), but they might be evidence of fighting for reasons other than cannibalism. In extant crocodylians, cannibalism is very common and has been recorded in many species, particularly those occurring at high densities (but there might be no causation since such species happen to be better studied). At a few months of age, crocodylians can already attack smaller individuals, at least in captivity (M. Roncas, pers. comm.), and scavenge on dead adults (VD, pers. obs.). In the best-studied species, the American alligator, cannibalism is well documented (Delany et al., 2011) and can be the main source of mortality, accounting for up to 50% (Rootes & Chabreck, 1993). Similarly high levels of juvenile mortality due to cannibalism in another species, the spectacled caiman, are density-dependent, suggesting that cannibalism might be important for population self-regulation (Staton & Dixon, 1975). As with many other aspects of crocodylian behavior, cannibalism is irregular and unpredictable: often a large adult would tolerate smaller conspecifics for months, then suddenly and for no obvious reasons kill and consume one of them, while continuing to tolerate others (VD, pers. obs.).

Consumption of eggs or, more rarely, juveniles is known in a few species of turtles and at least one tortoise, the desert tortoise (*Gopherus agassizii*), which is known to consume its own eggs (Ernst & Barbour, 1972). It appears to be rare, and its biological significance is unknown. Juvenile sea turtles, particularly Kemp's ridleys (*Lepidochelys kempii*), sometimes bite off pieces of each other's flippers in captivity (Marquez, 1994), but there is no evidence of such behavior in the wild.

Although predation on juveniles by adults has been recorded in the tuatara (*Sphenodon punctatus*), it is apparently rare and limited to populations with exceptionally high-density living in places with unusually scarce shelters (Cree et al., 1995).

In squamates, cannibalism is known in numerous taxa belonging to all major lineages. In a review by Polis and Myers (1985), cannibalism was reported for 45 squamate species; more recently, it has been documented in dozens of others but, unfortunately, no modern review is available. Polis and Myers believed squamate (and reptilian in general) cannibalism to be a rare, nonconsequential by-product of general predatory behavior. But it has now been recorded in numerous noncarnivorous species, and its biological functions appear to be remarkably diverse. As in crocodylians, it might be density-dependent and thus serve to regulate population size (Polis, 1981). This would explain why cannibalism appears to be particularly common on small predator-free islands (Žagar & Carretero, 2012), and why island lizards often exhibit gigantism, as cannibalism pressure would select for larger and faster-growing hatchlings (Mateo & Pleguezuelos, 2015). In addition to regulating population density, in some cases cannibalism might prevent disease transmission by forcing juveniles to find new hibernation sites rather than share them with adults, as suggested for the Dione rat snake (*Elaphe dione*) by Yu. M. Korotkov (unpubl. data), but in other cases, cannibalism is important or even essential for parasite circulation (Matuschka & Bannert, 1989). Cannibalism might also lower intraspecific competition and expand the species' ecological niche by forcing smaller individuals to use habitats different from those of larger individuals. Such forced habitat separation might occur between males and (usually larger) females in Lataste's vipers (*Vipera latastei*) (Freiria et al., 2006), between adults and juveniles in some *Sceloporus* lizards (Robbins et al., 2013), in Komodo dragons (D. R. King et al., 2002), and in common chameleons (*Chamaeleo chamaeleon*) (Keren-Rotem et al., 2006).

Predation of juveniles by adults (often only by the largest individuals of the larger sex) is the most frequently reported form of cannibalism in reptiles, and it is more common in species exhibiting postnatal parental care (J. C. Mitchell, 1986). Female Mexican lance-headed rattlesnakes (*Crotalus polystictus*) routinely consume nonviable offspring (Mociño-Deloya et al., 2009). W.-S. Huang (2008) found that female long-tailed skinks (*Mabuya longicaudata*) guarding their nests will sometimes consume their entire clutch in response to frequent predation attempts by Taiwanese kukri snakes (*Oligodon formosanus*), possibly because defending the nest becomes too energetically costly or too risky. Scavenging of dead conspecifics is common in many taxa, particularly in crocodylians (Pooley & Gans, 1976) and Komodo dragons (Auffenberg, 1981).

Siblicide among juveniles, which occurs regularly or even obligatorily in some species of fish, amphibians, and birds, seems to be extremely rare in lizards: it has been reported in just five species, and only in captivity (Petzold, 1971; D. R. King et al., 2002; Bonke et al., 2011; Robbins et al., 2013;). In snakes siblicide has been reported only in the best-studied species, the common garter snake, in captivity and very rarely in the wild (Fitch, 1965; J. C. Mitchell, 1986). It was once thought that all such cases in neonate gartersnakes and most, if not all, cases of cannibalism in snakes in general resulted from two snakes trying to swallow the same prey item at the same time (Polis & Myers, 1985). This does happen, particularly because neonates of gartersnakes and some other species have a strong tendency toward kleptoparasitism (Burghardt & Denny, 1983). However, there are now observations of neonate (Schwartz, 1985) and adult (Gray, 2007) gartersnakes attacking and eating each other in captivity in absence of other food items; in the case reported by Gray (2007), two males lived peacefully together from birth (to the same mother) for more than three years before the smaller one killed and partially consumed the larger one.

Recently, Maritz et al. (2019) made a surprising discovery. Cape cobras (*Naja nivea*) were found to frequently prey on smaller adults (with conspecifics making up 4% of all prey items), and in all observed cases, both the predator and the prey were male. It is possible that predation was the outcome of male-to-male combat, or that male bias was simply an artifact of small sample size. The biological significance of that bias (if it is real) is unknown.

10.3. Predator Avoidance

Many reptiles occur in aggregations, and it would be surprising if they did not evolve some socially enhanced forms of predator avoidance, but very little is known about whether they do. Alarm calls, so common in birds and mammals, are virtually unknown in reptiles; the only exception is distress calls used by juvenile crocodylians to elicit protection by adults (Mathevon et al., 2013). Rattling in rattlesnakes and rapid tail vibrating in many other species is often heard in response to disturbance or threat. Whether nearby conspecifics respond, or can even perceive such signals, is not known. It is also not known if adult crocodylians use some kind of alarm signals to warn their offspring of approaching predators, as claimed by some tribal hunters in Bolivia (Diego Llano, pers. comm.). It is known, however, that reptiles use chemical alarm signals: prairie rattlesnakes (*Crotalus viridis*) using communal dens release an alarm pheromone when disturbed, warning conspecifics of the presence of predators (Graves & Duvall, 1988).

In species practicing extended parental care, juveniles significantly minimize predation risk by remaining close to adults (see Chapter 7), and adults by nesting communally (Chapter 6), while in at least some other species hatchlings minimize predation by hatching synchronously (Chapter 8). But nonbreeding aggregations of adult reptiles often do not seem to have the antipredator function so commonly assumed for bird flocks, mammal herds, and fish schools: they frequently occur in species that are predation-free as adults, such as Galápagos tortoises (*Chelonoidis nigra* and related species) and marine iguanas (*Amblyrhynchus cristatus*), which live on islands without native predators sufficiently large to hunt them.

Some reptiles alter their antipredator response in presence of conspecifics, but it is often unclear why. Male upland calangos (*Tropidurus montanus*) run from predators for longer times when there are conspecifics nearby (Machado et al., 2007). On the other hand, predation risk often plays an important role in forming the social structure of a species: for example, local populations of cape spinytail iguanas (*Ctenosaura hemilopha*) cluster around rock crevices that provide good shelters; better access to such shelters might improve survival of dominant individuals (Carothers, 1981). Juvenile bronze anoles (*Anolis aeneus*) establish home territories around safe shelters and defend them (Stamps, 1983). Increasing predation risk imposes evolutionary costs on conspicuous social signaling: for example, losses to predation are higher in brightly colored male tawny (*Ctenophorus decresii*) and red-barred (*C. vadnappa*) crevice-dragons, as shown by a study using plastic models (Stuart-Fox et al., 2003).

Predator escaping behavior is contagious in many species occurring at high densities, so individuals likely benefit from increased vigilance by a group. A cape spiny-tailed iguana (*Ctenosaura hemilopha*) that spots an approaching predator and runs for cover often causes panicked retreat among other iguanas within sight (Carothers, 1981). Freshwater turtles basking on a log often drop into the water immediately if one of them sees an approaching human and dives, although one or two individuals sometimes stay behind (Moore & Seigel, 2006; VD, pers. obs.). On the other hand, some species seem to be oblivious to the behavior of conspecifics: communally basking European common vipers and meadow vipers (*V. ursinii*) usually stay in place after one of them spots an approaching human and escapes (VD, pers. obs.).

10.4. Habitat Choice, Thermoregulation, Roosting, and Hibernation

There are many examples of reptiles socially using various habitats, sometimes in unusual ways. For example, while green iguanas (*Iguana iguana*) are normally arboreal and live in trees, and even large cacti, some unusual locations have been noted: in the Llanos of Venezuela, iguanas make horizontal burrows into small cliffs left after human excavations. Hundreds of burrows at several horizontal levels have been recorded (Rodda & Burghardt, 1985). The results look like large versions of bee-eater cliff nests (Fig. 10.1). New Zealand common geckos (*Woodworthia maculata*) form aggregations of up to ~100 at denning sites (they are nocturnal and use dens during the day); these aggregations are not caused by lack of suitable sites (Hare & Hoare, 2005). In a few locations in North America and Eurasia, hundreds or even thousands of snakes form spectacular congregations at hibernation sites. Up to 5,000 snakes of five species hibernate in cracks in granite outcrops in a 2,500 m^2 area near Yanguletz River, Ukraine (Virlich, 1984). Much more famous is the gathering of over 75,000 common garter-snakes (*Thamnophis sirtalis*) near Narcisse, Manitoba (Corey, 1975). Fossils of aggregating snakes are known from the Oligocene (Breithaupt & Duvall, 1986). The first fossil record of aggregating in lizards is much older,

Fig. 10.1. Communal terrestrial nocturnal refuges used by green iguanas (*Iguana iguana*), showing scores (>100) of burrows, some with complex internal pathways, in a gravel borrow pit in the Llanos of Venezuela. Photograph by Gordon Burghardt.

dating back to the Early Cretaceous (S. E. Evans et al., 2007). The fossil record of aggregations in turtles and archosaurs is too extensive to cover even superficially, and the first certain record of social aggregation in reptiles is of early diapids, *Youngina capensis* (late Permian): 12 juveniles were found huddled together in an underground burrow, possibly to reduce water loss during a drought (R. M. Smith & Susan, 1995).

In most cases not obviously related to reproduction, it is unknown why reptiles aggregate. In places where freshwater turtles are still common, more than a dozen can sometimes be seen basking on a log while neighboring logs, seemingly identical, are vacant (VD, pers. obs.). Basking plays a role in thermoregulation and perhaps vitamin synthesis and parasite removal (E. O. Moll & Legler, 1971; Vogt, 1979); thermoregulation, in turn, can facilitate digestion and vitellogenesis. However, some incidences of basking-like behavior in turtles and crocodylians cannot be explained by thermoregulation, as they occur at night (Nordberg & McKnight, 2020; VD pers. obs.); such "nocturnal basking" might be purely social.

Interactions among basking turtles can include aggression, but in most cases groups of turtles bask together without agonistic interactions (reviewed in Lindeman, 1999). Too often, most of what we know about aquatic turtle social behavior is based on basking interactions. K. M. Davis has shown that much social behavior of a rather complex sort occurs underwater, and she observed social behavior in a mixed species exhibit of emydid turtles interacting in a large naturalistic aquarium at the Tennessee Aquarium in Chattanooga. She found a variety of competitive and aggressive behaviors as well as extensive titillation in both male and female turtles (K. M. Davis, 2009).

Similarly, pig-nosed turtles (*Carettochelys insculpta*) "bask" in groups in thermal springs on the river bottom for up to six months during the tropical winter, abandoning the springs as the river temperatures eclipse the thermal spring temperatures (Doody, Sims & Georges, 2001). Do they aggregate on purpose, or do they find that particular log or thermal spring more attractive for reasons obscure to a human observer? Such questions can often be answered experimentally, but the results of the few studies that have attempted to do that differ depending on study species. Hatchling lizards (*Varanus rosenbergi*) use their birth site termite mounds to thermoregulate and roost in for several weeks, and their field metabolic rates were significantly higher than controls in outdoor pens (Green et al., 1999). Burghardt (1983) found that captive juvenile common gartersnakes and brown snakes (*Storeria dekayi*) aggregated under some of many identical covers provided, and were more likely to aggregate with conspecifics. Similarly, Elfström and Zucker (1999) found that captive tree lizards (*Urosaurus ornatus*) aggregate for hibernation at particular sites even when numerous identical sites are provided. Formation of aggregations in major skinks (*Egernia frerei*) depended on relatedness rather than habitat pattern (Fuller et al., 2005). On the other hand, in a study of another species from the same genus by Michael et al. (2010), crevice skinks (*E. striolata*) apparently formed congregations only if particularly favorable sites were available, irrespective of social factors. Aggregations of neonate bearded dragons (*Pogona vitticeps*) were caused simply by mutual attraction to a common resource (Khan et al., 2010; see also Schutz et al., 2007). Similarly, yellow-footed tortoises (*Geochelone denticulata*) formed aggregations only when there was a shortage of suitable denning sites (Auffenberg, 1969). Regardless of underlying mechanisms, reptiles form groups more often than appreciated, but more data and observations are needed. For example, in how many species of lizards do individuals roost together while inactive? Broadley's flat lizards (*Platysaurus broadleyi*) roost together in the same crevices, and green iguanas excavate burrows in high densities within the same areas (Fig. 10.1) (Schutz et al., 2007).

There are some interesting studies on behavioral mechanisms of forming aggregations. Aragón et al. (2006) found that common lizards (*Zootoca vivipara*) use chemical cues from conspecifics to locate shelters; interestingly, the selection of shelter sites by juveniles was influenced by the odor of socially housed adult males, but not of isolated males, and also showed some dependence on the site of origin of the mother and the body condition of the juvenile. Juvenile tiger snakes (*Notechis scutatus*) also use chemical cues to form clusters at sites previously occupied by conspecifics, and benefit from clustering by being able to cool more slowly when air temperature drops (Aubret & Shine, 2010).

Gartersnakes selectively aggregate under rocks based on thermal characteristics (Huey et al., 1989). Aggregations of desert night lizards (*Xantusia vigilis*) improve reproductive success and survival, with thermal benefits of winter huddling disproportionately benefiting small juveniles, possibly encouraging delayed dispersal of offspring and formation of kin groups (Rabosky et al., 2012). Indeed, lizards of this species were found to disperse less and to form wintering aggregations more readily when related individuals were present (A. R. Davis, 2012); groups that contained nuclear family members were more stable than groups that contained less-related lizards (A. R. Davis et al., 2011). The authors of the latter study noted the similarities in kin-based group formation mechanisms between lizards and many bird and mammal species. Yeager and Burghardt (1991) found that plains gartersnakes (*T. radix*) avoided aggregating with individuals with which they previously competed over food. Patterson (1971) studied aggregations at hibernation sites by desert tortoises (*Gopherus agassizi*) and found that the presence of fresh fecal pellets prevented the tortoises from aggregating; as the temperature dropped and the tortoises stopped feeding, fresh pellets disappeared and aggregations could be formed. Marine iguanas huddle during cool weather and this has been shown to be functional (Boersma, 1982), especially for those in the middle of the huddle, a finding that parallels the work of J. R. Alberts and colleagues on newborn blind, hairless rat pups (J. R. Alberts, 1978).

Although many snakes congregate randomly with relatives and nonrelatives (Lyman-Henley & Burghardt, 1994; Lougheed et al., 1999), female rattlesnakes preferentially share communal hibernations sites with related females (Bushar et al., 1998), and such sites are very important in the complex social system of some rattlesnakes (see Chapter 3). It is likely that the role of congregations in the lives of reptiles is worthy of much more study.

11 |

Looking toward the Future

We have reviewed many aspects of the social lives of reptiles. The diversity is great and the burgeoning new findings are prompting far more consideration of the role of these animals in animal behavior research generally. Much of this is already occurring, especially among lizards. Lizards are the most diverse extant reptiles in terms of morphology, sensory reliance, reproduction, diet, ecological adaptations, and social organization. Many species, such as arboreal iguanas and anoles, are also, perhaps, among the easiest to observe and study; certainly, they have figured prominently in the work reported here. Turtles have advantages as study subjects, but the species are relatively few, they are long-lived, and, while they live in habitats from the open seas to mountain deserts, their social behavior variation may be less easy to study. Tortoises, along with terrestrial box and wood turtles, are the best studied behaviorally, being far more visible and also rather slow moving and more easily followed and observed than fast moving snakes and lizards. In the laboratory, however, pond turtles predominate. Crocodylians are often difficult to work with in experimental or controlled captive or laboratory settings, and there are fewer than 30 extant species. However, crocodylian-exhibit facilities and commercial farms do offer settings that can be exploited. A large proportion of both turtles and crocodylians are endangered. This makes their study both more urgent and more difficult. Snakes are almost as speciose as lizards, but they are less diverse in morphology, behavior, diet, sensory reliance, reproduction, and, apparently, social organization. The tuatara stands alone as a relict species for which comparative studies are rather impossible. Altogether, however, the fact remains that we have good behavioral data, from natural history to experiments, on only a small fraction, perhaps 1% or 2%, of the many thousands of reptiles. Much more information, even basic life history details, are needed to address comparative and evolutionary questions at the macroevolutionary levels. Gordon Rodda's compendium on lizard natural history (Rodda, 2020) is an example of what we need as resources to inspire and pose questions.

While all reptiles have much to investigate in terms of social behavior, we do know quite a bit now, as compared to where we were 40 years ago. We anticipate that intrepid, adventurous field workers around the world will collect observations and establish long-term field studies. Given the extensive effects of habitat destruction, poaching, and climate change, many young scientists and naturalists should be motivated to find out what they can while they can. Thus, more information should come available to extend, critique, and modify both the ideas expressed in this book and the data upon which they are founded. Of course, there will be many details of the existing literature that we may have missed, overlooked, or for space reasons could not discuss. We encourage readers to explore the many provided references to fascinating research on fascinating animals, which we could only cover briefly.

The emphasis in this chapter will be on some topics of timely interest, including phenomena typically studied in other animals, which push the boundaries of the questions to be asked and may be worth exploring. This is an interesting exercise, even if the future shows us that we were wrong (Burghardt, 2004). We will focus on some new questions, methods, and findings that highlight where we think the field may be heading. There is clearly overlap among these topics, as well as those discussed earlier, but the emphases differ. While some exceptional and novel phenomena and suggestions are highlighted in these brief sections, the need for comparative studies of species within and among lineages is also essential.

11.1. Mating Systems

Reptiles, as we have seen, present a wide range of mating systems. These range from polygamous systems to those nearly promiscuous to serial and long-term monogamy. The use of molecular genetics to uncover and assess multiple paternity, first clearly documented in common gartersnakes (*Thamnophis sirtalis*) (Schwartz et al., 1989), has shown it to be very widespread indeed, even in species that are considered widely dispersed and "nonsocial." Among the many fine papers on snakes, the long-term telemetry study by R. W. Clark et al. (2014) on western diamondback rattlesnakes (*Crotalus atrox*) stands out. One of the interesting issues needing more exploration, as shown in diamondbacks, is the demographic and social importance of having two mating seasons prior to female ovulation. Male spermatogenesis begins in spring and continues into late summer and early fall. The first mating season occurs in late summer. The second mating season occurs following emergence from brumation in spring. Females give birth in mid to late August in the Arizona populations studied. Sperm storage is involved and females only reproduce every two, perhaps three, years. Males only mate once with a given female, but both males and females mate with multiple partners, and multiple paternity is the norm. In this and similar systems, the operational sex ratio is thus highly male biased, and sexual selection is thus more intense on males than perhaps in species where both sexes reproduce annually. Male mate guarding also occurs. Such differences must have important consequences on the social lives of these species. In the western diamondback example, with known individuals where mating behavior was often observed, the authors "found little concordance between paternity and observations of courtship and mating behavior" (R. W. Clark et al., 2014, p. 1). Larger males did not sire more offspring than smaller males, but females with males staying in closer proximity did have greater reproductive success. Such studies indicate that we are far from understanding the mating systems in snakes, as compared to other reptiles as well as birds, many mammals, and even frogs and salamanders. Research on stable and generation-crossing social and family groups in some skinks provides exciting opportunities for comparative work on continents other than Australia (While et al., 2019).

Multiple paternity has now also been documented in other reptilian groups, and it has implications for sexual selection across a wide swath of taxa (Chapter 3). In lizards, especially, the social variation found within speciose groups, such as the skinks, warrants closer attention. The contrast with snakes, where polygynandry seems common and territoriality does not, should be studied. Could any of the major differences between snakes and lizards be rooted in the former evolving from a burrowing ancestor? We need far more data on social behavior in the blind and thread snakes (Scolecophidia) to compare with both the more advanced snakes (Alethinophidia) and the lizards most closely related to snakes, as well as lizards with traits convergent with a limbless burrowing life style. Burrowing animals are those we know the least about, yet they are critical for understanding social evolution in squamate reptiles. Parthenogenetic and clonal squamates also offer research opportunities not possible with natural populations of birds and mammals. Molecular genetic methods have helped elucidate mating patterns in reptiles, and these methods can also help in answering other questions about social grouping, dispersal, habitat and diet segregation, and many other phenomena described in this book.

11.2. Courtship

Courtship in reptiles is often a complex affair, as we have seen in Chapter 5. It can take many weeks, as in iguanas, or take place within days of emergence from dens, as in gartersnakes. The occurrence of socially induced adjustments in courtship due to the number and size of competitors is an intriguing area of future study (e.g., Shine, Langkilde & Mason, 2003b). Signals can be visual, chemical, tactile, and perhaps auditory. Visual signals involve movement displays, color, form, size, or combinations thereof. The role of territorial systems in the dynamics of courtship is of potential interest for more study, as is the male-male competition for mating access to females that seems to occur in at least some species in all the major reptile groups. The extent of differentiation and ritualization of courtship behavior and signals in areas of overlap of related species is clearly an area needing more attention, given the occurrence of, and evolutionary importance of, hybridization in speciation (Placyk et al., 2012). This is an area where a focus on multimodal signaling, sometimes dismissed as beyond reptilian capacities, can help in understanding how such signals are integrated and weighed in the brains of reptiles, and some model species could greatly inform the evolutionary neuroscience of terrestrial vertebrates. The comparative study across diverse crocodylian lineages by Dinets (2013c) is one of several studies that can be used as models. Sexual dimorphism, or lack thereof, might be usefully looked at phylogenetically in relation to mating systems. Monogamous geese have little dimorphism compared to polygamous ducks. Those lizards that are monogamous seem to also be less size and color dimorphic than more polygamous or lekking lizards, but this has not been formally analyzed, as far as we know, using modern methods.

11.3. Parental Care

The diversity in parental care is immense in reptiles, unlike in mammals and birds where it is virtually ubiquitous except for brood parasites. We are well beyond the time when a presumptive discovery of a nest of dinosaurs (Horner & Makela, 1979) supported their great remove from those "solitary" asocial reptiles, a myth then (Burghardt, 1977b), but still sometimes repeated. The recent documentation of parental care in snakes, other than many pitvipers (e.g., in pythons, a more ancient group), and other species is most exciting and also shows how postnatal parental care can evolve repeatedly and independently, just like leglessness in lizards. The recent report of paternal defense of a female rattlesnake with her brood (Hewlett & Schuett, 2019) is most exciting and shows how we know so little about even well-studied species.

Parental concern for offspring can also occur in unusual ways. For example, the oviparous tiger keelback snake (*Rhabdophis tigrinus*), not only has a toxic venom but also possesses poisonous nuchal glands that, accompanied by distinctive antipredator behaviors (such as neck arching and butting), release toxins (bufadienolides) that, if contacting the mouth or eyes, deter predators (Mori & Burghardt, 2008; Mori et al., 2012). This species primarily eats amphibians, including toads that contain toxic bufadienolides, and this toad diet is the source of the nuchal gland toxins (Hutchinson et al., 2007). Furthermore, offspring upon hatching may be fully provisioned with toxins in their nuchal glands if their mothers ate toads during gestation and immediately are capable of performing the unique antipredator displays. But, if the mothers have not eaten toads, the offspring have nontoxic glands. Hatchlings, however, readily eat toads and their glands become full of poison (Hutchinson et al., 2013). Island populations where the only amphibians available for *Rhabdophis* are nontoxic frogs do not have toxic glands (Hutchinson et al., 2007), and snakes from the island, including hatchlings, have reduced the performance of the distinctive displays, as well as striking and neck flattening, but they have an increased propensity to flee (Mori & Burghardt, 2000). However, when hatchlings from the toad-free island were fed toads, they increased the performance of the typical antipredator display as if they were aware that they now had a toxic defense (Mori & Burghardt, 2017). Moreover, maternal provisioning of offspring via the egg is not an incidental by-product of the maternal diet. Radio-tracking studies in the field showed that gravid females actively search out toad habitats and expend effort finding and capturing toads in preference to frogs. Laboratory prey trailing experiments confirmed that gravid females, but not nongravid females or adult males, were far more likely to trail toads than frogs (Kojima & Mori, 2015). When females became nongravid, they reverted to their normal frog preference. Thus, mothers, though not providing personal posthatching care, are clearly providing their offspring with unique resources to aid survival.

Parental care of a more traditional bent is increasingly being documented in lizards, and the first evidence for postnatal care in turtles, described in detail earlier, opens up new avenues and questions, especially for riverine species where areas inhabited by hatchlings during early life are distant from nesting areas. The comparison with sea turtle mothers, who do not return to the nest beaches for hatching, must mean that the costs and benefits differ. But how and why?

Parental care can be prolonged and complex in crocodylians, but the data available vary across species. However, differences among species clearly exist, although good comparative studies are rare. This is also an urgent area given that so many species, such as both gharial species, are endangered and persecuted, so that their behavior before human disturbance can only be surmised in many cases. Phylogenetic studies of courtship, parental care, and communication are needed.

There is also need for research on the behavioral consequences of shifts in many lizard (and some snake) species from oviparity to live-bearing (and possibly back, see Chapter 7, section 7.2).

11.4. Social Organization and Social Plasticity

The above topics on mating systems, courtship, and parental care are part of the social organization of reptile societies, but we also need to step back and look at the levels of the populations and groups within those populations. In nonhuman primate, canid, elephant, rodent, and other groups, the dispersal of juveniles or maturing animals from their natal areas is an important issue. Sex differentiated dispersal can also occur in reptiles (Doughty et al., 1994). The way animals array themselves in their habitats and interact with each other can be affected by whether the society is matrilineally or patrilineally organized, whether the society is based on territoriality or dominance, the type of mating system (including harem or lek systems), the role of parental care and kin recognition, and age differences in diet and habitat use. Habitat type—such as arboreal (shrub, trunk, canopy), fossorial, aquatic, terrestrial (desert, forest floor)—is also important. Animals with typically small home ranges, such as anoles, have different ways of organizing their communities as compared to animals that need to travel widely to obtain food or find brumation sites. Early attempts to classify animal societies and social organization, even in well-studied groups such as nonhuman primates, have floundered, though concepts such as fission-fusion and despotic societies are still used. But resource availability and variation within species in different habitats are important and may underlie differences within reptilian taxa, just as in primates, including, of course, humans. Brattstrom's pioneering paper (Brattstrom, 1974) on the role of space constraints in transforming social systems from territorial to hierarchical in the same saurian species needs replication and application to field populations (perhaps on small islands where population growth and density preclude long-distance dispersal) and may need to be included in phylogenetic studies of social evolution.

Another method that is being used to study social organization in reptiles and that is becoming increasingly sophisticated is social network analysis. For example, using network analysis, it was found that titillation displays in emydid turtles in a mixed species group are not emitted randomly, even within the same species or sex, and neither are spatial associations (K. M. Davis, 2009). Social network analysis is increasingly being used in social behavior studies of many reptiles and has great potential. For example, in a large multispecies exhibit at the Tennessee Aquarium in Chattanooga, titillation display interactions revealed interesting species, sex, and individual clusters (Plate 11.1) that remained largely stable across two years and showed species specificity. The extended family bonds, monogamy, and other phenomena being discovered in some skinks, but not others, opens research areas where applying network analysis may be profoundly important (Whiting & While, 2017).

11.5. Environmentally Induced Plasticity

One of the strikes against reptiles is the view, common even among herpetologists, that, unlike mammals, their behavior is rigid, instinct bound, and hard to alter. This view was countered in Chapter 9. Many studies are documenting how incubation temperature, diet, habitat structure, and social experiences can alter the behavior of reptiles. For example, studies transplanting anoles to islands with different perch diameters led to suggestions of rapid natural selection underlying changes in hind limb length by *Anolis* lizards (Losos et al., 1997), but experimental studies in the lab showed that such changes were due to posthatching plasticity in limb growth due to the altered habitats in which the animals grew up, not selection. Plasticity in the form of habituation, not selection, may also underlie population differences in recognition of predators that may differ across islands (Placyk & Burghardt, 2011). It has taken time for biologists to view plasticity as an alternative hypothesis for reptile behavior. Here is where the lessons of J. R. Platt (1964) need to be applied: science best progresses by testing alternative hypotheses. New revisions to the neo-Darwinian evolutionary synthesis are challenging many assumptions, as seen in the various chapters in Pigliucci and Müller (2010) and the overview in Laland et al. (2015). Epigenetic processes of gene switching and effects across generations are obvious areas of further research, as these may be enabled by environmental changes, urbanization, and human-facilitated invasions of alien species (as in the global population of some turtles, lizards, and snakes, especially rampant in Florida).

Reptiles, being ectothermic, often have metabolic, energetic, and other constraints that we do not fully appreciate, and we are often unaware we are applying anthropocentric biases (Rivas & Burghardt, 2002; Timberlake, 2007). Animals that act more slowly or with fewer overt signs that we can easily access can lead researchers to make significant conceptual errors upon which theoretical claims can founder. We also need to explore the many ways external conditions, including maternal conditions, alter the morphology, physiology, and behavior of reptiles, not just during the juvenile period but throughout their lives. Selection operates on the capacity for plastic responses as much as on the responses themselves. Thus, experiments must consider not only selection but also drift and plasticity in interpreting geographic, population, and species differences. As more reptiles become domesticated as pets, and others are farmed for skins, meat, medicine, and so on, the consequences on social behavior need investigation, as they could provide insightful "unnatural experiments." For example, many snakes and lizards are being bred as designer "candy" for the pet trade. What do all the color changes mean for mate selection and other behavior? While color may not be a factor in mate selection in most snakes, it certainly can be in many lizards. The studies showing the social, behavioral, and morphological consequences of selection on docility in foxes (Trut, 1999) suggest that there are opportunities for comparable studies in reptiles, as many species being bred are also being selected, intentionally or not, for behavior conducive to captive life with humans. Commercial reptile breeders sometimes refer to certain lineages as "docile" or "friendly," but no studies of the behavior of commercially bred (in fact, domesticated in some cases) reptiles compared to wild ones currently exist as far as we know.

11.6. Cognitive Processes, Social Recognition, and Social Learning

The claim that reptile brains are only about 10% the size of bird and mammal brains from the same sized animals is not only wrong (Font et al., 2019; Font, Burghardt & Leal, 2021) but misleading. Honeybees can do cognitive tasks that many animals with brains a thousand times larger cannot, and fish are solving problems that even nonhuman primates were thought incapable of doing 40 years ago (Bshary et al., 2002; Balcombe, 2016). The importance of the problem and context facing the animal is most significant (Burghardt, 2013). In terms of brain size in dinosaurs compared to reptiles, it seems that there is great diversity in dinosaurs, but that overall they might be considered intermediate between crocodylians and birds (Hopson, 1977; Brasier et al., 2017). What cognitive and learning abilities do reptiles possess?

The cognitive and decision-making processes shown by reptiles are perhaps best viewed as a subset of plasticity. This includes the learning studies of reptiles from habituation to problem solving discussed earlier in relation to social behavior (Chapter 9). Experimental studies of learning in reptiles go back to the early days of comparative psychology and have been extensively reviewed (Burghardt, 1977a). Since then, updates have appeared (Suboski, 1992; Wilkinson & Huber, 2012; Burghardt, 2013; Matsubara, Deeming & Wilkinson, 2017; Font, Burghardt & Leal, 2021; Szabo, et al., 2020). But far more needs to be done employing the innovative methods developed by comparative psychologists and ethologists studying mammals, birds, and fish. Bearded dragons (*Pogona vitticeps*) recognize visual illusions (Santacà et al., 2019) and tortoises recognize objects in photographs and visual displays (e.g., Wilkinson et al., 2013). Snakes that use distinctive tongue flicking in luring prey do this only in the presence of fish, but experiments with naïve neonates documented that chemical and visual stimuli using "underwater" video clips of swimming fish are also effective (Hansknecht & Burghardt, 2010). Thus, video presentations of stimuli, already shown effective in studying lizard headbob communication, may find increasing and innovative applications.

Social learning, including imitation, was long considered an advanced cognitive ability, perhaps not even shown in nonhuman primates. However, it has now been well documented in these and many other mammals, birds, and fishes. Although social learning in anoles was anticipated by N. Greenberg (1976), experimental evidence was first shown in turtles (K. M. Davis, 2009; Wilkinson et al., 2010; K. M. Davis & Burghardt, 2011) and some lizards (Wilkinson & Huber, 2012; D. W. A. Noble et al., 2014). Additional evidence for social learning was reported by Gutnick et al. (2020) in which giant tortoises at two different zoos were trained to approach and discriminate objects. Animals trained individually took many more trials to reach the criterion than animals able to observe other tortoises being trained. Another study showed that the invasive Italian wall lizard (*Podarcis sicula*) is able to use social information gleaned by observing either a conspecific or a congener (*Podarcis bocagei*) to solve a color discrimination foraging task (Damas-Moreira et al., 2018). Although most animals did not solve the task, more of those in the social treatment were successful and made fewer errors as well. Actual imitation of the actions of a conspecific in solving a problem has been demonstrated in bearded dragons (Kis et al., 2015). Social learning that influences gaze direction, tool use, foraging, mate selection, kleptoparasitism, and food competition is probably more widespread and behaviorally significant socially than we have imagined. More work in this area is needed, and it also needs to go beyond visually mediated tasks.

Consider conspecific recognition and the "dear enemy" phenomenon in which an animal recognizes members of its species that it has encountered in the past as different from strangers, as shown by Barash (1974) in two species of "solitary" mammals. Similar suggestive results were found using wild caught radio-tracked box turtles. These turtles reacted differently to turtles with overlapping home ranges than to those they never could have interacted with as adults (M. Davis, 1981). For example, head ducking behavior was shown more in stranger pairings than in neighbor pairings. This was true not only of wild caught adults but also captive hatchlings.

This book has repeatedly mentioned differences in behavior and temperament among different individuals. Are these stable and genetically based, or do even reptiles alter their responsiveness in social contexts by experience? Waters et al. (2017) review much of the literature on this topic, which primarily involves a few species of lizards and snakes. But we do know that social experience can affect individual behavior (e.g., Yeager & Burghardt, 1991), just as experience with predators can alter response to subsequent encounters with aversive stimuli and predators. Interestingly, differences in food amounts early in life can have lifelong effects on personality variation in identical clonal mourning geckos (*Lepidodactylus lugubris*), an obligate parthenogenetic lizard (Sakai, 2020). This fascinating work supports the conclusion at the end of section 11.1, above.

Chiszar et al. (1995) reviewed work showing that snakes, being chemically acute and relying on odors and vomodors, seem to be stressed, measured by increased activity and tongue flicking, when placed back into their home cage after it was cleaned as compared to when it was left soiled. Familiarity led to contentment. Other studies have shown that snakes also respond differently to cages containing cues from conspecifics as compared to themselves. This leads to the idea that snakes are aware of their chemical scent as compared to other conspecifics (Chiszar et al., 1980, 1991). Some comparable studies are found in lizards (De Fazio et al., 1977; Graves & Halpern, 1991; A. C. Alberts, 1992; Labra & Niemeyer, 1999). The chemical basis of self-recognition in one species has recently be uncovered (Mangiacotti et al., 2020).

These studies are intriguing because the classic work establishing that apes and some other "large-brained" animals are self-aware or self-conscious is based on the mirror mark test pioneered by Gordon Gallup (Gallup, 1970; Gallup et al., 2002). In this test, a chimpanzee, for example, will look into a mirror and remove something on his or her body that it could not see without using the mirror. Thus, it was aware it was peering at its own body and not mistaking the mirror image for a conspecific, as many animals do. But how does this really differ from a snake recognizing that the chemicals in its environment are its own? An experiment by Zuleyma Tang-Martinez on gartersnakes, *Thamnophis sirtalis*, also documented this (Halpin, 1990). A study from the GMB lab suggests that juvenile snakes from the same litter, thus genetically closely related, can discriminate their own chemical from not only clean controls but also same-sex littermates, even those fed on the same diets, as well as those fed on different diets, although there were sex differences as well. Such studies suggest that our visual-based anthropocentrism biases our view of such topics as self-awareness and consciousness. These are controversial subjects with nonhuman animals. The diverse views on these issues can be explored through the 57 relatively short chapters in *The Cognitive Animal* (Bekoff et al., 2002). Perhaps there are more lessons coming from serpents, beyond those from the Garden of Eden (Isbell, 2009).

11.7. Affective Processes

The emotional lives of reptiles, which include sentience, are something that has been little studied. But the claims of MacLean (1985, 1990) that positive emotions are only found in mammals and birds, as they have parental care, are clearly disproved by the many examples in this book of solicitude to offspring and conspecifics and of friendships, as indicated by social bonding, cooperative foraging, communal behavior, social play, and so on. The articles, commentaries, and debates in the online journal *Animal Sentience* offer detailed arguments and evidence on emotions (and even subjective feelings) across the animal kingdom. The fact that reptiles (unlike dogs) lack facial expressions and behavior we can empathize with, for example, does not preclude positive and negative emotions, regardless of how they are characterized (Mendl et al., 2010). An updated literature on reptile affect and sentience is now available (Lambert et al., 2019).

Attachments of reptiles to each other and often their human companions, as Komodo dragon keepers know well, do exist, but the responses may be subtle and unfamiliar to humans (B. B. Bowers & Burghardt, 1992). An adult female green iguana (*Iguana iguana*) in the lab of GMB would present her neck to people to stroke like a cat, but without the purring (Burghardt, 2000). There are, however, some spectacular examples of reptile–human bonds. An American crocodile (*Crocodylus acutus*) rescued and named Pocho by Gilberto "Chito" Shedden became a celebrity in Panama for being absolutely tame and very playful with its rescuer. Their unique relationship continued for 20 years, until the crocodile died, apparently of old age. The crocodile would swim with Shedden, rush at him with an open mouth in mock charges, sneak on him from behind as if to startle him, and accept being caressed, hugged, rotated in the water, and kissed on the snout (Dinets, 2015).

11.8. Play

For many decades, play was considered to be found only in mammals and some birds, apparent exceptions being—like Neill's (1971) views on crocodylian maternal behavior—considered misinterpretations and absurd (Bekoff & Byers, 1981; Fagen, 1981; Burghardt, 1984). Even apparent examples, such as those Burghardt (1978) briefly referred to, were based on brief observations and undocumented (no film or video). Certainly, reptiles would explore new habitats and investigate objects. GMB filmed wild hatchling green iguanas, hours after nest emergence, exploring his camera equipment in groups during his fieldwork in Panama in the early 1970s and newborn African chameleons wrestling in 1963. However, play was not something readily attributed to animals not already considered playful due both to our mammalocentric biases and the lack of a useful definition or criteria for identifying play. Burghardt attempted to develop objective criteria that could be applied to diverse animals and contexts in the absence of any prior bias to assuming the species was or was not playful. He came up with five criteria (Burghardt, 2005) that have been widely applied and have only had minor changes. A brief summary statement would be "play is repeated, seemingly non-functional behavior differing from more adaptive versions structurally, contextually, or developmentally, and initiated when the animal is in a relaxed, unstimulating, or low stress setting" (Burghardt, 2014, p. 91).

Using these criteria, play has now been found in numerous species other than birds and mammals, including reptiles, amphibians, fish, and some invertebrates (e.g., octopuses, spiders, and wasps) (Burghardt, 2005, 2014). Among reptiles, Nile soft-shelled turtles (*Trionyx triunguis*), geckos, monitor lizards, and various crocodylians engage in recognizable play activities (Fig. 11.1) (Burghardt et al., 1996, 1998a, 2002; Augustine et al., 2015; Barabanov et al., 2015; Burghardt, 2015; Dinets, 2015; Kane et al., 2019). Such activities can include pushing and manipulating balls and rings, playing tug of war and fetch with keepers, shaking old shoes like a dog, sliding down slopes, and other behavior that would readily be considered play if seen in a mammal. When sensitized to play as a possibility, some otherwise inexplicable findings can be possibly explained. For example, in the previously mentioned death roll study (see Chapter 3, section 3.2; Drumheller et al., 2019), some animals seemed more interested in playing with the bait than eating or fighting with it (Drumheller-Horton, pers. comm.). These examples have been discovered, at least in part, due to the increasing

Fig. 11.1. Male Cuban crocodile (*Crocodylus rhombifer*) giving its long-term female partner a ride around the pool at Zoo Miami, Florida. Photograph by Vladimir Dinets.

recognition of the importance of, and the increased use of, environmental enrichment for the well-being of reptiles as well as birds and mammals (e.g. Burghardt, 2013; Londoño et al., 2018).

The question arises, however, as to why, even when provisioned with toys and other objects, most reptiles do not engage in recognizable play, or do not play in as extended or complex manner as, for example, many mammals and birds. The reasons are several, but among the most important are the physiological constraints limiting sustained vigorous activity, the frequent lack of flexible and dexterous limbs used for much play (certainly pronounced in snakes), and the lack of parental care that buffers young reptiles from the demands of foraging, defense, and protection from inclement weather when young. Together, these factors underscore that any activities that are not of immediate survival value may not occur unless the animal has surplus resources of time, energy, and behavior (Burghardt, 1988). This "surplus resource theory" helps us understand that play emerges only when a constellation of life history, physiological, ecological, and behavioral factors are conducive to such behavior. Even in mammals, play is much more common in well-cared for captive animals than their wild counterparts, who have more demands for their survival (Burghardt, 1984). That being said, with the increased popularity of reptiles as companion animals and pets, many fascinating examples of reptile play with objects are proliferating on YouTube and other internet sites, published by the general public. A recent paper has modeled the conditions in which simple and complex play may evolve, and reptiles may be key species for study (Smaldino et al., 2019).

What about social play? Titillation displays with the long foreclaws of some North American emydid turtles, typically only found in adult males during courtship, are seen in hatchlings and juveniles of both sexes prior to sexual maturity or foreclaw elongation (Kramer & Burghardt, 1998) (Fig. 11.2). Headbob displays in juvenile lizards performed before sexual and territorial behavior emerge could also be play (Cooper, 1971; N. Greenberg & Hake, 1990; Burghardt, 2005). YouTube videos are replete with examples of lizards and turtles playing with cats and dogs, including tag games, ball competition, and tug of war. There are even reports of crocodylians

Fig. 11.2. Precocial courtship and titillation exchange in juvenile *Pseudemys nelsoni* prior to sexual maturity or foreclaw elongation: *top,* head-to-head, as in adult *Trachemys* and *Chrysemys; bottom,* swim above, as in adult *Pseudemys.* This may represent an example of the evolutionary origins of ritualized display recapitulated in juvenile behavior. Redrawn from Kramer & Burghardt, 1998.

playing with mammals such as river otters (*Lontra canadensis*) and humans (Dinets, 2015), and there is also some evidence of play-like behavior between juveniles and adults (Fig. 11.3). Thus, under favorable conditions, social play occurs, though like most play, it is usually observed in benign captive conditions. But even behavior in captivity may anticipate future observations in the field. For example, observations of a Komodo dragon repeatedly putting her head in boxes and pails and then walking around with them—seen in a video documentary called "The Power of Play" in episodes of *The Nature of Things* (CBC, 2019) and *Animals at Play* (BBC, 2019), and at https://vimeo.com/ondemand/thepowerofplay—were surprisingly similar to a subsequent YouTube video of a wild dragon emerging from the ocean onto a beach with its head inside a large empty sea

Fig. 11.3. A juvenile American alligator (*Alligator mississippiensis*) giving a ride to a younger crèchemate. Communal crèches of alligators often include juveniles of different ages. Big Cypress National Preserve, Florida. Photograph by Vladimir Dinets.

turtle shell. It is interesting to note that while play in many mammals occurs primarily in juveniles, a period which may not be that conducive to play in many reptiles, it is in older and adult long-lived reptiles that play is most often reported.

The functional role of play in the social lives of reptiles is not known. Given its spotty documentation to date, perhaps it is not an important topic to explore. However, parental care was also considered beyond the capacity of reptiles, and while now known to be widespread in crocodylians, it is also now being found, sporadically, and in various guises, in snakes, lizards, tuatara, and turtles. This variability is itself an opportunity to pose phylogenetic and functional questions. Molecular genetic studies may provide suggestions, but behavioral studies are ultimately necessary. In a similar vein, the factors that constrain and/or facilitate play in different reptile taxa are important to address. What we may tentatively conclude is that social play is less common than object play in reptiles, but observing young reptiles in social groups may result in more such behavior than we may have envisioned.

11.9. Neuroscience and Hormones

Both cognitive and affective processes involve the brain and nervous system. Play, being something animals voluntarily engage in, has neural underpinnings that overlap cognition and affect. However, motivation is also a major process underlying many social responses from courtship and mating to foraging, social hierarchies, territoriality, and fight-or-flight responses. Studies of reptiles in these areas have lagged behind mammals and birds, and even fish. One of the reasons is that reptiles are considered woefully mentally challenged and "pea-brained" compared to mammals and birds, due to claims that their brains average only 10% of the mass of similar sized birds and mammals. This conclusion is based on the work of Jerison (1973), who compiled the sparse available data. More recent analyzes, involving many more species, show that the difference in brain size between reptiles and mammals is not 10-fold, but 7-fold. This is still a sizable difference, but there are additional factors to consider, such as the heavy body armor of many reptiles (particularly crocodylians and turtles), as well as more effective use of brain volume in reptiles and birds compared to mammals (see Chapter 2). These new data and perspectives are discussed in detail in Font, Leal, and Burghardt (2021), along with other misconceptions about reptile behavior that echo many of the concerns raised here.

There is a growing number of studies on the brain mechanisms of social behavior in reptiles. Some of the early work was gathered in the symposium volume edited by N. Greenberg and MacLean (1978) that focused on lizards, but there is much newer work as well (reviews in Powers, 1990; Reiter et al., 2017). Roth et al. (2019) review brain and behavior studies in navigation, spatial learning, and other behavior in field and lab. Hormones are also highly important in reptile social behavior, and not just androgens and similar endocrinological products. Authoritative reviews can be found in Crews and Moore (2005) and Kabelik and Crews (2016). Some relevant questions are whether oxytocin influences social bonding in reptiles and what hormonal differences underlie the various color morphs of *Uta* and other species. We know that maternal stress has effects on lizard eggs and offspring (Anton et al., 2018; Ensminger et al., 2018); such studies are needed in almost all species facing changes in resources and climate.

How do nonhormonal drugs affect aggressive responses? Years ago, in studying Mexican black-bellied gartersnakes (*Thamnophis melanogaster*), we found that from birth on these animals were highly defensive in their antipredator behaviors, striking out at the slightest provocation, yet informal experiments with Valium in the GMB lab led to marked reduction in such proclivities. Can chemicals and epigenetics be used to modify social behavior in reptiles considering reptiles differ so greatly within and between species?

11.10. Social Evolution

Comparative social evolution in birds and mammals has been a major topic in animal behavior for decades. Consider two classic edited volumes by Crook (1970) and Rubenstein and Wrangham (1986). After decades, still nothing comparable exists for reptiles. The edited symposium by N. Greenberg and Crews (1977) is the closest we have, and even that treatment had no integrative scheme. We have premised this book on the view that nonavian reptiles are excellent taxa for exploring the patterns and processes of social evolution, yet as mentioned at the beginning, we have not presented any new phylogenetic analyzes or conceptual framework. Our main goal was to present, with many examples and extensive literature, the remarkable diversity and complexity reptiles possess and to stimulate both wonder and creative new explorations in field, zoo, and laboratory. There are increasing numbers of molecular phylogenies that now provide trees of the relationships among families, genera, species, and populations of many taxa. With more complete behavioral information, we can perform important studies. For example, within the Caribbean islands, the genus *Anolis* has repeatedly evolved ecomorphs independently. Thus, trunk and canopy species that seem closely related to similar living anoles on other islands may, in fact, have independently converged morphologically and behaviorally. Similar changes may have occurred in the evolution of forest and open grassland anoles. Such species provide ample scope for looking at the processes of social behavior shifts and sociality as well.

The repeated evolution of viviparity has been used to suggest how parental care may have evolved, including extended family life, as discussed earlier. The role of communal nesting may also have helped facilitate the sibling and juvenile bonding and gregariousness seen in many species, such as green iguanas and freshwater river turtles in the Amazon (and that most likely will be found in more species), opening more areas for comparative social evolution study.

There are many other examples where the interaction between and among animals can have far-reaching social consequences, including the processes of selection that may underlie the dances between predators and prey, mimics and aposematism (Weldon & Burghardt, 2015), and niche construction and multilevel selection (Laland et al., 2015). Attempts to unify social evolution theory (e.g., Lehtonen, 2020) may also underlie social evolution. Reptiles, especially squamates, may provide some of the most useful terrestrial vertebrate radiations to study.

11.11. Lessons from the Resilient

We are on the cusp of rapid, even radical, changes in how we view vertebrate social evolution, and non-avian reptiles are key players in this reassessment, which is already well underway in work on mammals, birds, and fishes. Much of the work on these species involves those that are easy to maintain and breed in captivity and have relatively short lifespans or age to sexual maturity (e.g., rats, mice, zebra finches, guppies, and zebra fish). Some lizards and snakes do have relatively fast maturity and hardiness in captivity, but they are not yet mainstream models. Anoles are now at the forefront of evolutionary studies, and gartersnakes and brown house snakes (*Lamprophis fuliginosus*) could also be important models.

The rapid urbanization of many habitats has led to some reptiles coexisting with humans and in altered habitats. With potentially dangerous animals, such as crocodylians, monitors, and large constricting or venomous snakes, the incidence of conflicts and injuries to humans and domestic animals needs to be documented and appropriate education or melioration methods implemented. The CrocBITE database (http://www.crocodile-attack.info) is one such tracking service. The consequences of urbanization, and of human encroachment on natural populations, on social behavior in reptiles are just starting to be explored, sometimes controversially. For example, Kamath and Losos (2017) reviewed the literature on territoriality in *Anolis* lizards, going back to the classic study by Evans (1938) and earlier, and claimed that all previous assertions of territoriality were flawed. This led to an exchange (Kamath & Losos, 2018) with Bush and Simberloff (2018) and Stamps (2018). Urbanization may have been an undercurrent in this controversy, as it may influence factors such as selection gradients, climate variables (temperature, humidity), diet, habitat structure (houses, utility poles, altered vegetation), predator risks, and other circumstances that could affect the social interactions and social organization of reptiles. Researchers on reptile social behavior in captivity, however, need to be continually mindful that behavior in captivity may be affected by the size, structure, and ecological salience of the provided environments (Bernheim et al., 2020). Genetic changes might also be occurring, and the roles of plasticity and selection, and how they interact, are worthy areas beginning to be explored in relation to processes of social behavior.

Related to this are the research opportunities that arise with invasive species. Many non-native reptiles are now thriving in numerous parts of the world due to human transport. These include red-eared sliders (*Trache-*

mys scripta) on all human-inhabited continents and many islands, numerous geckos in all warmer parts of the world, and no less than 56 species of alien reptiles in Florida (https://myfwc.com/wildlifehabitats/nonnatives/reptiles). Many of these animals are captured and killed. They could also be put to use in the service of science and conservation. We have one serious caution about research on reptile behavior that can compromise otherwise innovative work—one that applies to all research in animal behavior and otherwise, but which we have too often seen ignored in many of the studies reported in this book. This is the lack of attention to blind testing and interobserver reliability (Burghardt et al., 2012; Burghardt, 2020), especially when these methods can readily be implemented in laboratory and field studies. Video recording trials are also readily incorporated today. While increasing numbers of studies do all these things right, issues of reliability and confirmation bias should be something at least mentioned in all reported hypothesis-testing work, and we mention it here in closing as a reminder to everyone carrying out and evaluating research.

We are excited, however, that our changing world opens up many opportunities to study the five aims of ethology (Burghardt, 1997) in terms of sociality—mechanisms, development, function, evolution, and private experience—in fascinating animals. Such studies with reptiles promise to help us better understand the broad sweep of the evolution of social behavior in vertebrates. The great variation in some taxa, even closely related ones, contrasted with the conservatism in traits shown in others (such as the egg-laying number in *Anolis*) opens up great opportunities to study the evolutionary and ecological mechanisms of sociality in species where adaptation is both naturally occurring and where a changing climate is making it necessary. While we have documented, often only partially, much of the information acquired in recent years on non-avian reptile social lives, we have just scratched the surface of many mysteries and their illuminating, as well as delightful, solutions.

Abdala, V., & Diogo, R. (2010). Comparative anatomy, homologies and evolution of the pectoral and forelimb musculature of tetrapods with special attention to extant limbed amphibians and reptiles. *Journal of Anatomy, 217,* 536-573.

Abercrombie, C. L. (1978). Notes on West African crocodilians (Reptilia, Crocodilia). *Journal of Herpetology, 12,* 260-262.

Abraham, C., & Evans, R. (1999). The development of endothermy in American White Pelicans. *Condor, 101,* 832.

Ackerman, R. A., Seagrave, R. C., Dmi'el, R., & Ar, A. (1985). Water and heat exchange between parchment-shelled reptile eggs and their surroundings. *Copeia, 1985,* 703-711.

Agha, M., Lovich, J. E., Ennen, J. R., & Wilcox, E. (2013). Nest-guarding by female Agassiz's desert tortoise (*Gopherus agassizii*) at a wind-energy facility near Palm Springs, California. *Southwestern Naturalist, 58,* 254-257.

Aguirre, G., Adest, G., & Morafka, D. (1984). Home range and movement patterns of the Bolson tortoise, *Gopherus flavomarginatus. Acta Zoológica Mexicana Nueva Serie, 1,* 1-28.

Alberts, A. C. (1990). Chemical properties of femoral gland secretions in the desert iguana, *Dipsosaurus dorsalis. Journal of Chemical Ecology, 16,* 13-25.

Alberts, A. C. (1992). Pheromonal self-recognition in desert iguanas. *Copeia, 1992,* 229-232.

Alberts, A. C., Carter, R. L., Hayes, W. K., & Martins, E. P. (Eds.). (2004). *Iguanas: Biology and conservation,* Berkeley: University of California Press.

Alberts, A. C., Pratt, N. C., & Phillips, J. A. (1992). Seasonal productivity of lizard femoral glands: Relationship to social dominance and androgen levels. *Physiology & Behavior, 51,* 729-733.

Alberts, J. R. (1978). Huddling by rat pups: Group behavioral mechanisms of temperature regulation and energy conservation. *Journal of Comparative and Physiological Psychology, 92,* 231-245.

Alberts, J. R., & Cramer, C. P. (1988). Ecology and experience: Sources of means and meaning of developmental change. In E. M. Blass (Ed.), *Handbook of behavioral neurobiology* (Vol. 9, pp. 1-39). New York: Plenum.

Alderton, D. (1988). *Turtles and tortoises of the world.* London: Blandford Press.

Aldridge, R. D., & Brown, W. S. (1995). Male reproductive cycle, age at maturity, and cost of reproduction in the timber rattlesnake (*Crotalus horridus*). *Journal of Herpetology, 29,* 399-407.

Aldridge, R. D., & Sever, D. M. (2016). *Reproductive biology and phylogeny of snakes*. Boca Raton, FL: CRC Press.

Alexander, G. J. (2018). Reproductive biology and maternal care of neonates in southern African python (*Python natalensis*). *Journal of Zoology, 305*, 141–148.

Alexander, R. D. (1974). The evolution of social behavior. *Annual Review of Ecology and Systematics, 5*, 325–383.

Alexander, R. D. (1991). Social-learning and kin recognition-reply. *Ethology and Sociobiology, 12*, 387–399.

Alho, C. J., & Pádua, L. F. (1982). Reproductive parameters and nesting behavior of the Amazon turtle *Podocnemis expansa* (Testudinata: Pelomedusidae) in Brazil. *Canadian Journal of Zoology, 60*, 97–103.

Allaby, M. (2009). *Ecology: Plants, animals, and the environment*. New York: Infobase Publishing.

Allard, M. W., Miyamoto, M. M., Bjorndal, K. A., Bolten, A. B., & Bowen, B. W. (1994). Support for natal homing in green turtles from mitochondrial DNA sequences. *Copeia, 1994*, 34–41.

Allen, B. A., Burghardt, G. M., & York, D. S. (1984). Species and sex differences in substrate preference and tongue flick rate in three sympatric species of water snakes (*Nerodia*). *Journal of Comparative Psychology, 98*, 358–367.

Allen, L., Sanders, K. L., & Thomson, V. A. (2018). Molecular evidence for the first records of facultative parthenogenesis in elapid snakes. *Royal Society Open Science, 5*, 171–901.

Alonso, M., Sobern R. R., Ramos R., Thorbjarnarson, J. (2002). Mortality of eggs of *Crocodylus acutus* associated with the conduct of females in R. F. Monte Cabanigun, Cuba. Poster presented at the 16th Working Meeting of the Crocodile Specialist Group SSS/IUCN, October 2002, Gainesville, Florida.

Altig, R., & McDiarmid, R. W. (2007). Morphological diversity and evolution of egg and clutch structure in amphibians. *Herpetological Monographs, 21*, 1–32.

Amarello, M. (2012). Social Snakes? Non-random association patterns detected in a population of Arizona black rattlesnakes (*Crotalus cerberus*) [Master's thesis]. Arizona State University.

Amavet, P., Rosso, E., Markariani, R., & Piña, C. (2008). Microsatellite DNA markers applied to detection of multiple paternity in *Caiman latirostris* in Santa Fe, Argentina. *Journal of Experimental Zoology Part A: Ecological Genetics and Physiology, 309*, 637–642.

Amavet, P., Vilardi, J., Rueda, E., Larriera, A., & Saidman, B. (2012). Mating system and population analysis of the broad-snouted caiman (*Caiman latirostris*) using microsatellite markers. *Amphibia-Reptilia, 33*, 83–93.

Amdekar, M. S., & Thaker, M. (2019). Risk of social colours in an agamid lizard: Implications for the evolution of dynamic signals. *Biology Letters, 15*, 20190207.

AmphibiaWeb. (2019). AmphibiaWeb: Information on amphibian biology and conservation. [web application]. Retrieved March 10, 2020, from http://amphibiaweb.org.

Anan'eva, N. B., Borkin, L. Y., Darevskiy, I. S., & Orlov, N. L. (1998). *Amphibians and reptiles of Russia*. Moscow: ABF.

Ancel, A., Gilbert, C., Poulin, N., Beaulieu, M., & Thierry, B. (2015). New insights into the huddling dynamics of emperor penguins. *Animal Behaviour, 110,* 91–98.

Anderson, A., Russell, J., Booms, T., & Russell, D. (2015). Siblicide and cannibalism in Alaskan Boreal owls. *Journal of Raptor Research, 49,* 498–501.

Andersson, M. (1994). *Sexual selection.* Princeton, NJ: Princeton University Press.

Andrade, D., Nascimento, L., & Abe, A. (2006). Habits hidden underground: A review on the reproduction of the Amphisbaenia with notes on four neotropical species. *Amphibia-Reptilia, 27,* 207–217.

Andrén, C. (1986). Courtship, mating and agonistic behaviour in a free-living population of adders, *Vipera berus* (L.). *Amphibia-Reptilia, 7,* 353–383.

Andrews, H., & Whitaker, R. (1993). Captive breeding of freshwater turtles at the Centre for Herpetology/Madras Crocodile Bank. *Zoos Print, 8,* 12–14.

Angielczyk, K. D. (2009). *Dimetrodon* is not a dinosaur: Using tree thinking to understand the ancient relatives of mammals and their evolution. *Evolutionary Education and Outreach, 2,* 257–271.

Anton, A. J., Langkilde, T., Graham, S., & Fawcett, J. D. (2018). Effects of chronic corticosterone increases on the maternal behaviour of the Prairie Skink, *Plestiodon septentrionalis. Herpetological Journal, 28,* 123–126.

Apesteguía, S., Gómez, R. O., & Rougier, G. W. (2014). The youngest South American rhynchocephalian, a survivor of the K/Pg extinction. *Proceedings of the Royal Society B: Biological Sciences, 281,* 20140811.

Aragón, P., López, P., & Martín, J. (2000). Size-dependent chemosensory responses to familiar and unfamiliar conspecific faecal pellets by the Iberian rock-lizard, *Lacerta monticola. Ethology, 106,* 1115–1128.

Aragón, P., Massot, M., Gasparini, J., & Clobert, J. (2006). Socially acquired information from chemical cues in the common lizard, *Lacerta vivipara. Animal Behaviour, 72,* 965–974.

Arnold, K., & Neumeyer, C. (1987). Wavelength discrimination in the turtle *Pseudemys scripta elegans. Vision Research, 27,* 1501–1511.

Arnold, S. J. (2003). Too much natural history, or too little? *Animal Behaviour, 65,* 1056–1068.

Arnold, S. J., & Bennett, A. F. (1984). Behavioral variation in natural populations. III. Antipredator displays in the garter snake *Thamnophis radix. Animal Behaviour, 32,* 108–1118.

Arnold, S. J., & Wassersug, R. (1978). Differential predation on metamorphic anurans by garter snakes (Thamnophis): Social behavior as a possible defense. *Ecology, 59,* 1014–1022.

Aubret, F., Bignon, F., Kok, P., & Blanvillain, G. (2016). Only child syndrome in snakes: Eggs incubated alone produce asocial individuals. *Scientific Reports, 6,* 35752.

Aubret, F., Blanvillain, G., Bignon, F., & Kok, P. (2016). Heartbeat, embryo communication and hatching synchrony in snake eggs. *Scientific Reports, 6,* 23519.

Aubret, F., & Shine, R. (2009). Causes and consequences of aggregation by neonatal tiger snakes (*Notechis scutatus,* Elapidae). *Austral Ecology, 34,* 210–217.

Aubret, F., & Shine, R. (2010). Thermal plasticity in young snakes: How will climate change affect the thermoregulatory tactics of ectotherms? *Journal of Experimental Biology, 213*, 242–248.

Auffenberg, W. (1966). On the courtship of *Gopherus polyphemus*. *Herpetologica, 22*, 113–117.

Auffenberg, W. (1969). Social behavior of *Geochelone denticulata*. *Quarterly Journal of the Florida Academy of Sciences, 32*, 50–58.

Auffenberg, W. (1977). Display behavior in tortoises. *American Zoologist, 17*, 241–250.

Auffenberg, W. (1981). *The behavioral ecology of the Komodo monitor*. Gainesville: University of Florida Press.

Auffenberg, W. (1988). *Gray's monitor lizard*. Gainesville: University of Florida Press.

Augustine, L., Miller, K., & Burghardt, G. M. (2015). *Crocodylus rhombifer* (Cuban Crocodile). Play behavior. *Herpetological Review, 46*, 208–209.

Avanzini, M., Dalla Vecchia, F. M., Rigo, M., Mietto, P., & Roghi, G. (2007). A vertebrate nesting site in northeastern Italy reveals unexpectedly complex behavior for late Carnian reptiles. *Palaios, 22*, 465–475.

Avery, R. A. (1976). Thermoregulation, metabolism and social behaviour in Lacertidae. In A. d'A. Bellairs & C. B. Cox (Eds.), *Morphology and biology of reptiles* (pp. 245–259). Linnean Society of London Symposium Series No. 3.

Baeckens, S., De Meester, W., Tadić, Z., & Van Damme, R. (2019). Where to do number two: Lizards prefer to defecate on the largest rock in the territory. *Behavioural Processes, 167*, 103937.

Baicich, P. J., & Harrison, J. O. (1997). *A Guide to the nests, eggs, and nestlings of North American birds*. Princeton, NJ: Princeton University Press.

Bailey, I., Myatt, J. P., & Wilson, A. M. (2012). Group hunting within the Carnivora: Physiological, cognitive and environmental influences on strategy and cooperation. *Behavioral Ecology and Sociobiology, 67*, 1–17.

Baker, C. J., Franklin, C. E., Campbell, H. A., Irwin, T. R., & Dwyer, R. G. (2019). Ontogenetic shifts in the nesting behaviour of female crocodiles. *Oecologia, 189*, 891–904.

Baker, P., Costanzo, J., Iverson, J., & Lee, R., Jr. (2013). Seasonality and inter-specific and intraspecific asynchrony in emergence from the nest by hatchling freshwater turtles. *Canadian Journal of Zoology, 91*, 451–461.

Baker, R. E., & Gllingham, G. C. (1983). An analysis of courtship behavior in Blanding's turtle, *Emydoidea blandingi*. *Herpetologica, 39*, 166–173.

Balazs, G., & Ross, E. (1974). Observations on the pre-emergence behavior of the green turtle. *Copeia, 1974*, 986–988.

Balcombe, J. (2016). *What a fish knows*. New York: Scientific American / Farrar, Straus, and Giroux.

Ballen, C., Shine, R., & Olsson, M. (2014). Effects of early social isolation on the behaviour and performance of juvenile lizards, *Chamaeleo calyptratus*. *Animal Behaviour, 88*, 1–6.

Balme, D. M. (Ed.). (1991). *Aristotle. History of animals. Books VII–X*. Cambridge, MA: Harvard University Press.

Barabanov, V., Gulimova, V., Berdiev, R., & Saveliev, S. (2015). Object play in thick-toed geckos during a space experiment. *Journal of Ethology, 33*, 109–115.

Barash, D. P. (1974). Neighbor recognition in two "solitary" carnivores: The racoon (*Procyon lotor*) and red fox (*Vulpes fulva*). *Science, 185*, 794-796.

Barker, C. T., Naish, D., Newham, E., Katsamenis, O. L., & Dyke, G. (2017). Complex neuroanatomy in the rostrum of the Isle of Wight theropod *Neovenator salerii*. *Scientific Reports, 7*, 3749.

Barker, D. G., Murphy, J. B., & Smith, K. W. (1979). Social behavior in a captive group of Indian pythons, *Python molurus* (Serpentes, Boidae) with formation of a linear social hierarchy. *Copeia, 1979*, 466-471.

Barnett, K. E., Cocroft, R. B., & Fleishman, L. J. (1999). Possible communication by substrate vibration in a chameleon. *Copeia, 1999*, 225-228.

Barnosky, A. D., Koch, P. L., Feranec, R. S., Wing, S. L., & Shabel, A. B. (2004). Assessing the causes of late Pleistocene extinctions on the continents. *Science, 306*, 70-75.

Barnosky, A. D., Matzke, N., Tomiya, S., Wogan, G. O. U., Swartz, B., Quental, T. B., Marshall, C., McGuire, J. L., Lindsey, E. L., Maguire, K. C., Mersey, B., & Ferrer, E. A. (2011). Has the Earth's sixth mass extinction already arrived? *Nature, 471*, 51-57.

Barry, M., Shanas, U., & Brunton, D. H. (2014). Year-round mixed-age shelter aggregations in Duvaucel's geckos (*Hoplodactylus duvaucelii*). *Herpetologica, 70*, 395-406.

Bartram, W. (1791). *Travels through North and South Carolina, Georgia, east and west Florida*. Philadelphia, PA: James & Johnson.

Bartram, W. (1943). *Travels in Georgia and Florida, 1773-1774: A report to Dr. John Fothergill* (F. Harper, Ed.). *Trans. Amer. Philos. Soc. New Ser., 33*, 121-242.

Bass, A. H., & Chagnaud, B. P. (2012). Shared developmental and evolutionary origins for neural basis of vocal-acoustic and pectoral-gestural signaling. *Proceedings of the National Academy of Sciences, 109*, 10677-10684.

Batabyal, A., & Thaker, M. (2017). Signalling with physiological colours: High contrast for courtship but speed for competition. *Animal Behaviour, 129*, 229-236.

Bateson, G. (1966). Problems in cetacean and other mammalian communication. In K. E. Norris (Ed.), *Whales, dolphins and porpoises* (pp. 569-578). California: University of California Press.

Bauer, A. M., Cogger, H. G., & Zweifel, R. G. (Eds.). (1998). *Encyclopedia of reptiles and amphibians*. San Diego: Academic Press.

BBC (Producer). (2009). Sea krait hunt. *Planet Earth*. Retrieved from http://www.bbc.co.uk/programmes/p0038t09.

BBC (Producer). (2016). Galapagos racer hunt. *Planet Earth II*. Retrieved from https://www.youtube.com/watch?v=Rv9hn4IGofM.

BBC (Producer). (2019). The Power of Play. *Animals at Play*. https://www.bbc.co.uk/programmes/m00077gc.

Beck, D. D., & Ramirez-Bautista, A. (1991). Combat behavior of the beaded lizard, *Heloderma h. horridum*, in Jalisco, Mexico. *Journal of Herpetology, 25*, 481-484.

Bednarz, J. C. (1988). Cooperative hunting in Harris' hawks (*Parabuteo unicinctus*). *Science, 39*, 1525-1527.

Bekoff, M., Allen, C., & Burghardt, G. M. (Eds.). (2002). *The cognitive animal: Empirical and theoretical perspectives on animal cognition.* Cambridge, MA: MIT Press.

Bekoff, M., & Byers, J. A. 1981. A critical reanalysis of the ontogeny of mammalian social and locomotor play: An ethological hornet's nest. In K. Immelmann, G. W. Barlow, L. Petrinovich & M. Main (Eds.), *Behavioral development: The Bielefeld interdisciplinary project* (pp. 296–337). New York: Cambridge University Press.

Belkin, D. A. (1968). Aquatic respiration and underwater survival of two freshwater turtle species. *Respiration Physiology, 4,* 1–14.

Bellairs, A. 1970. *The life of reptiles.* New York: Universe Books.

Bellemain, E., Swenson, J. E., & Taberlet, P. (2006). Mating strategies in relation to sexually selected infanticide in a non-social carnivore: The brown bear. *Ethology, 112,* 238–246.

Bels, V. L., & Crama, Y. J.-M. (1994). Quantitative analysis of the courtship and mating behavior in the loggerhead musk turtle *Sternotherus minor* (Reptilia: Kinosternidae) with comments on courtship behavior in turtles. *Copeia, 1994,* 676–684.

Beltrán, I., Loiseleur, R., Durand, V., & Whiting, M. J. (2020). Effects of early thermal environment on the behavior and learning of a lizard with bimodal reproduction. *Behavioral Ecology and Sociobiology, 74,* 73. https://doi.org/10.1007/s00265-020-02849-6.

Belzer, B. (2002). A nine year study of eastern box turtle courtship with implications for reproductive success and conservation in a translocated population. *Turtle and Tortoise Newsletter, 6,* 17–26.

Bennett, P. M., & Owens, I. P. (2002). *Evolutionary ecology of birds: Life histories, mating systems and extinction.* Oxford: Oxford University Press.

Bennett, S. C. (1992). Sexual dimorphism of *Pteranodon* and other pterosaurs, with comments on cranial crests. *Journal of Vertebrate Paleontology, 12,* 422–434.

Benton, M. J. (2020). The origin of endothermy in synapsids and archosaurs and arms races in the Triassic. *Gondwana Research,* https://doi.org/10.1016/j.gr.2020.08.003.

Benton, M. J. (2000). *Vertebrate palaeontology* (2nd ed.). London: Blackwell Science.

Benton, M. J., Dhouailly, D., Jiang, B., & McNamara, M. (2019). The early origin of feathers. *Trends in Ecology & Evolution, 34,* 856–869.

Berg, C. (1898). Ueber die Eiablage, die Bruptflege und die Nahrung von *Amphisbaena darwinii. Verhandlungen Der Gesellschaft Deutscher Naturforscher Und Ärzte (Leipzig), 69,* 164–165.

Bernard, A., Lécuyer, C., Vincent, P., Amiot, R., Bardet, N., Buffetaut, E., . . . Prieur, A. (2010). Regulation of body temperature by some Mesozoic marine reptiles. *Science, 328,* 1379–1382.

Bernardo, J., & Plotkin, P. T. (2007). An evolutionary perspective on the arribada phenomenon and reproductive behavioral polymorphism of olive ridley sea turtles (*Lepidochelys olivacea*). In P. T. Plotkin (Ed.), *Biology and conservation of ridley sea turtles* (pp. 59–87). Baltimore: Johns Hopkins University Press.

Bernheim, M., Livne, S., & Shanas, U. (2020). Mediterranean spur-thighed tortoises (*Testudo graeca*) exhibit pre-copulatory behavior particularly under specific experimental setups. *Journal of Ethology.* https://doi.org/10.1007/s10164-020-00657-z.

Bernstein, J. M., Griffing, A. H., Daza, J. D., Gamble, T., & Bauer, A. M. (2016). Using alien resources: Caribbean dwark geckos nesting communally in invasive flora. *IRCF Reptiles and Amphibians, 23*, 40-43.

Berry, J. F., & Shine, R. (1980). Sexual size dimorphism and sexual selection in turtles (order Testudines). *Oecologia, 44*, 185-191.

Beruldsen, G. (1980). *A field guide to nests & eggs of Australian birds.* Adelaide: Rigby.

Besson, A. A., & Cree, A. (2010). A cold-adapted reptile becomes a more effective thermoregulator in a thermally challenging environment. *Oecologia, 163*, 571-581.

Besson, A. A., & Cree, A. (2011). Integrating physiology into conservation: An approach to help guide translocations of a rare reptile in a warming environment. *Animal Conservation, 14*, 28-37.

Bevan, E., Wibbels, T., Navarro, E., Rosas, M. Najera, B. M., Sarti, L., & Burchfield, P. (2016). Using unmanned aerial vehicle (UAV) technology for locating, identifying, and monitoring courtship and mating behaviour in the green turtle (*Chelonia mydas*). *Herpetological Review, 47*, 27-32.

Bever, G. S., Lyson, T. R., Field, D. J., & Bhullar, B. A. S. (2015). Evolutionary origin of the turtle skull. *Nature, 525*, 239-242.

Bezuijen, M. R., Webb, G. J. W., Hartoyo, P. S., Ramono, W. S., & Manolis, S. C. (1998). The false gharial (*Tomistoma schlegelii*) in Sumatra. In *Crocodiles: Proceedings of the 14th Working Meeting of the Crocodile Specialist Group, IUCN* (pp. 10-31). Gland, Switzerland, and Cambridge, UK: The World Conservation Union.

Bezy, R. L., & Cole, C. J. (2014). Amphibians and reptiles of the Madrean Archipelago of Arizona and New Mexico. *American Museum Novitates, 2014*, 1-25.

Birchard, G. F, Black, C. P., Schuett, G. W., & Black, V. D. (1984). Influence of pregnancy on oxygen consumption, heart rate, and hematology in the garter snake: Implications for the "cost of reproduction" in live-bearing reptiles. *Journal of Comparative Biochemistry and Physiology, 77A*, 519-523.

Birchard, G. F., & Marcellini, D. (1996). Incubation time in reptilian eggs. *Journal of Zoology, London, 240*, 621-635.

Bishop, D. C., & Echternacht, A. C. (2004). Emergence behavior and movements of winter-aggregated Green Anoles (*Anolis carolinensis*) and the thermal characteristics of their crevices in Tennessee. *Herpetologica, 60*, 168-177.

Bjorndal, K. A. (1979). Cellulose digestion and volatile fatty acid production in the green turtle, *Chelonia mydas. Comparative Biochemistry and Physiology— Part A: Physiology, 63*, 127-133.

Bjorndal, K. A. (1980). Nutrition and grazing behavior of the green turtle *Chelonia mydas. Marine Biology, 56*, 147-154.

Bjorndal, K. A. (1982). The consequences of herbivory for the life history pattern of the Caribbean green turtle, *Chelonia mydas*. In K. A. Bjorndal (Ed.), *Biology and conservation of sea turtles* (pp. 111-116). Washington, DC: Smithsonian Institution Press.

Black, C. P, Birchard, G. F., Schuett, G. W., & Black, V. D. (1984). Influence of incubation substrate water content on oxygen uptake in embryos of the Burmese python (*Python molurus bivittatus*). In R. S. Seymour (Ed.), *Respiration and metabolism in reptiles* (pp. 137–145). New York: Plenum Press.

Blackburn, D. G. (1985). Evolutionary origins of vivipary in the Reptilia. II. Serpentes, Amphisbaenia and Ichthyosauria. *Amphibia-Reptilia, 5*, 259–291.

Blackburn, D. G. (2015). Evolution of viviparity in squamate reptiles: Reversibility reconsidered. *Journal of Experimental Zoology, 324*, 473–486.

Blackburn, D. G., & Flemming, A. F. (2012). Invasive implantation and intimate placental associations in a placentotrophic African lizard, *Trachylepis ivensi* (Scincidae). *Journal of Morphology, 273*, 137–159.

Blackburn, D. G., & Sidor, C. A. (2015). Evolution of viviparous reproduction in Paleozoic and Mesozoic reptiles. *International Journal of Developmental Biology, 58*, 935–948.

Blouin-Demers, G., & Weatherhead, P. J. (2000). A novel association between a beetle and a snake: Parasitism of *Elaphe obsoleta* by *Nicrophorus pustulatus*. *Ecoscience, 7*, 395–397.

Blouin-Demers, G., Weatherhead, P. J., & Row, J. R. (2004). Phenotypic consequences of nest-site selection in black rat snakes (*Elaphe obsoleta*). *Canadian Journal of Zoology, 82*, 449–456.

Bock, B. C., & McCracken, G. F. (1988). Genetic structure and variability in the green iguana (*Iguana iguana*). *Journal of Herpetology, 22*, 316–322.

Bock, B. C., & Rand, A. S. (1989). Factors influencing nesting synchrony and hatching success at a green iguana nesting aggregation in Panama. *Copeia, 1989*, 978–986.

Bock, B. C., Rand, A. S., & Burghardt, G. M. (1985). Seasonal migration and nesting site fidelity in the iguana. In M. A. R. Rankin (Ed.), *Migration: Mechanisms and adaptive significance*. Austin: University of Texas Press.

Bock, B. C., Rand, A. S., & Burghardt, G. M. (1989). Nesting season movements of female green iguanas (*Iguana iguana*) in Panama. *Copeia, 1989*, 214–216.

Boersma, P. D. (1982). The benefits of sleeping aggregations in marine iguanas, *Amblyrhynchus cristatus*. In G. M. Burghardt & A. S. Rand (Eds.), *Iguanas of the world: Their behavior, ecology, and conservation* (pp. 292–299). Park Ridge, NJ: Noyes Publications.

Böhme, W., & Ziegler, T. (2009). A review of iguanian and anguimorph lizard genitalia (Squamata: Chamaeleonidae; Varanoidea, Shinisauridae, Xenosauridae, Anguidae) and their phylogenetic significance: Comparisons with molecular data sets. *Journal of Zoological Systematics and Evolutionary Research, 47*, 189–202.

Boice, R. (1970). Competitive feeding behaviours in captive *Terrapene c. carolina*. *Animal Behaviour, 18*, 703–710.

Bolhuis, J., & Van Kampen, H. (1992). An evaluation of auditory learning in filial imprinting. *Behaviour, 122*, 195.

Bonke, R., Böhme, W., Opiela, K., & Rödder, D. (2011). A remarkable case of cannibalism in juvenile leopard geckos, *Eublepharis macularius* (Blyth, 1854) (Squamata: Eublepharidae). *Herpetology Notes, 4*, 211–212.

Bonke, R., Whitaker, N., Roedder, D., & Boehme, W. (2015). Vocalizations in two rare crocodilian species: A comparative analysis of distress calls of *Tomistoma schlegelii* (Müller, 1838) and *Gavialis gangeticus* (Gmelin, 1789). *North-Western Journal of Zoology*, *11*, 151–162.

Bonnet, X., Lagarde, F., Henen, B. T., Corbin, J., Nagy, K. A., Naulleau, G., . . . Cambag, R. (2001). Sexual dimorphism in steppe tortoises (*Testudo horsfieldii*): Influence of the environment and sexual selection on body shape and mobility. *Biological Journal of the Linnean Society*, *72*, 357–372.

Bonnet, X., Shine, R., & Lourdais, O. (2002). Taxonomic chauvinism. *Trends in Ecology and Evolution*, *17*, 1–3.

Boomsma, J. (2007). Kin selection versus sexual selection: Why the ends do not meet. *Current Biology*, *17*, R673-R683.

Boomsma, J. (2009). Lifetime monogamy and the evolution of eusociality. *Philosophical Transactions of the Royal Society B: Biological Sciences*, *364*, 3191–3207.

Booth, D., & Thompson, M. (1991). A comparison of reptilian eggs with those of megapode birds. In D. C. Deeming & M. W. J. Ferguson (Eds.), *Egg incubation: Its effects on embryonic development in birds and reptiles* (pp. 325–344). Cambridge and New York: Cambridge University Press.

Booth, J., & Peters, J. A. (1972). Behavioural studies on the green turtle (*Chelonia mydas*) in the sea. *Animal Behaviour*, *20*, 808–812.

Booth, W., & Schuett, G. W. (2011). Molecular genetic evidence for alternative reproductive strategies in North American pitvipers (Serpentes: Viperidae): Long-term sperm storage and facultative parthenogenesis. *Biological Journal of the Linnean Society*, *101*, 934–942.

Booth, W., & Schuett, G. W. (2016). The emerging phylogenetic pattern of parthenogenesis in snakes. *Biological Journal of the Linnean Society*, *118*, 172–186.

Booth, W., Smith, C. F., Eskridge, P. H., Hoss, S. K., Mendelson, J. R., & Schuett, G. W. (2012). Facultative parthenogenesis discovered in wild vertebrates. *Biology Letters*, *8*, 983–985.

Borden J. H. (1985). Aggregation pheromones. In G. A. Kerkut & L. I. Gilbert (Eds.), *Comprehensive insect physiology biochemistry and pharmacology* (Vol. 9, pp. 257–285). Elmsford, NY: Pergamon Press.

Borteiro, C., Kolenc, F., & Verdes, J. M. (2013). Aggregative behaviour in the fossorial lizard *Amphisbaena darwinii* (Squamata, Amphisbaenidae). *Cuadernos de Herpetologia*, *27*, 57–58.

Bortolotti, G. R., Wiebe, K. L., & Iko, W. M. (1991). Cannibalism of nestling American kestrels by their parents and siblings. *Canadian Journal of Zoology*, *69*, 1447–1453.

Botha-Brink, J., & Angielczyk, K. D. (2010). Do extraordinarily high growth rates in Permo-Triassic dicynodonts (Therapsida, Anomodontia) explain their success before and after the end-Permian extinction? *Zoological Journal of the Linnean Society*, *160*, 341–365.

Botha-Brink, J., & Modesto, S. P. (2007). A mixed-age classed "pelycosaur" aggregation from South Africa: Earliest evidence of parental care in amniotes? *Proceedings of the Royal Society B: Biological Sciences*, *274*, 2829–2834.

Boucher, M., Tellez, M., & Anderson, J. T. (2020). Differences in distress: Variance and production of American Crocodile (*Crocodylus acutus*) distress calls in Belize. *Ecology and Evolution, 10*, 9624–9634.

Bowers, B. B., & Burghardt, G. M. (1992). The scientist and the snake: Relationships with reptiles. In H. Davis & D. Balfour (Eds.), *The inevitable bond: Examining scientist-animal interactions* (pp. 250–263). Cambridge, UK: Cambridge University Press.

Bowers, R. I. (2018). A common heritage of behaviour systems. *Behaviour, 155*, 415–442.

Bowmaker, J. K. (1998). Evolution of colour vision in vertebrates. *Eye, 12*, 541–547.

Bowman, R. (2003). Apparent cooperative hunting in Florida scrub-jays. *Wilson Bulletin, 115*, 197–199.

Branch, B. (1989). Alternative life-history styles in reptiles. In M. N. Bruton (Ed.), *Alternative life-history styles in animals* (pp. 127–151). Norwell, MA: Kluwer Academic.

Branch, B. (1998). *Field guide to the snakes and other reptiles of southern Africa.* Sanibel Island, FL: Ralph Curtis Books.

Branch, W. R. (1989). Alternative life-history styles in reptiles. In M. N. Bruton (Ed.), *Alternative life-history styles of animals* (pp. 127–151). Amsterdam: Springer Netherlands.

Brandley, M. C., Young, R. L., Warren, D. L., Thompson, M. B., & Wagner, G. P. (2012). Uterine gene expression in the live-bearing lizard, *Chalcides ocellatus*, reveals convergence of squamate reptile and mammalian pregnancy mechanisms. *Genome Biology and Evolution, 4*, 394–411.

Brasier, M. D., Norman, D. R., Liu, A. G., Cotton, L. J., Hiskocks, J. E. H., Garwood, R. J., Antcliffe, J. B., & Wacey, D. (2017). Remarkable preservation of brain tissues in an Early Cretaceous iguanodontian dinosaur. *Geological Society of London, Special Publications, 448*, 383–398.

Brattstrom, B. H. (1971). Social and thermoregulatory behavior of the bearded dragon, *Amphibolurus barbatus. Copeia, 1971*, 484–497.

Brattstrom, B. H. (1973). Social and maintenance behavior of the echidna, *Tachyglossus aculeatus. Journal of Mammalogy, 54*, 50–71.

Brattstrom, B. H. (1974). The evolution of reptilian social behavior. *American Zoologist, 14*, 35–49.

Brattstrom, B. H. (1978). Learning studies in lizards. In N. Greenberg & P. D. MacLean (Eds.), *Behavior and neurology of lizards: An interdisciplinary colloquium* (pp. 173–181). Rockville, MD: National Institute of Mental Health.

Braz, H., & Manço, D. G. (2011). Natural nests of the false-coral snake *Oxyrhopus guibei* in southeastern Brazil. *Herpetology Notes, 4*, 187–189.

Brazaitis, P., & Watanabe, M. E. (2011). Crocodilian behaviour: A window to dinosaur behaviour? *Historical Biology, 23*, 73–90.

Breithaupt, B. H., & Duvall, D. (1986). The oldest record of serpent aggregation. *Lethaia, 19*(2), 181–185.

Brien, M. L., Lang, J. W., Webb, G. J., Stevenson, C., & Christian, K. A. (2013). The good, the bad, and the ugly: Agonistic behavior in juvenile crocodilians. *PLoS ONE, 8*, e80872. https://doi.org/10.1371/journal.pone.0080872.

Brien, M. L., Webb, G. J., Lang, J. W., & Christian, K. A. (2013). Intra- and interspecific agonistic behaviour in hatchling Australian freshwater crocodiles (*Crocodylus johnstoni*) and saltwater crocodiles (*Crocodylus porosus*). *Australian Journal of Zoology*, *61*, 196. https://doi.org/10.1071/zo13035.

Brien, M. L., Webb, G. J., Lang, J. W., McGuinness, K. A., & Christian, K. A. (2013). Born to be bad: Agonistic behaviour in hatchling saltwater crocodiles (*Crocodylus porosus*). *Behaviour*, *150*, 737-762.

Brillet, C., & Pailette, M. (1991). Acoustic signals of the nocturnal lizard *Gekko gecko*: Analysis of the "long complex sequence." *Bioacoustics*, *3*, 33-34.

Briskie, J. V., & Montgomerie, R. (1997). Sexual selection and the intromittent organ of birds. *Journal of Avian Biology*, *28*, 73-86.

Brito, E. S., Strussmann, C., & Baicere-Silva, C. M. (2009). Courtship behavior of *Mesoclemmys vanderhaegei* (Bour, 1973) (Testudines: Chelidae) under natural conditions in the Brazilian Cerrado. *Herpetology Notes*, *2*, 67-72.

Britton, A. R. C. (2001). Review and classification of call types of juvenile crocodilians and factors affecting distress calls. In G. C. Grigg, F. Seebacher & C. E. Franklin (Eds.), *Crocodilian biology and evolution* (pp. 364-377). Chipping Norton, Australia: Surrey Beatty & Sons.

Britton, A. R. C., Whitaker, R., & Whitaker, N. (2012). Here be a dragon: Exceptional size in a saltwater crocodile (*Crocodylus porosus*) from the Philippines. *Herpetological Review*, *44*, 541-546.

Brochu, C. A. (2000). A digitally-rendered endocast for *Tyrannosaurus rex*. *Journal of Vertebrate Paleontology*, *20*, 1-6.

Brochu, C. A. (2003). Phylogenetic approaches towards crocodilian history. *Annual Review of Earth and Planetary Sciences*, *31*, 357-397.

Brodie, E. D., Jr., Nussbaum, R. A., & Storm, R. M. (1969). An egg-laying aggregation of five species of Oregon reptiles. *Herpetologica*, *25*, 223-227.

Brown, A. M. (1984). Ultrasound in gecko distress calls (Reptilia: Gekkonidae). *Israel Journal of Zoology*, *33*, 95-101.

Brown, C. R., & Brown, M. B. (1996). *Coloniality in the cliff swallow: The effect of group size on social behavior*. Chicago: University of Chicago Press.

Brown, C. R., & Brown, M. B. (2001). Avian coloniality. In V. Nolan and C. F. Thompson (Eds.), *Current ornithology* (Vol. 16, pp. 1-82). Boston, MA: Springer.

Brown, G. P., & Shine, R. (2005). Female phenotype, life history, and reproductive success in free-ranging snakes (*Tropidonophis mairii*). *Ecology*, *86*, 2763-2770.

Brown, J. L. (1966). Types of group selection. *Nature*, *211*, 870.

Brown, J. L., & Orians, G. (1970). Spacing patterns in mobile animals. *Annual Review of Ecology and Systematics*, *1*, 239-262.

Brown, L., & Macdonald, D. W. (1995). Predation on green turtle *Chelonia mydas* nests by wild canids at Akyatan Beach, Turkey. *Biological Conservation*, *71*, 55-60.

Brown, S. G., & Duffy, P. K. (1992). The effects of egg-laying site, temperature, and salt water on incubation time and hatching success in the gecko *Lepidodactylus lugubris*. *Journal of Herpetology*, *26*, 510-513.

Brown, W. (1993). Biology, status and management of the timber rattlesnake (*Crotalus horridus*): A guide for conservation. *Herpetological Circulars*, *22*, 1-78.

Brua, R. (2002). Parent-embryo interactions. *Oxford Ornithology Series, 13*, 88-99.

Brueggen, J. (2001). Parental care in Siamese crocodiles (*Crocodylus siamensis*). *Crocodile Specialist Group Newsletter, 20*, 66-67.

Bruinjé, A. C., Coelho, F. E. A., Maggi, B. S., & Costa, G. C. (2020). Chemical signalling behaviour in intrasexual communication of lizards lacking femoral pores. *Ethology, 126*, 772-779. https://doi.org/10.1111/eth.13021.

Brumm, H., & Zollinger, S. A. (2017). Vocal plasticity in a reptile. *Proceedings of the Royal Society B, 284*, 20170451.

Bruner, G., Fernández-Marín, H., Touchon, J. C., & Wcislo, W. T. (2012). Eggs of the blind snake, *Liotyphlops albirostris*, are incubated in a nest of the lower fungus-growing ant, *Apterostigma* cf. *goniodes*. *Psyche, 212*, 1-5.

Brusatte, S. L. (2012). *Dinosaur paleobiology*. Hoboken, NJ: John Wiley & Sons, Ltd.

Bshary, R., Wickler, W., & Fricke, H. (2002). Fish cognition: A primate's eye view. *Animal Cognition, 5*, 1-13.

Budd, G. E. (2008). The earliest fossil record of the animals and its significance. *Philosophical Transactions of the Royal Society B: Biological Sciences, 363*, 1425-1434.

Buffetaut, E., & Taquet, P. (1977). The giant crocodilian *Sarcosuchus* in the Early Cretaceous of Brazil and Niger. *Palaeontology, 20*, 203-208.

Bugden, S., & Evans, R. (1999). The development of a vocal thermoregulatory response to temperature in embryos of the domestic chicken. *The Wilson Bulletin, 111*, 188-194.

Bull, C. M. (1988). Mate fidelity in an Australian lizard *Trachydosaurus rugosus*. *Behavioral Ecology and Sociobiology, 23*, 45-49.

Bull, C. M. (1994). Population dynamics and pair fidelity in sleepy lizards. In L. J. Vitt & E. R. Pianka (Eds.), *Lizard ecology: Historical and experimental perspectives* (pp. 159-174). Princeton, NJ: Princeton University Press.

Bull, C. M. (2000). Monogamy in lizards. *Behavioural Processes, 51*, 7-20.

Bull, C. M., & Baghurst, B. C. (1998). Home range overlap of mothers and their offspring in the sleepy lizard, *Tiliqua rugosa*. *Behavioral Ecology and Sociobiology, 42*, 357-362.

Bull, C. M., & Burzacott, D. A. (2006). The influence of parasites on the retention of long-term partnerships in the Australian sleepy lizard, *Tiliqua rugosa*. *Oecologia, 146*, 675-680.

Bull, C. M., Cooper, S. J. B., & Baghurst, B. C. (1998). Social monogamy and extra-pair fertilization in an Australian lizard, *Tiliqua rugosa*. *Behavioral Ecology and Sociobiology, 44*, 63-72.

Bull, C. M., Griffin, C., Bonnett, M., Gardner, M., & Cooper, S. (2001). Discrimination between related and unrelated individuals in the Australian lizard *Egernia striolata*. *Behavioral Ecology and Sociobiology, 50*, 173-179.

Bull, C. M., Griffin, C. L., & Johnston, G. R. (1999). Olfactory discrimination in scat-piling lizards. *Behavioral Ecology, 10*, 136-140.

Bull, J. J. (1980). Sex determination in reptiles. *Quarterly Review of Biology, 55*, 3-21.

Bull, J. J., & Shine, R. (1979). Iteroparous animals that skip opportunities for reproduction. *The American Naturalist, 114*, 296-303.

Bulté, G., Chlebak, R. J., Dawson, J. W., & Blouin-Demers, G. (2018). Studying mate choice in the wild using 3D printed decoys and action cameras: A case of study of male choice in the northern map turtle. *Animal Behaviour, 138*, 141–143.

Burger, J. (1976). Behavior of hatchling diamondback terrapins (*Malaclemys terrapin*) in the field. *Copeia, 1976*, 742–748.

Burger, J. (1977). Determinants of hatching success in diamondback terrapin, *Malaclemys terrapin*. *American Midland Naturalist, 97*, 444–464.

Burger, J. (1991a). Effects of incubation temperature on behavior of hatchling pine snakes: Implications for reptilian distribution. *Behavioral Ecology and Sociobiology, 28*, 297–303.

Burger, J. (1991b). Response to prey chemical cues by hatchling pine snakes (*Pituophis melanoleucus*): Effects of temperature and experience. *Journal of Chemical Ecology, 17*, 1069–1078.

Burger, J. (1998). Effects of incubation temperature on hatchling pine snakes: Implications for survival. *Behavioral Ecology and Sociobiology, 43*, 11–18.

Burger, J., & Zappalorti, R. T. (1991). Nesting behavior of pine snakes (*Pituophis m. melanoleucus*) in the New Jersey Pine Barrens. *Journal of Herpetology, 25*, 152–160.

Burger, J., & Zappalorti, R. T. (1992). Philopatry and nesting phenology of pine snakes *Pituophis melanoleucus* in the New Jersey Pine Barrens. *Behavioral Ecology and Sociobiology, 30*, 331–336.

Burghardt, G. M. (1970a). Chemical perception in reptiles. In J. W. Johnston, Jr., D. G. Moulton & A. Turk (Eds.), *Communication by chemical signals* (pp. 241–308). New York: Appleton-Century-Crofts.

Burghardt, G. M. (1970b). Defining "communication." In J. W. Johnston, Jr., D. G. Moulton & A. Turk (Eds.), *Communication by chemical signals* (p. 5–18). New York: Appleton-Century-Crofts.

Burghardt, G. M. (1973). Instinct and innate behavior: Toward an ethological psychology. In J. A. Nevin & G. S. Reynolds (Eds.), *The study of behavior: Learning, motivation, emotion, and instinct* (pp. 322–400). Glenview, IL: Scott, Foresman.

Burghardt, G. M. (1975). Chemical prey preference polymorphism in newborn garter snakes, *Thamnophis sirtalis*. *Behaviour, 52*, 202–225.

Burghardt, G. M. (1977a). Learning processes in reptiles. *Biology of the Reptilia, 7*, 555–681.

Burghardt, G. M. (1977b). Of iguanas and dinosaurs: Social behavior and communication in neonate reptiles. *American Zoologist, 17*, 177–190.

Burghardt, G. M. (1978). Behavioral ontogeny in reptiles: Whence, whither, and why. In G. M. Burghardt & M. Bekoff (Eds.), *The development of behavior: Comparative and evolutionary aspects* (pp. 149–174). New York: Garland STPM Press.

Burghardt, G. M. (1980). Behavioral and stimulus correlates of vomeronasal functioning in reptiles: Feeding, grouping, sex, and tongue use. In D. Müller-Schwarze & R. M. Silverstein (Eds.), *Chemical signals: Vertebrates and aquatic invertebrates* (pp. 275–301). Boston, MA: Springer.

Burghardt, G. M. (1983). Aggregation and species discrimination in newborn snakes. *Zeitschrift für Tierpsychologie, 61,* 89-101.

Burghardt, G. M. (1984). On the origins of play. In P. K. Smith (Ed.), *Play in animals and humans* (pp. 5-41). London: Basil Blackwell.

Burghardt, G. M. (Ed.) (1985). *Foundations of comparative ethology.* New York: Van Nostrand Reinhold.

Burghardt, G. M. (1988). Precocity, play, and the ectotherm-endotherm transition: Superficial adaptation or profound reorganization? In E. M. Blass (Ed.), *Handbook of behavioral neurobiology* (Vol. 9, pp. 107-148). New York: Plenum.

Burghardt, G. M. (1990). Chemically mediated predation in vertebrates: Diversity, ontogeny, and information. In D. W. McDonald, D. Müller-Schwarze & S. E. Natynczuk (Eds.), *Chemical signals in vertebrates* (Vol. 5, pp. 475-499). Oxford: Oxford University Press.

Burghardt, G. M. (1997). Amending Tinbergen: A fifth aim for ethology. In R. W. Mitchell, N. S. Thompson & H. L. Miles (Eds.), *Anthropomorphism, anecdotes, and animals* (pp. 254-276). Albany, NY: SUNY Press.

Burghardt, G. M. (1998a). The evolutionary origins of play revisited: Lessons from turtles. In M. Bekoff & J. A. Byers (Eds.), *Animal play: Evolutionary, comparative, and ecological perspectives* (pp. 1-26). New York: Cambridge University Press.

Burghardt, G. M. (1998b). Snake stories: From the additive model to ethology's fifth aim. In L. Hart (Ed.), *Responsible conduct of research in animal behavior* (pp. 77-95). Oxford: Oxford University Press.

Burghardt, G. M. (2000). Staying close. In M. Bekoff (Ed.), *The smile of a dolphin: Remarkable accounts of animal emotions* (pp. 162-165). New York: Random House/Discovery Books.

Burghardt, G. M. (2004). Iguana research: Looking back and looking forward. In A. C. Alberts, R. L. Carter, W. K. Hayes & E. P. Martins (Eds.), *Iguanas: Biology and conservation* (pp. 1-12). Berkeley: University of California Press.

Burghardt, G. M. (2005). *The genesis of animal play: Testing the limits.* Cambridge, MA: MIT Press.

Burghardt, G. M. (2013). Environmental enrichment and cognitive complexity in reptiles and amphibians: Concepts, review and implications for captive populations. *Applied Animal Behaviour Science, 147,* 286-298.

Burghardt, G. M. (2014). A brief glimpse at the long evolutionary history of play. *Animal Behavior and Cognition, 1,* 90-98.

Burghardt, G. M. (2015). Play in fishes, frogs, and reptiles. *Current Biology, 25,* R9-R10.

Burghardt, G. M. (2020). Insights found in century-old writings on animal behaviour and some cautions for today. *Animal Behaviour.* https://doi.org/10.1016/j.anbehav.2020.02.010.

Burghardt, G. M., Allen, B. A., & Frank, H. (1986). Exploratory tongue flicking by green iguanas in laboratory and field. In D. Duvall, D. Müller-Schwarze & R. M. Silverstein (Eds.), *Chemical signals in vertebrates* (Vol. 4, pp. 305-321). New York: Plenum Press.

Burghardt, G. M., Bartmess-LeVasseur, J. N., Browning, S. A., Morrison, K. E., Stec, C. L., Zachau, C. E., & Freeberg, T. M. (2012). Minimizing observer bias in behavioral studies: A review and recommendations. *Ethology*, *118*, 511–517.

Burghardt, G. M., & Bowers, R. I. (2017). From instinct to behavior systems: An integrated approach to ethological psychology. In J. Call, G. M. Burghardt, I. M. Pepperberg, C. T. Snowdon, & T. Zentall (Eds.), *APA handbook of comparative psychology: Vol 1. Basic Concepts, Methods, Neural Substrate, and Behavior* (pp. 333–364). Washington, DC: American Psychological Association.

Burghardt, G. M., Chiszar, D., Murphy, J., Romano, J., Walsh, T., & Manrod, J. (2002). Behavioral diversity, complexity, and play in Komodo dragons. In J. Murphy, C. Ciofi, C. de La Panouse & T. Walsh (Eds.), *Komodo dragons: Biology and conservation* (pp. 78–117). Washington, DC: Smithsonian Press.

Burghardt, G. M., & Denny, D. (1983). Effects of prey movement and prey odor on feeding in garter snakes. *Zeitschrift für Tierpsychologie*, *62*, 329–347.

Burghardt, G. M., Greene, H. W., & Rand, A. S. (1977). Social behavior in hatchling green iguanas: Life at a reptile rookery. *Science*, *195*, 689–691.

Burghardt, G. M., & Layne, D. G. (1995). Effects of ontogenetic processes and rearing conditions. In C. Warwick, F. L. Frye, & J. B. Murphy (Eds.), *Health and welfare of captive reptiles* (pp. 165–185). London: Chapman & Hall.

Burghardt, G. M., & Layne-Colon, D. G. (2021). Effects of ontogeny, rearing conditions, and individual differences on reptile behaviour: Welfare, conservation, and invasive species implications. In C. Warwick, P. C. Arena, & G. M. Burghardt (Eds.), *Health and Welfare of Captive Reptiles* (2nd ed.). Cham, Switzerland: Springer Nature Switzerland AG.

Burghardt, G. M., & Milostan, M. (1995). Ethological studies on reptiles and amphibians: Lessons for species survival. In J. Demarest, B. Durrant & E. Gibbons (Eds.), *Captive conservation of endangered species: An interdisciplinary approach* (pp. 187–203). New York: SUNY Press.

Burghardt, G. M., & Rand, A. S. (1980). *Dragon of the trees: The green iguana.* Photography and script for 28 min. color sound 16 mm film. Office of Telecommunications, Smithsonian Institution.

Burghardt, G. M., & Rand, A. S. (Eds.). (1982). *Iguanas of the world: Their behavior, ecology, and conservation.* Park Ridge, NJ: Noyes Publications.

Burghardt, G. M., & Rand, A. S. (1985). Group size and growth rate in hatchling green iguanas (*Iguana iguana*). *Behavioral Ecology and Sociobiology*, *18*, 101–104.

Burghardt, G. M., & Schwartz, J. M. (1999). Geographic variations on methodological themes from comparative ethology: A Natricine snake perspective. In S. A. Foster & J. A. Endler (Eds.), *Geographic variation in behavior: Perspectives on evolutionary mechanisms* (pp. 69–94). New York: Oxford University Press.

Burghardt, G. M., Ward, B., & Rosscoe, R. (1996). Problem of reptile play: Environmental enrichment and play behavior in a captive Nile soft-shelled turtle, *Trionyx triunguis*. *Zoo Biology*, *15*, 223–238.

Burke, V. J., Rathbun, S. L., Bodie, J. R., & Gibbons, J. W. (1998). Effect of density on predation rate for turtle nests in a complex landscape. *Oikos*, *83*, 3–11.

Burkhardt, R. W., Jr. (2005). *Patterns of behavior.* Chicago: University of Chicago Press.

Bury, R. B. (1979). Population ecology of freshwater turtles. In M. Harless & H. Morlock (Eds.), *Turtles: Perspectives and research* (pp. 561–572). New York: John Wiley & Sons.

Bury, R. B., & Wolfheim, J. H. (1973). Aggression in free-living pond turtles (*Clemmys marmorata*). *BioScience, 23,* 659–662.

Bush, J. M., Quinn, M. K. M., Balreira, E. C., & Johnson, M. A. (2016). How do lizards determine dominance? Applying ranking algorithms to animal social behaviour. *Animal Behaviour, 118,* 65–74.

Bush, J. M., & Simberloff, D. (2018). A case for anole territoriality. *Behavioral Ecology and Sociobiology, 72,* 111. https://doi.org/10.1007/s00265-018-2522-6.

Bushar, L. M., Reinert, H. K., & Gelbert, L. (1998). Genetic variation and gene flow within and between local populations of the timber rattlesnake, *Crotalus horridus. Copeia, 1998,* 411–422.

Bustard, H. R. (1965). Observations on Australian geckos. *Herpetologica, 21,* 294–302.

Bustard, H. R. (1967). Mechanism of nocturnal emergence from the nest in green turtle hatchlings. *Nature, 214,* 317.

Bustard, H. R. (1969). Defensive display behavior in the bandy-bandy, *Vermicella annulata* (Serpentes: Elapidae). *Herpetologica, 25,* 319–320.

Bustard, H. R. (1972). *Sea turtles: Natural history and conservation.* New York: Taplinger.

Bustard, H. R. (1979). Defensive mechanisms and predation in gekkonid lizards. *British Journal of Herpetology, 6,* 9–11.

Butler, B. O., & Graham, T. E. (1995). Early post-emergent behavior and habitat selection in hatchling Blanding's turtles, *Emydoidea blandingii,* in Massachusetts. *Chelonian Conservation Biology, 1,* 187–196.

Calderon-Espinosa, M., Jerez, A., & Rangel, G. F. M. (2018). Living in the extremes: The thermal ecology of communal nests of the highland Andean lizard *Anadia bogotensis* (Squamata: Gymnophthalmidae). *Current Herpetology, 37,* 158–171.

Caldwell, J. P. (1986). Selection of egg deposition sites: A seasonal shift in the southern leopard frog, *Rana sphenocephala. Copeia, 1986,* 249–253.

Calsbeek, R., & Bonneaud, C. (2008). Postcopulatory fertilization bias as a form of cryptic sexual selection. *Evolution, 62,* 1137–1148.

Campbell, A. L., Naik, R. R., Sowards, L., & Stone, M. O. (2002). Biological infrared imaging and sensing. *Micron, 33,* 211–225.

Campbell, H. W. (1973). Observations on the acoustic behaviour of crocodilians. *Zoologica, 58,* 1–11.

Campos, Z., Sanaiotti, T., Muniz, F., Farias, I., & Magnusson, W. E. (2012). Parental care in the dwarf caiman, *Paleosuchus palpebrosus* Cuvier, 1807 (Reptilia: Crocodilia: Alligatoridae). *Journal of Natural History, 46,* 2979–2984.

Cannon, M., Carpenter, R., & Ackerman, R. (1986). Synchronous hatching and oxygen consumption of Darwin's rhea eggs (*Pterocnemia pennata*). *Physiological Zoology, 59,* 95–108.

Cantwell, L. R., & Forrest, T. G. (2013). Response of *Anolis sagrei* to acoustic calls from predatory and nonpredatory birds. *Journal of Herpetology, 47,* 293–298.

Capula, M., & Luiselli, L. (1997). A tentative review of sexual behavior and alternative reproductive strategies of the Italian colubrid snakes. *Herpetozoa, 10*, 107–119.

Carey, M. J., Phillips, R. A., Silk, J. R., & Shaffer, S. A. (2014). Trans-equatorial migration of short-tailed shearwaters revealed by geolocators. *Emu-Austral Ornithology, 114*, 352–359.

Carothers, J. H. (1981). Dominance and competition in an herbivorous lizard. *Behavioral Ecology and Sociobiology, 8*, 261–266.

Carpenter, C. C. (1966). Miscellaneous notes on Galapagos lava lizards (*Tropidurus*: Iguanidae). *Herpetologica, 26*, 377–386.

Carpenter, C. C. (1977). Communication and displays of snakes. *American Zoologist, 17*, 217–223.

Carpenter, C. C. (1978). Comparative display behavior in the genus *Sceloporus* (Iguanidae). *Contributions in Biology and Geology to the Milwaukee Public Museum, 18*, 1–71.

Carpenter, C. C. (1982). The aggressive displays of iguanine lizards. In G. M. Burghardt & A. S. Rand (Eds.), *Iguanas of the world: Their behavior, ecology, and conservation* (pp. 215–331). Park Ridge, NJ: Noyes Publications.

Carpenter, C. C. (1984). Dominance in snakes. *University of Kansas Museum of Natural History Special Publications, 10*, 195–202.

Carpenter, C. C. (1986). An inventory of the display-action-patterns in lizards. *Smithsonian Herpetological Information Service, 68*, 1–18.

Carpenter, C. C., Badham, J. A., & Kimble, B. (1970). Behavior patterns of three species of *Amphibolurus* (Agamidae), *Copeia, 1970*, 497–505.

Carpenter, C. C., & Ferguson, G. W. (1977). Variation and evolution of stereotyped behavior in reptiles. *Biology of the Reptilia, 7*, 335–554.

Carpenter, C. C., Murphy, J. B., & Mitchell, L. A. (1978). Combat bouts with spur use in the Madagascan boa (*Sanzinia madagascariensis*). *Herpetologica, 34*, 207–212.

Carpenter, K. (1999). *Eggs, nests, and baby dinosaurs: A look at dinosaur reproduction*. Bloomington: Indiana University Press.

Carpenter, K. & Alf, K. (1994). Global distribution of dinosaur eggs, nests, and babies. In K. Carpenter, K. F. Hirsch & J. R. Horner (Eds.), *Dinosaur eggs and babies* (pp. 15–30). Cambridge, UK: Cambridge University Press.

Carr, A. (1967). *So excellent a fishe: A natural history of sea turtles*. Garden City, NY: American Museum of Natural History.

Carr, A., & Hirth, H. (1961). Social facilitation in green turtle siblings. *Animal Behaviour, 9*, 68–70.

Carter, D. B. (1999). Nesting and evidence of parental care by the lace monitor *Varanus varius*. *Mertensiella, 11*, 137–147.

Cartland, L. K., Cree, A., Sutherland, W. H. F., Grimmond, N. M., & Skeaff, C. M. (1994). Plasma concentrations of total cholesterol and triacylglycerol in wild and captive juvenile tuatara (*Sphenodon punctatus*). *New Zealand Journal of Zoology, 21*, 399–406.

Carvajal-Orcampo, V., Angel-Villejo, M. C., Gutierrez-Cardenas, P. D., Ospina-Bautista, F., & Varon, J. V. (2019). A case of communal egg-laying of *Gonatodes albogularis* (Sauria, Sphaerodactylidae) in bromeliads (Poales, Bromiliaceae). *Herpetozoa, 32*, 45–49.

Case, T. J. (1982). Ecology and evolution of the insular gigantic chuckwallas, *Sauromalus hispidus* and *Sauromalus varius*. In G. M. Burghardt & A. S. Rand (Eds.), *Iguanas of the world: Their behavior, ecology, and conservation* (pp. 184–212). Park Ridge, NJ: Noyes Publications.

Castanet, J., Newman, D. G., & Saint Girons, H. (1988). Skeletochronological data on the growth, age, and population structure of the tuatara, *Sphenodon punctatus*, on Stephens and Lady Alice Islands, New Zealand. *Herpetologica, 44*, 25–37.

CBC. (2019). The power of play. *The nature of things*. https://www.cbc.ca /natureofthings/episodes/the-power-of-play.

Chabert, T., Colin, A., Aubin, T., Shacks, V., Bourquin, S. L., Elsey, R. M., Acosta, J. G., & Mathevon, N. (2015). Size does matter: Crocodile mothers react more to the voice of smaller offspring. *Scientific Reports, 5*, 15547.

Chance, M., & Jolly, C. (1970). *Social groups of monkeys, apes and men*. New York: E. P. Dutton.

Chapple, D. G. (2003). Ecology, life-history, and behavior in the Australian scincid genus *Egernia*, with comments on the evolution of complex sociality in lizards. *Herpetological Monographs, 17*, 145–180.

Chapple, D. G., & Keogh, J. S. (2005). Complex mating system and dispersal patterns in a social lizard, *Egernia whitii*. *Molecular Ecology, 14*, 1215–1227.

Chapple, D. G., & Keogh, J. S. (2006). Group structure and stability in social aggregations of White's skink, *Egernia whitii*. *Ethology, 112*, 247–257.

Charruau, P., & Hénaut, Y. (2012). Nest attendance and hatchling care in wild American crocodiles (*Crocodylus acutus*) in Quintana Roo, Mexico. *Animal Biology, 62*, 29–51.

Cheetham, E., Doody, J. S., Stewart, B., & Harlow, P. (2011). Embryonic mortality as a cost of communal nesting in the delicate skink. *Journal of Zoology, London, 283*, 234–242.

Chelazzi, G., & Delfino, G. (1986). A field test on the use of olfaction in homing by *Testudo hermanni* (Reptilia: Testudinidae). *Journal of Herpetology, 20*, 451–455.

Chen, J., Jono, T., Cui, J., Yue, X., & Tang, Y. (2016). The acoustic properties of low intensity vocalizations match hearing sensitivity in the webbed-toed gecko, *Gekko subpalmatus*. *PLoS ONE, 11*, e0146677.

Chen, Q., Liu, Y., Brauth, S. E., Fang, G., & Tang, Y. (2017). The thermal background determines how the infrared and visual systems interact in pit vipers. *Journal of Experimental Biology, 220*, 3103–3109.

Chinsamy, A., Chiappe, L. M., Marugán-Lobón, J., Gao, C., & Zhang, F. (2013). Gender identification of the Mesozoic bird *Confuciusornis sanctus*. *Nature Communications, 4*, 1381.

Chiszar, D., Simonsen, L., Radcliffe, C., & Smith, H. M. (1979). Rate of tongue flicking by cottonmouths (*Agkistrodon piscivorus*) during prolonged exposure to various food odors, and strike-induced chemosensory searching by the cantil (*Agkistrodon bilineatus*). *Transactions of the Kansas Academy of Science, 1903*, 49–54.

Chiszar, D., Smith, H. M., Bogert, C. M., & Vidaurri, J. (1991). A chemical sense of self in timber and prairie rattlesnakes. *Bulletin of the Psychonomic Society, 29*, 153-154.

Chiszar, D., Tomlinson, W. T., Smith, H. M., Murphy, J. B., & Radcliffe, C. W. (1995). Behavioural consequences of husbandry manipulations: Indicators of arousal, quiescence and environmental awareness. In C. Warwick, F. L. Frye & J. B. Murphy (Eds.), *Health and welfare of captive reptiles* (pp. 186-204). London: Chapman & Hall.

Chiszar, D., Welborn, S., Wand, M. A., Scudder, K. M., & Smith, H. M. (1980). Investigatory behavior in snakes, II: Cage cleaning and the induction of defecation in snakes. *Animal Learning and Behavior, 8*, 505-510.

Christens, E., & Bider, J. (1987). Nesting activity and hatching success of the painted turtle (*Chrysemys picta marginata*) in southwestern Quebec. *Herpetologica, 75*, 55-65.

Christensen-Dalsgaard, J., & Carr, C. E. (2008). Evolution of a sensory novelty: Tympanic ears and the associated neural processing. *Brain Research Bulletin, 75*, 365-370.

Christian, K. A., Griffiths, A. D., Bedford, G., & Jenkin, G. (1999). Androgen concentrations and behavior of frillneck lizards (*Chlamydosaurus kingii*). *Journal of Herpetology, 33*, 12-17.

Christian, K. A., & Tracy, C. (1981). The effect of the thermal environment on the ability of hatchling Galapagos land iguanas to avoid predation during dispersal. *Oecologia, 49*, 218-223.

Clark, H., Florio, B., & Hurowitz, R. (1955). Embryonic growth of *Thamnophis s. sirtalis* in relation to fertilization date and placental function. *Copeia, 1955*, 9-13.

Clark, R. W. (2004). Kin recognition in rattlesnakes. *Proceedings of the Royal Society of London. Series B: Biological Sciences, 271*, S243-S245.

Clark, R. W. (2007). Public information for solitary foragers: Timber rattlesnakes use conspecific chemical cues to select ambush sites. *Behavioral Ecology, 18*, 487-490.

Clark, R. W., Brown, W., Stechert, R., & Greene, H. (2012). Cryptic sociality in rattlesnakes (*Crotalus horridus*) detected by kinship analysis. *Biology Letters, 8*, 523-525.

Clark, R. W., Schuett, G. W., Repp, R. A., Amarello, M., Smith, C. F., & Herrmann, H.-W. (2014). Mating systems, reproductive success, and sexual selection in secretive species: A case study of the western diamond-backed rattlesnake, *Crotalus atrox. PLoS ONE, 9*, e90616.

Cloudsley-Thompson, J. (1970). On the biology of the desert tortoise *Testudo sulcata* in Sudan. *Journal of Zoology, 160*, 17-33.

Clusella Trullas, S., Van Wyk, J. H., & Spotila, J. R. (2007). Thermal melanism in ectotherms. *Journal of Thermal Biology, 32*, 235-245.

Cody, M. L. (1966). A general theory of clutch size. *Evolution, 20*, 174-184.

Colbert, E. H., Morales, M., & Minkoff, E. C. (2001). Colbert's evolution of the vertebrates: A history of the backboned animals through time (5th ed.). New York: Wiley-Liss.

Colbert, P., Spencer, R., & Janzen, F. (2010). Mechanism and cost of synchronous hatching. *Functional Ecology*, *24*, 112–121.

Congdon, J., D. Breitenbach, G., van Loben Sels, R., & Tinkle, D. (1987). Reproduction and nesting ecology of snapping turtles (*Chelydra serpentina*) in southeastern Michigan. *Herpetologica*, *43*, 39–54.

Congdon, J. D., & Gatten, R. E. (1989). Movements and energetics of nesting *Chrysemys picta*. *Herpetologica*, *45*, 94–100.

Congdon, J. D., Gatten, R. E., & Morreale, S. J. (1989). Overwintering activity of box turtles (*Terrapene carolina*) in South Carolina. *Journal of Herpetology*, *23*, 179–181.

Congdon, J. D., Tinkle, D., Breitenbach, G., & van Loben Sels, R. (1983). Nesting ecology and hatching success in the turtle *Emydoidea blandingi*. *Herpetologica*, *49*, 417–429.

Constanzo, J. P. (1989). Conspecific scent trailing by garter snakes (*Thamnophis sirtalis*) during autumn: Further evidence for use of pheromones in den location. *Journal of Chemical Ecology*, *15*, 2531–2538.

Cooper, W. E., Jr. (1971). Display behavior of hatchling *Anolis carolinensis*. *Herpetologica*, *27*, 498–500.

Cooper, W. E., Jr., (1993). Tree selection by the broad-headed skink, *Eumeces laticeps*: Size, holes, and cover. *Amphibia-Reptilia*, *14*, 285–294.

Cooper, W. E., Jr., & Burghardt, G. M. (1990a). A comparative analysis of scoring methods for chemical discrimination of prey by squamate reptiles. *Journal of Chemical Ecology*, *16*, 45–65.

Cooper, W. E., Jr., & Burghardt, G. M. (1990b). Vomerolfaction and vomodor. *Journal of Chemical Ecology*, *16*, 103–105.

Cooper, W. E., Jr., Caffrey, C., & Vitt, L. J. (1985). Aggregation in the banded gecko, *Coleonyx variegatus*. *Herpetologica*, *41*, 342–350.

Cooper, W. E., Jr., & Greenberg, N. (1992). Reptilian coloration and behavior. *Biology of the Reptilia*, *18*, 298–422.

Cooper, W. E., Jr., López, P., & Salvador, A. (1994). Pheromone detection by an amphisbaenian. *Animal Behaviour*, *47*, 1401–1411.

Cooper, W. E., & Vitt, L. J. (1997). Maximizing male reproductive success in the broad-headed skink (*Eumeces laticeps*): Preliminary evidence for mate guarding, size-assortative pairing, and opportunistic extra-pair mating. *Amphibia-Reptilia*, *18*, 59–73.

Coppinger, R. P., & Smith, C. K. (1989). A model for understanding the evolution of mammalian behavior. In H. Genoways (Ed.), *Current mammalogy* (Vol. 2, pp. 335–374). New York: Plenum.

Corey, A. C. (1975). *Snake dens of Narcisse*. Winnipeg: Manitoba Fish and Wildlife Enhancement Fund.

Cornelis, G., Funk, M., Vernochet, C., Leal, F., Tarazona, O. A., Meurice, G., . . . Heidmann, T. (2017). An endogenous retroviral envelope syncytin and its cognate receptor identified in the viviparous placental *Mabuya* lizard. *Proceedings of the National Academy of Sciences*, *114*, E10991–E11000.

Cornelius, S. E. (1986). *The sea turtles of Santa Rosa National Park*. San Jose, Costa Rica: Fundacion de Parques Nacionales.

Cornelius, S. E., Ulloa, M. A., Castro, J. C., Mata del Valle, M., & Robinson, D. C. (1991). Management of olive ridley sea turtles (*Lepidochelys olivacea*) nesting at Playas Nancite and Ostional, Costa Rica. *Neotropical Wildlife Use and Conservation, 1*, 111-135.

Cossette, A. P., & Brochu, C. A. (2020). A systematic review of the giant alligatoroid *Deinosuchus* from the Campanian of North America and its implications for the relationships at the root of Crocodylia. *Journal of Vertebrate Paleontology, 40*. https://doi.org/10.1080/02724634.2020.1767638.

Costa, C., de Oliveira, H., Drummond, L., Tonini, R., Filipe, J., & Zaldivar-Rae, J. (2013). *Kentropyx calcarata* (Squamata: Teiidae): Mating behavior in the wild. *North-Western Journal of Zoology, 9*, 198-200.

Costanzo, J. P. (1989). Conspecific scent trailing by garter snakes (*Thamnophis sirtalis*) during autumn: Further evidence for use of pheromones in den location. *Journal of Chemical Ecology, 15*, 2531-2538.

Cott, H. B. (1961). Scientific results of an inquiry into the ecology and economic status of the Nile crocodile (*Crocodilus niloticus*) in Uganda and Northern Rhodesia. *Transactions of the Zoological Society of London, 29*, 211-356.

Couper, P. J. (1996). *Nephrurus asper* (Squamata: Gekkonidae): Sperm storage and other reproductive data. *Memoirs of the Queensland Museum, 39*, 487-487.

Couper, P. J., & Ingram, G. J. (1992). A new species of skink of *Lerista* from Queensland and a re-appraisal of *L. allanae*. *Memoirs of the Queensland Museum, 32*, 55-59.

Covacevich, J., & Limpus, C. (1972). Observations on community egg-laying by the yellow-faced whip snake, *Demansia psammophis* (Schlegel) 1837 (Squamata: Elapidae). *Herpetologica, 3*, 208-210.

Cree A. (1994). Low annual reproductive output in female reptiles from New Zealand. *New Zealand Journal of Zoology, 21*, 351-372.

Cree, A. (2002). Tuatara. In T. Halliday & K. Adler (Eds.), *The new encyclopedia of reptiles and amphibians* (pp. 210-211). Oxford, UK: Oxford University Press.

Cree, A. (2014). *Tuatara: Biology and conservation of a venerable survivor*. Christchurch, NZ: Canterbury University Press.

Cree, A., Daugherty, C. H., & Hay, J. M. (1995). Reproduction of a rare New Zealand reptile, the tuatara *Sphenodon punctatus*, on rat-free and rat-inhabited islands. *Conservation Biology, 9*, 373-383.

Crespi, B. (2001). The evolution of social behavior in microorganisms. *Trends in Ecology & Evolution, 16*, 178-183.

Crews, D., & Fitzgerald, K. T. (1980). "Sexual" behavior in parthenogenetic lizards (*Cnemidophorus*). *Proceedings of the National Academy of Sciences, 77*, 499-502.

Crews, D., Grassman, M., & Lindzey, J. (1986). Behavioral facilitation of reproduction in sexual and unisexual whiptail lizards. *Proceedings of the National Academy of Sciences, 83*, 9547-9550.

Crews, D., & Greenberg, N. (1981). Function and causation of social signals in lizards. *American Zoologist, 21*, 273-294.

Crews, D., & Moore, M. C. (2005). Historical contributions of research on reptiles to behavioral neuroendocrinology. *Hormones and Behavior, 48*, 384-394.

Crews, D., Sanderson, N., & Dias, B. G. (2009). Hormones, brain, and behavior in reptiles. In D. W. Pfaff, A. Arnold, A. Etgen, S. Fahrbach & R. Rubin (Eds.), *Hormones, brain and behavior* (2nd ed., Vol. 2, pp. 771–816). San Diego: Academic Press.

Crivelli, A. (1981). The Dalmatian pelican, *Pelecanus crispus* Bruch 1832, a recently world-endangered bird species. *Biological Conservation, 20*, 297–310.

Crompton, A. W., Owerkowicz, T., Bhullar, B. A., & Musinsky, C. (2017). Structure of the nasal region of non-mammalian cynodonts and mammaliaforms: Speculations on the evolution of mammalian endothermy. *Journal of Vertebrate Paleontology, 37*, e1269116.

Crook, J. H. (Ed.). (1970). *Social behaviour in birds and mammals.* London: Academic Press.

Crowe-Riddell, J. M., Simões, B. F., Partridge, J. C., Hunt, D. M., Delean, S., Schwerdt, J. G., . . . Sanders, K. L. (2019). Phototactic tails: Evolution and molecular basis of a novel sensory trait in sea snakes. *Molecular Ecology, 28*, 2013–2028.

Crump, M. L. (1995). Parental care. *Amphibian Biology, 2*, 518–567.

Cuéllar, O., & Kluge, A. G. (1972). Natural parthenogenesis in the gekkonid lizard *Lepidodactylus lugubris. Journal of Genetics, 61*, 14–26.

Cullen, E. (1957). Adaptations in the kittiwake to cliff-nesting. *Ibis, 99*, 275–302.

Cunnington, G. M., & Cebek, J. E. (2005). Mating and nesting behavior of the eastern hognose snake (*Heterodon platirhinos*) in the northern portion of its range. *American Midland Naturalist, 154*, 474–479.

Currie, P. J. (1998). Possible evidence of gregarious behavior in Tyrannosaurids. *Gaia, 15*, 271–277.

Czech, B., Krausman, P. R., & Borkhataria, R. (1998). Social construction, political power, and the allocation of benefits to endangered species. *Conservation Biology, 12*, 1103–1112.

Czech-Damal, N. U., Liebschner, A., Miersch, L., Klauer, G., Hanke, F. D., Marshall, C., & Hanke, W. (2012). Electroreception in the Guiana dolphin (*Sotalia guianensis*). *Proceedings of the Royal Society of London B: Biological Sciences, 279*, 663–668.

Damas-Moreira, I., Oliveira, D., Santos, J. L., Riley, J. L., Harris, D. J., & Whiting, M. J. (2018). Learning from others: An invasive lizard uses social information from both conspecifics and heterospecifics. *Biology Letters, 14*, 20180532. https://doi.org/10.1098/rsbl.2018.0532.

Danchin, E., & Wagner, R. H. (1997). The evolution of coloniality: The emergence of new perspectives. *Trends in Ecology & Evolution, 12*, 342–347.

Danforth, B. N. (1991). Female foraging and intranest behavior of a communal bee, *Perdita portalis* (Hymenoptera: Andrenidae). *Annals of the Entomological Society of America, 84*, 537–548.

Danielyan, F., Arakelyan, M., & Stepanyan, I. (2008). Hybrids of *Darevskia valentini, D. armeniaca* and *D. unisexualis* from a sympatric population in Armenia. *Amphibia-Reptilia, 29*, 487–504.

Darwin, C. (1839). *Voyages of the Adventure and Beagle*, Volume III. Journal and Remarks. 1832–1836. London: Henry Coburn.

Darwin, C. (1845). *Voyage of the Beagle round the world*. London: Murray.

Darwin, C. (1859). *On the origin of species by means of natural selection*. London: Murray.

Darwin, C. (1871). *The descent of man, and selection in relation to sex*. London: Murray.

Davenport, M. (1995). Evidence of possible sperm storage in the caiman, *Paleosuchus palpebrosus*. *Herpetological Review, 26*, 14–15.

Davis, A. R. (2012). Kin presence drives philopatry and social aggregation in juvenile Desert Night Lizards (*Xantusia vigilis*). *Behavioral Ecology, 23*, 18–24.

Davis, A. R., Corl, A., Surget-Groba, Y., & Sinervo, B. (2011). Convergent evolution of kin-based sociality in a lizard. *Proceedings of the Royal Society of London, 278*, 1507–1514.

Davis, K. M. (2009). *Sociality, cognition and social learning in turtles (Emydidae)*. Unpublished doctoral dissertation. Knoxville: University of Tennessee.

Davis, K. M., & Burghardt, G. M. (2011). Turtles (*Pseudemys nelsoni*) learn about visual cues indicating food from experienced turtles. *Journal of Comparative Psychology, 125*, 404–410.

Davis, L., Glenn, T., Elsey, R., Dessauer, H., & Sawyer, R. (2001). Multiple paternity and mating patterns in the American alligator, *Alligator mississippiensis*. *Molecular Ecology, 10*, 1011–1024.

Davis, M. (1981). Aspects of the social and spatial experience of eastern box turtles *Terrapene carolina carolina*. Unpublished doctoral dissertation. Knoxville: University of Tennessee.

Dayananda, B., & Webb, J. K. (2017). Incubation under climate warming affects learning ability and survival in hatchling lizards. *Biology Letters, 13*, 20170002.

De Braga, M., & Reisz, R. R. (1995). A new diapsid reptile from the uppermost Carboniferous (Stephanian) of Kansas. *Palaeontology, 38*, 199–199.

Deeming, D. C. (1991). Reasons for the dichotomy in egg turning in birds and reptiles. In D. C. Deeming & M. W. Ferguson, M. W. (Eds.), *Egg incubation: Its effects on embryonic development in birds and reptiles* (pp. 307–323). Cambridge, UK: Cambridge University Press.

Deeming, D. C. (Ed.). (2004). *Reptilian incubation: Environment, evolution and behaviour*. Nottingham: Nottingham University Press.

De Fazio, A., Simon, C. A., Middendorf, G. A., & Romano, D. (1977). Iguanid substrate licking: A response to novel situations in *Sceloporus jarrovi*. *Copeia, 1977*, 706–709.

De Haan, C. C. (1999). *Malpolon monspessulanus*. In W. Böhme (Ed.), *Handbuch der Reptilien und Amphibien Europas* (Vol. 3/IIA: Schlangen 2, pp. 661–756). Wiesbaden: Aula-Verlag.

Dehn, M. (1990). Vigilance for predators: Detection and dilution effects. *Behavioral Ecology and Sociobiology, 26*, 337–342.

Deitz, D. C., & Jackson, D. R. (1979). Use of American alligator nests by nesting turtles. *Journal of Herpetology, 13*, 510–512.

Delany, M. F., Woodward, A. R., Kiltie, R. A., & Moore, C. T. (2011). Mortality of American alligators attributed to cannibalism. *Herpetologica, 67*, 174–185.

Delean, S., & Harvey, C. (1984). Notes on the reproduction of *Nephrurus deleani* (Reptilia: Gekkonidae). *Transactions of the Royal Society of South Australia, 108*, 221–222.

del Hoyo, J., Elliott, A., Christie, D. A., & Sargatal, J. E. (Eds.) (1992–2013). *Handbook of the birds of the world*. Barcelona, Spain: Lynx Edicions.

De Lisle, H. F. (1996). *The natural history of monitor lizards*. Malabar, FL: Krieger Publishing Company.

DeNardo, D., Moeller, K. T., Seward, M., & Repp, R. (2018). Evidence for atypical nest overwintering by hatchling lizards, *Heloderma suspectum*. *Proceedings of the Royal Society, B, 285.* https://doi.org/10.1098/rspb.2018.0632.

Deraniyagala, P. (1939). *The tetrapod reptiles of Ceylon*. Colombo: The Colombo Museum.

de Silva, A., Bauer, A. M., Austin, C. C., Goonewardene, S., Hawke, Z., Vanneck, V., . . . Goonasekera, M. M. (2004). Distribution and natural history of *Calodactylodes illingworthorum* (Reptilia: Gekkonidae) in Sri Lanka: Preliminary findings. *Lyriocephalus, 5,* 192–198.

de Silva, A., Bauer, A. M., Austin, C.C., Perera, B. J. K, Jayaratne, R. L., Goonasekera, M. M., . . . Drion, A. (2004). Some cultural traits of the inhabitants of the Nilgala fire savannah, Sri Lanka, towards animals: With special reference to the herpetofauna. *Lyriocephalus, 5,* 183–191.

de Sousa, P., & Freire, E. (2010). Communal nests of *Hemidactylus mabouia* (Moreau de Jonnès, 1818) (Squamata: Gekkonidae) in a remnant of Atlantic Forest in northeastern Brazil. *Biotemas, 23,* 231–234.

Dewsbury, D. (1984a). *Comparative psychology in the 20th century*. Stroudsburg, PA: Hutchinson Ross.

Dewsbury, D. (Ed.). (1984b). *Foundations of comparative psychology*. New York: Van Nostrand Reinhold.

Diamond, A. (1976). Breeding biology and conservation of hawksbill turtles, *Eretmochelys imbricata* L., on Cousin Island, Seychelles. *Biological Conservation, 9,* 199–215.

Dias, W. A., Iori, F. V., Ghilardi, A. M., & Fernandes, M. A. (2020). The pterygoid region and cranial airways of *Caipirasuchus paulistanus* and *Caipirasuchus montealtensis* (Crocodyliformes, Sphagesauridae), from the Upper Cretaceous Adamantina Formation, Bauru Basin, Brazil. *Cretaceous Research, 106,* 104192.

Díaz, V. D., Suberbiola, X. P., & Sanz, J. L. (2012). Juvenile and adult teeth of the titanosaurian dinosaur *Lirainosaurus* (Sauropoda) from the Late Cretaceous of Iberia. *Geobios, 45,* 265–274.

Diegert, C. F., & Williamson, T. E. (1998). A digital acoustic model of the lambeosaurine hadrosaur *Parasaurolophus tubicen*. *Journal of Vertebrate Paleontology, 18* (3, Suppl.), 38A.

Dieter, C. T. (2000). *The ultimate guide to crocodilians in captivity*. Darwin: Crocodile Encounter.

Dinets, V. (2010). Nocturnal behavior of the American alligator (*Alligator mississippiensis*) in the wild during the mating season. *Herpetological Bulletin, 111,* 4–11.

Dinets, V. (2011a). *Crocodylus palustris* (mugger crocodile) signaling behavior. *Herpetological Review, 42,* 424.

Dinets, V. (2011b). *Crocodylus rhombifer* (Cuban crocodile) courtship behavior. *Herpetological Review, 42,* 232.

Dinets, V. (2011c). Effects of aquatic habitat continuity on signal composition in crocodilians. *Animal Behaviour, 82*, 191-201.

Dinets, V. (2011d). *Melanosuchus niger* (black caiman) signaling behavior. *Herpetological Review, 42*, 424.

Dinets, V. (2013a). Do individual crocodilians adjust their signaling to habitat structure? *Ethology Ecology & Evolution, 25*, 174-184.

Dinets, V. (2013b). *Dragon songs: Love and adventure among crocodiles, alligators and other dinosaur relations.* New York: Arcade.

Dinets, V. (2013c). Long-distance signaling in extant crocodilians. *Copeia, 2013*, 517-526.

Dinets, V. (2013d). Underwater sound locating ability in the American alligator (*Alligator mississippiensis*). *Journal of Herpetology, 47*, 521-523.

Dinets, V. (2014). Why avian systematics are no longer scientific. *South Dakota Bird Notes, 66*, 26-29.

Dinets, V. (2015). Apparent coordination and collaboration in cooperatively hunting crocodilians. *Ethology Ecology & Evolution, 27*, 244-250.

Dinets, V. (2017). Coordinated hunting by Cuban boas. *Animal Behavior and Cognition, 4*, 24-29.

Dinets, V., Brueggen, J. C., & Brueggen, J. D. (2014). Crocodilians use tools for hunting. *Ethology Ecology & Evolution, 27*, 74-78.

Ditmars, R. L. (1907). *The reptile book: A comprehensive, popularised work on the structure and habits of the turtles, tortoises, crocodilians, lizards and snakes which inhabit the United States and Northern Mexico.* New York: Doubleday.

Donihue, C. M., Herrel, A., Martín, J., Foufopoulos, J., Pafilis, P., & Baeckens, S. (2020). Rapid and repeated divergence of animal chemical signals in an island introduction experiment. *Journal of Animal Ecology, 89*, 1458-1467.

Doody, J. S. (1995). A comparative nesting study of two syntopic species of softshell turtles (*Apalone mutica* and *Apalone spinifera*) in southcentral Louisiana [Unpublished Master's thesis], Southeastern Louisiana University.

Doody, J. S. (2011). Environmentally cued hatching in reptiles. *Integrative and Comparative Biology, 51*, 49-61.

Doody, J. S. (2019). *Intellagama lesueurii* (Water Dragon), *Acritoscincus duperreyi* (Three-lined Skink), *Lampropholis guichenoti* (Garden Skink). Communal nesting. *Herpetological Review, 50*, 141.

Doody, J. S., Burghardt, G. M., & Dinets, V. (2013). Breaking the social-nonsocial dichotomy: A role for reptiles in vertebrate social behavior research? *Ethology, 119*, 95-103.

Doody, J. S., Castellano, C., Green, B., Rhind, D., Sims, R., & Robinson, T. (2009). Population declines in Australian predators caused by an invasive species. *Animal Conservation, 12*, 46-53.

Doody, J. S., Castellano, C., Rakotondrainy, R., Ronto, W., Rakotondriamanga, T., Duchene, J., & Randria, Z. (2011). Aggregated drinking behavior of radiated tortoises (*Astrochelys radiata*) in arid southwestern Madagascar. *Chelonian Conservation and Biology, 10*, 145-146.

Doody, J. S., Castellano, C. M., Rhind, D., & Green, B. (2013). Indirect facilitation of a native mesopredator by an invasive species: Are cane toads re-shaping tropical riparian communities? *Biological Invasions, 15*, 559–568.

Doody, J. S., Coleman, K. E., Coleman, L., & Stephens, G. (2017). *Anolis sagrei* (Brown Anole). Communal nesting. *Herpetological Review, 48*, 841–842.

Doody, J. S., Coleman, K. E., Coleman, L., & Stephens, G. (2019). *Anolis sagrei.* Environmentally cued hatching. *Herpetological Review, 50*, 136–137.

Doody, J. S., Ellis, R., & Rhind, D. (2015). *Gehyra australis* (tree dtella) and *Gehyra pilbara* (Pilbara dtella): Environmentally cued hatching. *Herpetological Review, 46*, 257–258.

Doody, J. S, Freedberg, S., & Keogh, J. S. (2009). Communal egg-laying in reptiles and amphibians: Evolutionary patterns and hypotheses. *The Quarterly Review of Biology, 84*, 229–252.

Doody, J. S., Georges, A., & Young, J. E. (2003). Twice every second year: Reproduction in the pig-nosed turtle, *Carettochelys insculpta*, in the wet-dry tropics of Australia. *Journal of Zoology, London, 259*, 179–188.

Doody, J. S., Georges, A., & Young, J. E. (2004). Determinants of reproductive success and offspring sex in a turtle with environmental sex determination. *Biological Journal of the Linnean Society, 81*, 1–16.

Doody, J. S., Georges, A., Young, J. E., Pauza, M. D., Pepper, A. L., Alderman, R. L., & Welsh, M. A. (2001). Embryonic aestivation and emergence behaviour in the pig-nosed turtle, *Carettochelys insculpta. Canadian Journal of Zoology, 79*, 1062–1072.

Doody, J. S., Green, B., Rhind, D., Castellano, C. M., Sims, R., & Robinson, T. (2009). Population-level declines in Australian predators caused by an invasive species. *Animal Conservation, 12*, 46–53.

Doody, J. S., James, H., Colyvas, K., McHenry, C., & Clulow, S. (2015). Deep nesting in a lizard déjà vu devil's corkscrews: First helical reptile burrow and deepest reptile nest. *Biological Journal of the Linnean Society, 116*, 13–26.

Doody, J. S., James, H., Ellis, R., Gibson, N., Raven, M., Mahoney, S., . . . McHenry, C. R. (2014). Cryptic and complex nesting in the yellow-spotted monitor, *Varanus panoptes. Journal of Herpetology, 48*, 363–370.

Doody, J. S., McHenry, C., Brown, M., Canning, G., Vas, G., & Clulow, S. (2018). Deep, helical, communal nesting and emergence in the sand monitor: Ecology informing paleoecology? *Journal of Zoology, 305*, 88–95.

Doody, J. S., McHenry, C. R., Durkin, L., Brown, M., Simms, A., Coleman, L., & Clulow, S. (2018). Deep communal nesting by yellow-spotted monitors in a desert ecosystem: Indirect evidence for a response to extreme dry conditions. *Herpetologica, 74*, 306–310.

Doody, J. S., & Paull, P. (2013). Hitting the ground running: Environmentally cued hatching in a lizard. *Copeia, 2013*, 160–165.

Doody, J. S., Rhind, D., Green, B., Castellano, C., McHenry, C., & Clulow, S. (2017). Chronic effects of an invasive species on an animal community. *Ecology, 98*, 2093–2101.

Doody, J. S., & Schembri, B. (2014a). *Acritoscincus platynota* (red-throated skink): Environmentally cued hatching. *Herpetological Review, 45*, 693.

Doody, J. S., & Schembri, B. (2014b). *Carlia schmeltzii* (robust rainbow skink): Environmentally cued hatching. *Herpetological Review, 45,* 494.

Doody, J. S., Schembri, B., & Sweet, S. (2015). *Varanus pilbarensis* (Pilbara rock monitor) and *V. glauerti* (Kimberley rock monitor). Tail display behavior. *Herpetological Review, 46,* 439-440.

Doody, J. S., Sims, R., & Georges, A. (2001). Thermal spring use by pig-nosed turtles (*Carettochelys insculpta*) in tropical Australia. *Chelonian Conservation and Biology, 4,* 81-87.

Doody, J. S., Sims, R. A., & Georges, A. (2003). Gregarious behavior of nesting turtles (*Carettochelys insculpta*) does not reduce nest predation risk. *Copeia, 2003,* 894-898.

Doody, J. S., Soanes, R., Castellano, C. M., Rhind, D., Green, B., McHenry, C. R., & Clulow, S. (2015). Invasive toads shift predator-prey densities in animal communities by removing top predators. *Ecology, 96,* 2544-2554.

Doody, J. S., Stewart, B., Camacho, C., & Christian, K. (2012). Good vibrations? Sibling embryos expedite hatching in a turtle. *Animal Behaviour, 83,* 645-651.

Doody, J. S., Trembath, D., Davies, C. Freier, D., Stewart, B., & Kay G. (2010). A mating aggregation of olive pythons, *Liasis olivaceous,* with comments on mating systems and sexual dimorphism in snakes. *Herpetofauna, 39,* 92-94.

Doody, J. S., & Welsh, M. (2005). First glimpses into the ecology of the red-faced turtle, *Emydura victoriae,* in tropical Australia. *Herpetofauna, 35,* 11.

Doody, J. S., & Young, J. E. (1995). Temporal variation in reproduction and clutch mortality of southern leopard frogs (*Rana utricularia*) in south Mississippi. *Journal of Herpetology, 29,* 614-616.

Doody, J. S., Zavala, G. E., Osterman, E., Coleman, L., and Clulow, S. (2018). *Dendrophidion clarkii* and *Atractus dunni.* Communal nesting. *Herpetological Review, 49,* 339-340.

Dooling, R. J., & Popper, A. N. (2000). Hearing in birds and reptiles: An overview. In R. J. Dooling & R. R. Fay (Eds.), *Comparative hearing: Birds and reptiles* (pp. 1-12). New York: Springer.

Dorcas, M. E., & Peterson, C. R. (1998). Daily body temperature variation in free-ranging rubber boas. *Herpetologica, 54,* 88-103.

Doughty, P., Sinervo, B., & Burghardt, G. M. (1994). Sex-biased dispersal in a polygynous lizard, *Uta stansburiana. Animal Behaviour, 47,* 227-229.

Drumheller, S. K., Darlington, J., & Vliet, K. A. (2019). Surveying the death roll behavior across Crocodylia. *Ethology, Ecology & Evolution, 31,* 329-347.

Drumheller, S. K., McHugh, J. B., Kane, M., Riedel, A., & D'Amore, D. C. (2020). High frequencies of theropod bite marks provide evidence for feeding, scavenging, and possible cannibalism in a stressed Late Jurassic ecosystem. *PLoS ONE, 15,* e0233115.

Drummond, H., & Burghardt, G. M. (1982). Orientation in dispersing hatchling green iguanas (*Iguana iguana*). In G. M. Burghardt & A. S. Rand (Eds.) *Iguanas of the world: Their behavior, ecology, and conservation* (pp. 271-291). Park Ridge, NJ: Noyes Publications.

Drummond, H., & Burghardt, G. M. (1983). Nocturnal and diurnal nest emergence in green iguanas. *Journal of Herpetology, 17,* 290-292.

Dubach, J., Sajewicz, A., & Pawley, R. (1997). Parthenogenesis in the Arafuran file snake (*Acrochordus arafurae*). *Herpetological Natural History*, *5*, 11-18.

Duda, P. L., & Gupta, V. K. (1981). Courtship and mating behaviour of the Indian soft shell turtle, *Lissemys punctata punctata*. *Proceedings: Animal Sciences*, *90*, 453-461.

Duffield, G., & Bull, M. (2002). Stable social aggregations in an Australian lizard, *Egernia stokesii*. *Naturwissenschaften*, *89*, 424-427.

Dugan, B. A. (1982a). A field study of the headbob displays of male green iguanas (*Iguana iguana*): Variation in form and context. *Animal Behaviour*, *30*, 327-338.

Dugan, B. A. (1982b). The mating behavior of the green iguana, *Iguana iguana*. In G. M. Burghardt & A. S. Rand (Eds.), *Iguanas of the world: Their behavior, ecology, and conservation* (pp. 320-341). Park Ridge, NJ: Noyes Publications.

Dugan, B. A., & Wiewandt, T. (1982). Socio-ecological determinants of mating strategies in iguanine lizards. In G. M. Burghardt & A. S. Rand (Eds.), *Iguanas of the world: Their behavior, ecology, and conservation* (pp. 303-319). Park Ridge, NJ: Noyes Publications.

Dugas-Ford, J., Rowell, J. J., & Ragsdale, C. W. (2012). Cell-type homologies and the origins of the neocortex. *Proceedings of the National Academy of Sciences*, *109*, 16974-16979.

Dundee, H. A., & Miller, M. C. (1968). Aggregative behavior and habitat conditioning by the prairie ringneck snake, *Diadophis punctatus arnyi*. *Tulane Studies in Zoology*, *15*, 41-58.

Dunson, W. A. (1982). Low water vapor conductance of hard-shelled eggs of the gecko lizards *Hemidactylus* and *Lepidodactylus*. *Journal of Experimental Zoology*, *219*, 377-379.

Dunson, W. A., & Bramham, C. R. (1981). Evaporative water loss and oxygen consumption of three small lizards from the Florida Keys: *Sphaerodactylus cinereus*, *S. notatus*, and *Anolis sagrei*. *Physiological Zoology*, *54*, 253-259.

Dunson, W. A., & Ehlert, G. (1971). Effects of temperature, salinity and surface water flow on distribution of the sea snake *Pelamis*. *Limnology and Oceanography*, *16*, 845-853.

Duvall, D. (1982), Western fence lizard (*Sceloporus occidentalis*) chemical signals. III. An experimental ethogram of conspecific body licking. *Journal of Experimental Zoology*, *221*, 23-26.

Duvall, D., Graves, B. M., & Carpenter, G. C. (1987). Visual and chemical composite signaling effects of *Sceloporus* lizard fecal boli. *Copeia*, *1987*, 1028-1031.

Duvall, D., King, M. B., & Graves, B. M. (1983). Fossil and comparative evidence for possible chemical signaling in the mammal-like reptiles. In D. Müller-Schwarze & R. M. Silverstein (Eds.), *Chemical signals in vertebrates* (Vol. 3, 25-36). New York: Plenum Press.

Duvall, D., Schuett, G. W., & Arnold, S. J. (1993). Ecology and evolution of snake mating systems. In R. A. Seigel and J. T. Collins (Eds.), *Snakes: Ecology and behavior* (pp. 165-200). McGraw-Hill, New York.

Dyke, G., Vremir, M., Kaiser, G., & Naish, D. (2012). A drowned Mesozoic bird breeding colony from the Late Cretaceous of Transylvania. *Naturwissenschaften*, *99*, 435-442.

Dyson, M. L., Klump, G. M., & Gauger, B. (1998). Absolute hearing thresholds and critical masking ratios in the European barn owl: A comparison with other owls. *Journal of Comparative Physiology A*, *182*, 695-702.

Earley, R. L., Attum, O., & Eason, P. (2002). Varanid combat: Perspectives from game theory. *Amphibia-Reptilia*, *23*, 469-486.

Eckert, S. A. (2002). Swim speed and movement patterns of gravid leatherback sea turtles (*Dermochelys coriacea*) at St Croix, US Virgin Islands. *Journal of Experimental Biology*, *205*, 3689-3697.

Eckrich, C. E., & Owens, D. W. (1995). Solitary versus arribada nesting in the olive ridley sea turtles (*Lepidochelys olivacea*): A test of the predator-satiation hypothesis. *Herpetologica*, *51*, 349-354.

Eglis, A. (1962). Tortoise behavior: A taxonomic adjunct. *Herpetologica*, *18*, 1-8.

Ehrenfeld, D. W. (1979). Behavior associated with nesting. In M. Harless & H. Morlock (Eds.), *Turtles: Perspectives and research* (pp. 417-434). New York: Wiley and Sons.

Ehrenfeld, J. G., & Ehrenfeld, D. W. (1973). Externally secreting glands of freshwater and sea turtles. *Copeia*, *1973*, 305-314.

Eifler, D. (2001). *Egernia cunninghami* (Cunningham's skink). Escape behavior. *Herpetological Review*, *32*, 40.

Eifler, D., Eifler, M., Malela, K., & Childers, J. (2016). Social networks in the Little Scrub Island ground lizard (*Ameiva corax*). *Journal of Ethology*, *34*, 343-348.

Elfström, B. E. O., & Zucker, N. (1999). Winter aggregation and its relationship to social status in the tree lizard, *Urosaurus ornatus*. *Journal of Herpetology*, *33*, 240-248.

Elphick, M. J., Pike, D. A., Bezzina, C., & Shine, R. (2013). Cues for communal egg-laying in lizards (*Bassiana duperreyi*, Scincidae). *Biological Journal of the Linnean Society*, *110*, 839-842.

Emlen, S. (1982a). The evolution of helping. I. An ecological constraints model. *The American Naturalist*, *119*, 29-39.

Emlen, S. (1982b). The evolution of helping. II. The role of behavioral conflict. *The American Naturalist*, *119*, 40-53.

Emslie, S. D. (1982). *Movements and behavior of the Sonora mud turtle (Kinosternon sonoriense) in Central Arizona* [Doctoral dissertation]. Northern Arizona University.

Engelstaedter, J. (2008). Constraints on the evolution of asexual reproduction. *BioEssays*, *30*, 1138-1150.

Ensminger, D., Langkilde, T., Owen, D. MacLeod, K., & Sheriff, M. (2018). Maternal stress alters the phenotype of the mother, her eggs, and her offspring in a wild caught lizard. *Journal of Animal Ecology*, *87*, 1685-1697.

Eraud, C., Dorie, A., Jacquet, A., & Faivre, B. (2008). The crop milk: A potential new route for carotenoid-mediated parental effects. *Journal of Avian Biology*, *39*, 247-251.

Erickson, G. M., Rogers, K. C., & Yerby, S. A. (2001). Dinosaurian growth patterns and rapid avian growth rates. *Nature*, *412*, 429-433.

Ernst, C. H., & Barbour, R. W. (1972). *Turtles of the United States*. Lexington: University of Kentucky.

Ernst, C. H., & Lovich, J. E. (2009). *Turtles of the United States and Canada.* Baltimore: Johns Hopkins University Press.

Espinoza, R. E., & Lobo, F. (1996). Possible communal nesting in two species of *Liolaemus* lizards (Iguania: Tropiduridae) from northern Argentina. *Herpetological Natural History, 4,* 65–68.

Espinoza, R. E., Wiens, J. J., & Tracy, C. R. (2004). Recurrent evolution of herbivory in small, cold-climate lizards: Breaking the ecophysiological rules of reptilian herbivory. *Proceedings of the National Academy of Sciences of the United States of America, 101,* 16819–16824.

Etkin, W. (Ed.). (1964). *Social behavior and organization among vertebrates.* Chicago, IL: University of Chicago Press.

Evans, L. T. (1938). Courtship behavior and sexual selection of *Anolis. Journal of Comparative Psychology, 26,* 475–497.

Evans, L. T. (1951). Field study of the social behavior of the black lizard, *Ctenosaura pectinata. American Museum Novitates, 1943,* 1–26.

Evans, L. T. (1953). The courtship pattern of the box turtle, *Terrapene c. carolina. Herpetologica, 9,* 189–192.

Evans, L. T. (1959). A motion picture study of maternal behavior of the lizard, *Eumeces obsoletus. Copeia, 1959,* 103–110.

Evans, S. E., & Klembara, J. (2005). A choristoderan reptile (Reptilia: Diapsida) from the Lower Miocene of northwest Bohemia (Czech Republic). *Journal of Vertebrate Paleontology, 25,* 171–184.

Evans, S. E., Wang, Y., & Jones, M. E. (2007). An aggregation of lizard skeletons from the Lower Cretaceous of China. *Senckenbergiana Lethaea, 87,* 109–118.

Ewert, M. (1979). The embryo and its egg: Development and natural history. In M. Harless & H. Morlock (Eds.), *Turtles: Perspectives and research* (pp. 333–413). New York: John Wiley and Sons.

Ewert, M. (1985). Embryology of turtles. In C. Gans, F. Billet & P. F. A. Maderson (Eds.), *Biology of the Reptilia* (Vol. 14, pp. 75–267). New York: John Wiley & Sons.

Ewert, M. (1991). Cold torpor, diapause, delayed hatching and aestivation in reptiles and birds. In D. L. Deeming & M. W. J. Ferguson (Eds.), *Egg incubation: Its effects on embryonic development in birds and reptiles* (pp. 173–191). New York: Cambridge University Press.

Fagen, R. (1981) *Animal play behavior.* Oxford: Oxford University Press.

Falcon-Lang, H. J., Benton, M. J., & Stimson, M. (2007). Ecology of early reptiles inferred from Lower Pennsylvanian trackways. *Journal of the Geological Society, London, 164,* 1113–1118.

Fantin, C., Ferreira, J., Magalhães, M., da Silva Damasseno, T., de Melo Pereira, D. I., & Vogt, R. C. (2017). Kinship analysis of offspring of the giant South American river turtle (*Podocnemis expansa*) using microsatellite DNA markers. *Chelonian Conservation and Biology, 16,* 123–127.

Faria, C., Zarza, E., Reynoso, V., & Emerson, B. (2010). Predominance of single paternity in the black spiny-tailed iguana: Conservation genetic concerns for female-biased hunting. *Conservation Genetics, 11,* 1645–1652.

Fassett, J., Zielinski, R. A., & Budahn, J. R. (2002). Dinosaurs that did not die: Evidence for Paleocene dinosaurs in the Ojo Alamo Sandstone, San Juan Basin, New Mexico. In C. Koeberl & K. MacLeod (Eds.), *Catastrophic events and mass extinctions: Impacts and beyond* (pp. 307-336). Boulder, CO: The Geological Society of America.

Fathinia, B., Anderson, S. C., Rastegar-Pouyani, N., Jahani, H., & Mohamadi, H. (2009). Notes on the natural history of *Pseudocerastes urarachnoides* (Squamata: Viperidae). *Russian Journal of Herpetology, 16*, 134-138.

Fathinia, B., Rastegar-Pouyani, N., Rastegar-Pouyani, E., Todehdehghan, F., & Amiri, F. (2015). Avian deception using an elaborate caudal lure in *Pseudocerastes urarachnoides* (Serpentes: Viperidae). *Amphibia-Reptilia, 36*, 223-231.

Feiner, N., Souza-Lima, S. D., Jorge, F., Naem, S., Aubret, F., Uller, T., & Nadler, S. A. (2020). Vertical transmission of a nematode from female lizards to the brains of their offspring. *The American Naturalist.* https://doi.org/10.1086/708188.

Fenner, A., & Bull, C. (2011a). The use of scats as social signals in a solitary, endangered scincid lizard, *Tiliqua adelaidensis*. *Wildlife Research, 37*, 582-587.

Fenner, A., & Bull, C. (2011b). Responses of the endangered pygmy bluetongue lizard to conspecific scats. *Journal of Ethology, 29*, 69-77.

Fenwick, A. M., Greene, H. W., & Parkinson, C. L. (2011). The serpent and the egg: Unidirectional evolution of reproductive mode in vipers? *Journal of Zoological Systematics and Evolutionary Research, 50*, 59-66.

Ferguson, G. W. (1966). Releasers of courtship and territorial behaviour in the side blotched lizard *Uta stansburiana*. *Animal Behaviour, 14*, 89-92.

Fernández, J. B., Medina, M., Kubisch, E. L., Manero, A. A., Scolaro, J. A., & Ibargüengoytía, N. R. (2015). Female reproductive biology of the lizards *Liolaemus sarmientoi* and *L. magellanicus* from the southern end of the world. *Herpetological Journal, 25*, 101-108.

Ferrara, C. R., Mortimer, J. A., & Vogt, R. C. (2014). First evidence that hatchlings of *Chelonia mydas* emit sounds. *Copeia, 2014*, 245-247.

Ferrara, C. R., Vogt, R. C., Eisemberg, C. C., & Doody, J. S. (2017). First evidence of the pig-nosed turtle (*Carettochelys insculpta*) vocalizing underwater. *Copeia, 2017*, 29-32.

Ferrara, C. R., Vogt, R. C., Giles, J. C., & Kuchling, G. (2014). Chelonian vocal communication. In G. Witzany (Ed.), *Biocommunication of animals* (pp. 261-274). Dordrecht, Netherlands: Springer.

Ferrara, C. R., Vogt, R. C., Harfush, M. R., Sousa-Lima, R. S., Albavera, E., & Tavera, A. (2014). First evidence of leatherback turtle (*Dermochelys coriacea*) embryos and hatchlings emitting sounds. *Chelonian Conservation and Biology, 13*, 110-114.

Ferrara, C. R., Vogt, R. C., & Sousa-Lima, R. S. (2013). Turtle vocalizations as the first evidence of posthatching parental care in chelonians. *Journal of Comparative Psychology, 127*, 24-32.

Ferrara, C. R., Vogt, R. C., Sousa-Lima, R. S., Tardio, B. M., & Bernardes, V. C. D. (2014). Sound communication and social behavior in an Amazonian river turtle (*Podocnemis expansa*). *Herpetologica, 70*, 149-156.

Fields, S., & Johnston, M. (2005). Whither model organism research? *Science, 307*, 1885–1886.

Figueroa, A., McKelvy, A. D., Grismer, L. L., Bell, C. D., & Lailvaux, S. P. (2016). A species-level phylogeny of extant snakes with description of a new colubrid subfamily and genus. *PLoS ONE, 11*, e0161070.

Filadelfo, T., Danta, P. T., & Ledo, R. M. D. (2013). Evidence of a communal nest of *Kentropyx calcarata* (Squamata: Teiidae) in the Atlantic Forest of northeastern Brazil. *Phyllomedusa, 12*, 143–146.

Fitch, H. S. (1954). Life history and ecology of the five-lined skink, *Eumeces fasciatus*. *University of Kansas Publications of the Museum of Natural History, 8*, 1–156.

Fitch, H. S. (1965). An ecological study of the garter snake, *Thamnophis sirtalis*. *University of Kansas Publications of the Museum of Natural History, 15*, 493–564.

Fitch, H. S., & Hillis, D. M. (1984). The *Anolis* dewlap: Interspecific variability and morphological associations with habitat. *Copeia, 1984*, 315–323.

FitzGibbon, S., & Franklin, C. (2010). The importance of the cloacal bursae as the primary site of aquatic respiration in the freshwater turtle, *Elseya albagula*. *Australian Zoologist, 35*, 276–282.

Flannery, T. (2002). *The future eaters: An ecological history of the Australasian lands and people*. New York: Grove Press.

Font, E., Burghardt, G. M., and Leal, M. (2021). Reptile brains, behavior, and cognition: Multiple misconceptions. In C. Warwick, P. C. Arena & G. M. Burghardt (Eds.), *Health and welfare of captive reptiles* (2nd ed.). Cham, Switzerland: Springer. Nature Switzerland AG.

Font, E., & Carazo, P. (2010). Animals in translation: Why there is meaning (but probably no message) in animal communication. *Animal Behaviour, 80*, e1–e6.

Font, E., Carazo, P., Pérez I de Lanuza, G., & Kramer, M. (2012). Predator-elicited foot shakes in wall lizards (*Podarcis muralis*): Evidence for a pursuit-deterrent function. *Journal of Comparative Psychology, 126*, 87–96.

Font, E., García-Roa, R., Pincheira-Donoso, D., & Carazo, P. (2019). Rethinking the effects of body size on the study of brain size evolution. *Brain, Behavior and Evolution, 22*, 1–14.

Font, E., Pérez i de Lanuza, G., & Sampedro, C. (2009). Ultraviolet reflectance and cryptic sexual dichromatism in the ocellated lizard, *Lacerta* (*Timon*) *lepida* (Squamata: Lacertidae). *Biological Journal of the Linnean Society, 97*, 766–780.

Fontanarrosa, G., & Abdala, V. (2016). Bone indicators of grasping hands in lizards. *PeerJ, 4*, e1978.

Ford, D. P., & Benson, R. B. (2019). A redescription of *Orovenator mayorum* (Sauropsida, Diapsida) using high-resolution μCT, and the consequences for early amniote phylogeny. *Papers in Palaeontology, 5*, 197–239.

Ford, N. B. (1986). The role of pheromone trails in the sociobiology of snakes. In D. Duvall, D. Müller-Schwarze & R. M. Silverstein (Eds.), *Chemical signals in vertebrates* (Vol. 4, pp. 261–278). New York: Plenum Press.

Ford, N. B., & Burghardt, G. M. (1993). Perceptual mechanisms and the behavioral ecology of snakes. In R. A. Seigel & J. T. Collins (Eds.), *Snakes: Ecology and behavior* (pp. 117–164). New York: McGraw-Hill Inc.

Ford, N. B., & O'Bleness, M. L. (1986). Species and sexual specificity of pheromone trails of the garter snake, *Thamnophis marcianus*. *Journal of Herpetology*,
20, 259–262.

Foster, W. A., & Treherne, J. E. (1981). Evidence for the dilution effect in the
selfish herd from fish predation on a marine insect. *Nature*, *293*, 466–467.

Foureaux, G., Egami, M. I., Jared, C., Antoniazzi, M. M., Gutierre, R. C., & Smith,
R. L. (2010). Rudimentary eyes of squamate fossorial reptiles (Amphisbaenia
and Serpentes). *Anatomical Record*, *293*, 351–357.

Fowler, J. A. (1966). A communal nesting site for the smooth green snake in
Michigan. *Herpetologica*, *22*, 231.

Fowler, L. E. (1979). Hatchling success and nest predation in the green sea turtle,
Chelonia mydas, at Tortuguero, Costa Rica. *Ecology*, *60*, 946–955.

Fox, S., McCoy, K., & Baird, T. (Eds.). (2003). *Lizard social behavior*. Baltimore:
John Hopkins University Press.

Frankenberg, E. (1982a). Social behaviour of the parthenogenetic Indo-Pacific
gecko, *Hemidactylus garnotii*. *Zeitschrift für Tierpsychologie*, *59*, 19–28.

Frankenberg, E. (1982b). Vocal behavior of the Mediterranean house gecko,
Hemidactylus turcicus. *Copeia*, *1982*, 770–775.

Fredriksson, G. M. (2005). Predation on sun bears by reticulated python in East
Kalimantan, Indonesian Borneo. *Raffles Bulletin of Zoology*, *53*, 165–168.

Freedberg S., Ewert M. A., Ridenhour B. J., Neiman M., Nelson C. E. (2005).
Nesting fidelity and molecular evidence for natal homing in the freshwater
turtle, *Graptemys kohnii*. *Proceedings of the Royal Society of London, Series B:
Biological Sciences*, *272*, 1345–1350.

Freeman, A. R., & Hare, J. F. (2015). Infrasound in mating displays: A peacock's
tale. *Animal Behaviour*, *102*, 241–250.

Freiria, F. M., Brito, J. C., & Avia, M. L. (2006). Ophiophagy and cannibalism
in *Vipera latastei* Boscá, 1878 (Reptilia, Viperidae). *Herpetological Bulletin*, *96*,
26.

Friedel, P., Young, B. A., & van Hemmen, J. L. (2008). Auditory localization of
ground-borne vibrations in snakes. *Physical Review Letters*, *100*, 048701.

Friesen, C. R., Shine, R., Krohmer, R. W., & Mason, R. T. (2013). Not just a
chastity belt: The functional significance of mating plugs in garter snakes,
revisited. *Biological Journal of the Linnean Society*, *109*, 893–907.

Friesen, C. R., Uhrig, E. J., Squire, M. K., Mason, R. T., & Brennan, P. L. (2014).
Sexual conflict over mating in red-sided garter snakes (*Thamnophis sirtalis*) as
indicated by experimental manipulation of genitalia. *Proceedings of the Royal
Society B: Biological Sciences*, *281*, 20132694.

Froese, A. D., & Burghardt, G. M. (1974). Food competition in captive juvenile
snapping turtles, *Chelydra serpentina*. *Animal Behaviour*, *22*, 734–739.

Fry, B. G., Vidal, N., Norman, J. A., Vonk, F. J., Scheib, H., Ramjan, S. R., . . .
Hodgson, W. C. (2006). Early evolution of the venom system in lizards and
snakes. *Nature*, *439*, 584.

Fuller, S. J., Bull, C. M., Murray, K., & Spencer, R. J. (2005). Clustering of related
individuals in a population of the Australian lizard, *Egernia frerei*. *Molecular
Ecology*, *14*, 1207–1213.

Funston, G. F., Currie, P. J., Ryan, M. J., & Dong, Z. M. (2019). Birdlike growth and mixed-age flocks in avimimids (Theropoda, Oviraptorosauria). *Scientific Reports*, *9*, 1–21.

Furry, K., Swain, T., & Chiszar, D. (1991). Strike-induced chemosensory searching and trail following by prairie rattlesnakes (*Crotalus viridis*) preying upon deer mice (*Peromyscus maniculatus*): Chemical discrimination among individual mice. *Herpetologica*, *47*, 69–78.

Futumya, D. J. (1998). Wherefore and whither the naturalist? *The American Naturalist*, *151*, 1–6.

Gachet, H. (1833). Observations sur l'accouplement du Lézard des murailles (*Lac. muralis*). *Actes de la Société linnéenne de Bordeaux*, *VI*, 106–113.

Gage, S. H., & Gage, S. P. (1886). Aquatic respiration in soft-shelled turtles: A contribution to the physiology of respiration in vertebrates. *The American Naturalist*, *20*, 233–236.

Gaither, N. S., & Stein, B. E. (1979). Reptiles and mammals use similar sensory organizations in the midbrain. *Science*, *205*, 595–597.

Galbraith, D., Chandler, M., & Brooks, R. (1987). The fine structure of home ranges of male *Chelydra serpentina*: Are snapping turtles territorial? *Canadian Journal of Zoology*, *65*, 2623–2629.

Galef, B. G., Jr., & Giraldeau, L. A. (2001). Social influences on foraging in vertebrates: Causal mechanisms and adaptive functions. *Animal Behaviour*, *61*, 3–15.

Galef, B. G., Jr., & Laland, K. N. (2005). Social learning in animals: Empirical studies and theoretical models. *BioScience*, *55*, 489–499.

Galeotti, P., Sacchi, R., Fasola, M., & Ballasina, D. (2005). Do mounting vocalizations in tortoises have a communication function? A comparative analysis. *The Herpetological Journal*, *15*, 61–71.

Gallup, G. G., Jr. (1970). Chimpanzees: Self recognition. *Science*, *167*, 86–87.

Gallup, G. G., Jr., Anderson, J. R., & Shillito, D. J. (2002). The mirror test. In M. Bekoff, C. Allen & G. M. Burghardt (Eds.), *The cognitive animal: Empirical and theoretical perspectives on animal cognition* (pp. 325–333). Cambridge, MA: MIT Press.

Galoyan, E., Moskalenko, V., Gabelaia, M., Tarkhnishvili, D., Spangenberg, V., Chamkina, A., & Arakelyan, M. (2020). Syntopy of two species of rock lizards (*Darevskia raddei* and *D. portschinskii*) may not lead to hybridization between them. *Zoologischer Anzeiger*, *288*, 43–52.

Galoyan, E. A., Tsellarius, E. Y., & Arakelyan, M. S. (2019). Friend-or-foe? Behavioural evidence suggests interspecific discrimination leading to low probability of hybridization in two coexisting rock lizard species (Lacertidae, *Darevskia*). *Behavioral Ecology and Sociobiology*, *73*, 46.

Gamble, T., Coryell, J., Ezaz, T., Lynch, J., Scantlebury, D. P., & Zarkower, D. (2015). Restriction site-associated DNA sequencing (RAD-seq) reveals an extraordinary number of transitions among gecko sex-determining systems. *Molecular Biology and Evolution*, *32*, 1296–1309.

Gans, C., Gillingham, J. C., & Clark, D. L. (1984). Mating and male combat in tuatara, *Sphenodon punctatus*. *Journal of Herpetology*, *18*, 194–197.

Garberoglio, F. F., Apesteguía, S., Simões, T. R., Palci, A., Gómez, R. O., Nydam, R. L., . . . Caldwell, M. W. (2019). New skulls and skeletons of the Cretaceous legged snake *Najash*, and the evolution of the modern snake body plan. *Science Advances, 5,* eaax5833.

García, R. A. (2007). An "egg-tooth"-like structure in titanosaurian sauropod embryos. *Journal of Vertebrate Paleontology, 27,* 247-252.

García-Roa, R., Iglesias, M., Gosá, A., & Cabido, C. (2015). *Podarcis muralis* (common wall lizard). Communal nesting. *Herpetological Review, 46,* 435-436.

Gardner, M., Bull, C., & Cooper, S. (2002). High levels of genetic monogamy in the group-living Australian lizard *Egernia stokesii. Molecular Ecology, 11,* 1787-1794.

Gardner, M., Bull, C., Cooper, S., & Duffield, G. (2001). Genetic evidence for a family structure in stable social aggregations of the Australian lizard *Egernia stokesii. Molecular Ecology, 10,* 175-183.

Gardner, M., Bull, C., Fenner, A., Murray, K., & Donnellan, S. (2007). Consistent social structure within aggregations of the Australian lizard, *Egernia stokesii* across seven disconnected rocky outcrops. *Journal of Ethology, 25,* 263-270.

Gardner, M., Pearson, S., Johnston, G., & Schwarz, M. (2016). Group living in squamate reptiles: A review of evidence for stable aggregations. *Biological Reviews, 91,* 925-936.

Garnett, S. T. (2009). Metabolism and survival of fasting estuarine crocodiles. *Journal of Zoology, 208,* 493-502.

Garrick, L. D., & Lang, J. W. (1977). Social signals and behaviors of adult alligators and crocodiles. *American Zoologist, 17,* 225-239.

Garrick, L. D., Lang, J. W., & Herzog, H. A. (1978). Social signals of adult American alligators. *Bulletin of the American Museum of Natural History, 160,* 153-192.

Garrison, G., Phillips, M., Eifler, M., & Eifler, D. (2016). Intraspecific variation in opportunistic use of trophic resources by the lizard *Ameiva corax* (Squamata: Teiidae). *Amphibia-Reptilia, 37,* 331-334.

Gaston, K. J., & May, R. M. (1992). Taxonomy of taxonomists. *Nature, 356,* 281-282.

Gauthier, J. A. (1994). The diversification of the amniotes. In D. Prothero (Ed.), *Major features of vertebrate evolution: Short courses in paleontology* (pp. 129-159). Boulder, CO: The Paleontological Society.

Gaze, P. (2001). Tuatara recovery plan: 2001-2011. *Threatened species recovery plan series* (Vol. 47). Wellington, New Zealand: Department of Conservation.

Gehlbach, F. R., Watkins, J. F., & Kroll, J. C. (1971). Pheromone trail-following studies of typhlopid, leptotyphlopid, and colubrid snakes. *Behaviour, 40,* 282-294.

Gehlbach, F. R., Watkins, J. F., & Reno, H. W. (1968). Blind snake defensive behavior elicited by ant attacks. *Biological Sciences, 18,* 784-785.

Gelfand, D. L., & McCracken, G. F. (1986). Individual variation in the isolation calls of Mexican free-tailed bat pups (*Tadarida brasiliensis mexicana*). *Animal Behaviour, 34,* 1078-1086.

Georges, A., Limpus, C., & Stoutjesdijk, R. (1994). Hatchling sex in the marine turtle *Caretta caretta* is determined by proportion of development at a temperature, not daily duration of exposure. *Journal of Experimental Zoology, 270*, 432–444.

Gerber, G. (1997). Nesting behavior of the Little Cayman rock iguana, *Cyclura nubila caymanensis. Joint Annual Meeting, American Society of Ichthyologists and Herpetologists/Herpetologists League/Society for the Study of Amphibians and Reptiles*. Seattle, Washington.

Germano, D. J., & Smith, P. T. (2010). Molecular evidence for parthenogenesis in the Sierra garter snake, *Thamnophis couchii* (Colubridae). *Southwestern Naturalist, 55*, 280–282.

Gerum, R. C., Fabry, B., Metzner, C., Beaulieu, M., Ancel, A., & Zitterbart, D. P. (2013). The origin of traveling waves in an emperor penguin huddle. *New Journal of Physics, 15*, 125022.

Gienger, C. M., & Tracy, C. R. (2008). Ecological interactions between Gila monsters (*Heloderma suspectum*) and desert tortoises (*Gopherus agassizii*). *Southwestern Naturalist, 53*, 265–268.

Giles, J. C., Davis, J. A., McCauley, R. D., & Kuchling, G. (2009). Voice of the turtle: The underwater acoustic repertoire of the long-necked freshwater turtle, *Chelodina oblonga. Journal of the Acoustical Society of America, 126*, 434–443.

Gill, J. A. (2007). Approaches to measuring the effects of human disturbance on birds. *Ibis, 149*, 9–14.

Gillingham, J. C. (1987). Social behavior. In R. A. Siegel, J. T. Collins & S. S. Novak (Eds.), *Snakes: Ecology and evolutionary biology* (pp. 184–209). New York: Macmillan.

Gillingham, J. C. (1995). Normal behaviour. In C. Warwick, F. L. Frye & J. B. Murphy (Eds.), *Health and welfare of captive reptiles* (pp. 131–164). London: Chapman & Hall.

Gillingham, J. C., Carmichael, C., & Miller, T. (1995). Social behavior of the tuatara, *Sphenodon punctatus. Herpetological Monographs, 9*, 5–16.

Gilmore, R. G. (1993). Reproductive biology of lamnoid sharks. *Environmental Biology of Fishes, 38*, 95–114.

Giraldeau, L.-A. (1997). The ecology of information use. In J. R. Krebs & N. B. Davies (Eds.), *Behavioural ecology: An evolutionary approach* (pp. 42–68). Hoboken, NJ: Wiley-Blackwell.

Giraldeau, L.-A., Valone, T. J., & Templeton, J. J. (2002). Potential disadvantages of using socially acquired information. *Philosophical Transactions of the Royal Society of London. Series B: Biological Sciences, 357*, 1559–1566.

Girondot, M., Tucker, A. D., Rivalan, P., Godfrey, M. H., & Chevalier, J. (2002). Density-dependent nest destruction and population fluctuations of Guianan leatherback turtles. *Animal Conservation, 5*, 75–84.

Gist, D., Bagwill, A., Lance, V., Sever, D., & Elsey, R. (2008). Sperm storage in the oviduct of the American alligator. *Journal of Experimental Zoology Part A: Ecological Genetics and Physiology, 309*, 581–587.

Glaw, F., Köhler, J. R., Townsend, T. M., & Vences, M. (2012). Rivaling the world's smallest reptiles: Discovery of miniaturized and microendemic new species of leaf chameleons (*Brookesia*) from Northern Madagascar. *PLoS ONE, 7*, e31314.

Glen, F., Broderick, A., Godley, B., & Hays, G. (2005). Patterns in the emergence of green (*Chelonia mydas*) and loggerhead (*Caretta caretta*) turtle hatchlings from their nests. *Marine Biology, 146,* 1039–1049.

Godfrey, S., Ansari, T., Gardner, M., Farine, D., & Bull, C. (2014). A contact-based social network of lizards is defined by low genetic relatedness among strongly connected individuals. *Animal Behaviour, 97,* 35–43.

Godfrey, S., Bull, C., James, R., & Murray, K. (2009). Network structure and parasite transmission in a group living lizard, the gidgee skink, *Egernia stokesii. Behavioral Ecology and Sociobiology, 63,* 1045–1056.

Godfrey, S., Bull, C., Murray, K., & Gardner, M. (2006). Transmission mode and distribution of parasites among groups of the social lizard *Egernia stokesii. Parasitology Research, 99,* 223–230.

Godwin, C. D. (2017). Softshell turtle nesting ecology with emphasis on ORV impacts and flooding [Doctoral dissertation]. Louisiana State University, ProQuest Dissertations Publishing.

Golinski, A., John-Alderb, H., & Kratochvílc, L. (2011). Male sexual behavior does not require elevated testosterone in a lizard (*Coleonyx elegans*, Eublepharidae). *Hormones and Behavior, 59,* 144–150.

Gomez-Mestre, I., Wiens, J., & Warkentin, K. (2008). Evolution of adaptive plasticity: Risk-sensitive hatching in neotropical leaf-breeding treefrogs. *Ecological Monographs, 78,* 205–224.

Goonatilake, W., and Peries, A.L. (2001). Range extension of *Calodacty-lodes illingworthorum* Deraniyagala, 1953. (Gekkonidae: Reptilia). *Loris, 22,* 23–26.

Goonewardene, S., Hawke, Z., Vanneck, V., Drion, A., de Silva, A., Jayarathne, R., & Perera, J. (2003). Diversity of Nilgala Fire Savannah, Sri Lanka: With special reference to its herpetofauna. *Report of Project Hoona,* Rajarata University, Sri Lanka.

Gottlieb, G. (1991). Experiential canalization of behavioral development: Theory. *Developmental Psychology, 27,* 4–13.

Graham, J. B. (1974). Aquatic respiration in the sea snake *Pelamis platurus. Respiration Physiology, 21,* 1–7.

Grant, C. (1971). Snake eggs. *Science, 172,* 792.

Grant, P. B., & Samways, M. J. (2015). Acoustic prey and a listening predator: Interaction between calling katydids and the bat-eared fox. *Bioacoustics, 24,* 49–61.

Graves, B. M. (1989). Defensive behavior of female prairie rattlesnakes (*Crotalus viridis*) changes after parturition. *Copeia, 1989,* 791–794.

Graves, B. M., & Duvall, D. (1987). An experimental study of aggregation and thermoregulation in prairie rattlesnakes (*Crotalus viridis viridis*). *Herpetologica, 1987,* 259–264.

Graves, B. M., & Duvall, D. (1988). Evidence of an alarm pheromone from the cloacal sacs of prairie rattlesnakes. *Southwestern Naturalist, 33,* 339–345.

Graves, B. M., & Duvall, D. (1995). Aggregation of squamate reptiles associated with gestation, oviposition, and parturition. *Herpetological Monographs, 9,* 102–119.

Graves, B. M., & Halpern, M. (1991). Discrimination of self from conspecific chemical cues in *Tiliqua scincoides* (Sauria: Scincidae). *Journal of Herpetology*, 25, 125–6.

Gray, B. S. (2007). A note on cannibalism in the common garter snake, *Thamnophis sirtalis sirtalis*. *Chicago Herpetological Society*, 42, 6.

Greeff, J., & Whiting, M. (2000). Foraging-mode plasticity in the lizard *Platysaurus broadleyi*. *Herpetologica*, 56, 402–407.

Green, B., McKelvey, M., & Rismiller, P. (1999). The behaviour and energetics of hatchling *Varanus rosenbergi*. *Mertensiella*, 11, 105–112.

Greenberg, B. (1943). Social behavior of the western banded gecko, *Coleonyx variegatus* Baird. *Physiological Zoology*, 16, 110–122.

Greenberg, B., & Noble, G. K. (1944). Social behavior of the American chameleon (*Anolis carolinensis* Voigt). *Physiology and Zoology*, 17, 392–439.

Greenberg, N. (1976). Observations on social feeding in lizards. *Herpetologica*, 32, 348–352.

Greenberg, N., & Crews, D. (1977). Introduction to the symposium: Social behavior in reptiles. *American Zoologist*, 17, 153–154.

Greenberg, N., & Crews, D. (1990). Endocrine and behavioral responses to aggression and social dominance in the green anole lizard, *Anolis carolinensis*. *General and Comparative Endocrinology*, 77, 246–255.

Greenberg, N., & Hake, L. (1990). Hatching and neonatal behavior of the lizard, *Anolis carolinensis*. *Journal of Herpetology*, 24, 402–405.

Greenberg, N., & Jenssen, T. A. (1982) Displays of captive banded iguanas. In G. M. Burghardt & A. S. Rand (Eds.), *Iguanas of the world: Their behavior, ecology, and conservation* (pp. 232–251). Park Ridge, NJ: Noyes Publications.

Greenberg, N., & MacLean, P. D. (Eds.) (1978). *Behavior and neurology of lizards: An interdisciplinary colloquium*. Rockville, MD: National Institute of Mental Health.

Greene, H. W. (1997). *Snakes: The evolution of mystery in nature*. Berkeley: University of California Press.

Greene, H. W. (2005). Organisms in nature as a central focus for biology. *Trends in Ecology and Evolution*, 20, 23–27.

Greene, H. W. (2013). *Tracks and shadows: Field biology as art*. Berkeley: University of California Press.

Greene, H. W., Burghardt, G. M., Dugan, B. A., & Rand, A. S. (1978). Predation and the defensive behavior of green iguanas (Reptilia, Lacertilia, Iguanidae). *Journal of Herpetology*, 12, 169–176.

Greene, H. W., May, P. G., Hardy, D. L., Sr., Sciturro, J. M., & Farrell, T. M. (2002). Parental behavior by vipers. In G. W. Schuett, M. Höggren, M. E. Douglas & H. W. Greene (Eds.), *Biology of the vipers* (pp. 179–206). Eagle Mountain, UT: Eagle Mountain Publishing.

Greene, H. W., Rodríguez, J. J. S., & Powell, B. J. (2006). Parental behavior in anguid lizards. *South American Journal of Herpetology*, 1, 9–16.

Greene, M. J., & Mason, R. T. (2000). Courtship, mating, and male combat of the brown tree snake, *Boiga irregularis*. *Herpetologica*, 56, 166–175.

Greene, M. J., Stark, S. L., & Mason, R. T. (2001). Pheromone trailing behavior of the brown tree snake, *Boiga irregularis*. *Journal of Chemical Ecology*, 27, 2193–2201.

Greer, A. E. (1989). *The biology and evolution of Australian lizards.* Chipping Norton, Australia: Surrey Beatty and Sons.

Gregory, M. S., Silvers, A., and Sutch, D. (1978). *Sociobiology and human nature.* San Francisco: Jossey-Bass.

Gregory, P. T. (1974). Patterns of spring emergence of the red-sided garter snake (*Thamnophis sirtalis parietalis*) in the Interlake region of Manitoba. *Canadian Journal of Zoology, 52,* 1063-1069.

Gregory, P. T. (1975). Aggregations of gravid snakes in Manitoba, Canada. *Copeia, 1975,* 185-186.

Gregory, P. T. (1984). Habitat, diet, and composition of assemblages of garter snakes (*Thamnophis*) at eight sites on Vancouver Island. *Canadian Journal of Zoology, 62,* 2013-2022.

Gregory, P. T., Macartney, J. M., & Larsen, K. W. (1987). Spatial patterns and movements. In R. A. Seigel, J. T. Collins & S. S. Novak (Eds.), *Snakes: Ecology and evolutionary biology* (pp. 366-395). New York: Macmillan.

Gregory, P. T., & Stewart, K. W. (1975). Long-distance dispersal and feeding strategy of the red-sided garter snake (*Thamnophis sirtalis parietalis*) in the Interlake of Manitoba. *Canadian Journal of Zoology, 53,* 238-245.

Grigg, G. C., & Kirshner, D. (2015). *Biology and evolution of crocodylians.* Clayton, Australia: CSIRO.

Groen, J., Kaastra-Berga, G., & Kaastra, S. (2020). The first documented case of arboreal foraging by two male adders (*Vipera berus*) raiding the nest of a blue tit (*Cyanistes caeruleus*). *Herpetology Notes, 13,* 583-586.

Grosse, A. M., Buhlmann, K. A., Harris, B. B., DeGregorio, B. A., Moule, B. M., Horan, R. V., III, & Tuberville, T. D. (2012). Nest guarding in the gopher tortoise (*Gopherus polyphemus*). *Chelonian Conservation Biology, 11,* 148-151.

Groves, J. D. (1981). Observations and comments on the post-parturient behavior of some tropical boas of the genus *Epicrates*. *British Journal of Herpetology, 6,* 89-91.

Guillette, L. J., Jr., Spielvogel, S., & Moore, F. L. (1981). Luteal development, placentation, and plasma progesterone concentration in the viviparous lizard *Sceloporus jarrovi*. *General and Comparative Endocrinology, 43,* 20-29.

Gutnick, T., Weissenbacher, A., & Kuba, M. J. (2020). The underestimated giants: Operant conditioning, visual discrimination and long-term memory in giant tortoises. *Animal Cognition, 23,* 159-167.

Guyétant, R. (1966). Observations écologiques sur les pontes de *Rana temporaria* L. dans la région de Besançon. *Physiologie et Biologie Animale, 2,* 13-18.

Hagelin, J. C., & Jones, I. L. (2007). Bird odors and other chemical substances: A defense mechanism or overlooked mode of intraspecific communication? *Auk, 124,* 741-761.

Haines, R. W. (1946). A revision of the movements of the forearm in tetrapods. *Journal of Anatomy, 80,* 1-11.

Håkansson, P., & Loman, J. (2004). Communal spawning in the common frog *Rana temporaria*-egg temperature and predation consequences. *Ethology, 110,* 665-680.

Hall, D., & Steidl, R. (2007). Movements, activity, and spacing of Sonoran mud turtles (*Kinosternon sonoriense*) in interrupted mountain streams. *Copeia*, *2007*, 403–412.

Halliday, T., & Verrell, P. (1986). Sexual selection and body size in amphibians. *Herpetological Journal*, *1*, 86–92.

Halliwell, B., Uller, T., Chapple, D. G., Gardner, M. G., Wapstra, E., & While, G. M. (2017). Habitat saturation promotes delayed dispersal in a social reptile. *Behavioral Ecology*, *28*, 515–522.

Halliwell, B., Uller, T., Holland, B. R., & While, G. M. (2017). Live bearing promotes the evolution of sociality in reptiles. *Nature Communications*, *8*, 1–8.

Halliwell, B., Uller, T., Wapstra, E., & While, G. M. (2017). Resource distribution mediates social and mating behavior in a family living lizard. *Behavioral Ecology*, *28*, 145–153.

Halloy, M., Boretto, J. M., & Ibargüengoytía, N. R. (2007). Signs of parental behavior in *Liolaemus elongatus* (Sauria: Liolaemidae) of Neuquén, Argentina. *South American Journal of Herpetology*, *2*, 141–147.

Halloy, M., & Halloy, S. (1997). An indirect form of parental care in a high altitude viviparous lizard, *Liolaemus huacahuasicus* (Tropiduridae). *Bulletin of the Maryland Herpetological Society*, *33*, 139–155.

Halpern, M. (1992). Nasal chemical senses in reptiles: Structure and function. *Biology of the Reptilia*, *18*, 423–523.

Halpern, M., & Martinez-Marcos, A. (2003). Structure and function of the vomeronasal system: An update. *Progress in Neurobiology*, *70*, 245–318.

Halpin, Z. T. (1990). Responses of juvenile eastern garter snakes (*Thamnophis sirtalis sirtalis*) to own, conspecific, and clean odors. *Copeia*, *1990*, 1157–60.

Hamilton, W. D. (1964a). The genetical evolution of social behaviour. *Journal of Theoretical Biology*, *7*, 1–17.

Hamilton, W. D. (1964b). The genetical evolution of social behaviour, II. *Journal of Theoretical Biology*, *7*, 17–52

Hansknecht, K. A., & Burghardt, G. M. (2010). Stimulus control of lingual predatory luring and related foraging tactics of Mangrove Saltmarsh Snakes (*Nerodia clarkii compressicauda*). *Journal of Comparative Psychology*, *124*, 159–165.

Hardy, J. D. (1957). Bat predation by the Cuban boa, *Epicrates angulifer* Bibron. *Copeia*, *1957*, 151–152.

Hare, K. M., & Hoare, J. M. (2005). *Hoplodactylus maculatus* (Common Gecko). Aggregations. *Herpetological Review*, *36*, 179.

Harless, M., Walde, A., Delaney, D., Pater, L., & Hayes, W. (2009). Home range, spatial overlap, and burrow use of the desert tortoise in the West Mojave Desert. *Copeia*, *2009*, 378–389.

Harrell, T. L., Jr., Pérez-Huerta, A., & Suarez, C. A. (2016). Endothermic mosasaurs? Possible thermoregulation of Late Cretaceous mosasaurs (Reptilia, Squamata) indicated by stable oxygen isotopes in fossil bioapatite in comparison with coeval marine fish and pelagic seabirds. *Palaeontology*, *59*, 351–363.

Harrington, S. M., & Reeder, T. W. (2017). Phylogenetic inference and divergence dating of snakes using molecules, morphology and fossils: New insights into convergent evolution of feeding morphology and limb reduction. *Biological Journal of the Linnean Society, 121*, 379–394.

Harrison, A. (2013). Size-assortative pairing and social monogamy in a neotropical lizard, *Anolis limifrons* (Squamata: Polychrotidae). *Breviora, 534*, 1–9.

Harrison, C. (1975). *Field guide to the nests, eggs and nestlings of European birds.* London: Collins.

Harshman, J. (2007). Classification and phylogeny of birds. In B. G. M. Jamieson (Ed.), *Reproductive biology and phylogeny of birds* (pp. 1–35). Enfield, NH: Science Publishers.

Hasson, O. (1991). Pursuit-deterrent signals: Communication between prey and predator. *Trends in Ecology & Evolution, 6*, 325–329.

Hastings, A. K., & Hellmund, M. (2015). Rare in situ preservation of adult Crocodylian with eggs from the middle Eocene of Geiseltal, Germany. *Palaios, 30*, 446–461.

Hayes, L. D. (2000). To nest communally or not to nest communally: A review of rodent communal nesting and nursing. *Animal Behaviour, 59*, 677–688.

Hayes, W. K., Carter, R. L., Cyril, S., Jr., & Thornton, B. (2004). Conservation of an endangered Bahamian rock iguana, I. In A. C. Alberts, R. L. Carter, W. K. Hayes & E. P. Martins (Eds.), *Iguanas: Biology and conservation* (pp. 232–257). Berkeley: University of California Press.

Hays, G., Speakman, J., & Hayes, J. (1992). The pattern of emergence by loggerhead turtle (*Caretta caretta*) hatchlings on Cephalonia, Greece. *Herpetologica, 48*, 396–401.

Hayward, M. W., & Somers, M. J. (2009). *Reintroduction of top-order predators.* Oxford: Wiley-Blackwell.

Head, J. J. (2015). Fossil calibration dates for molecular phylogenetic analysis of snakes 1: Serpentes, Alethinophidia, Boidae, Pythonidae. *Palaeontologia Electronica, 18*, 1–17.

Head, J. J., Mahlow, K., & Müller, J. (2016). Fossil calibration dates for molecular phylogenetic analysis of snakes 2: Caenophidia, Colubroidea, Elapoidea, Colubridae. *Palaeontologia Electronica, 19*, 1–21.

Heatwole, H., & Sullivan, B. K. (1995). *Amphibian Biology. Volume 2: Social Behaviour.* Chipping Norton, Australia: Surrey Beatty & Sons.

Hedges, S. B. (2008). At the lower size limit in snakes: Two new species of threadsnakes (Squamata, Leptotyphlopidae, *Leptotyphlops*) from the Lesser Antilles. *Zootaxa, 1841*, 1–30.

Hedges, S. B., & Thomas, R. (2011). At the lower size limit in amniote vertebrates: A new diminutive lizard from the West Indies. *Caribbean Journal of Science, 37*, 168–173.

Heideman, J. L. (1993). Social organization and behaviour of *Agama aculeata aculeata* and *Agama planiceps planiceps* (Reptilia: Agamidae) during the breeding season. *Journal of the Herpetological Association of Africa, 42*, 28–31.

Heilmann, G. (1927). *The origin of birds.* New York: Dover Publications.

Heinrich, B. (1988). Winter foraging at carcasses by three sympatric corvids, with emphasis on recruitment by the raven, *Corvus corax. Behavioral Ecology and Sociobiology, 23*, 141–156.

Hendrickson, J. (1958). The green sea-turtle, *Chelonia mydas* (Linn.) in Malaya and Sarawak. *Proceedings of the Zoological Society of London, 130*, 455–535.

Henzell, R. (1972). Adaptation to aridity in lizards of the *Egernia whitei* species-group [Doctoral dissertation]. University of Adelaide.

Hepper, P. G., Scott, D., & Shahidullah, S. (1993). Newborn and fetal response to maternal voice. *Journal of Reproductive and Infant Psychology, 11*, 147–153.

Hernandez, A., Villavicencio, W., Ljustina, O., & Doody, J. S. (2017). *Anolis equestris*. Environmentally cued hatching. *Herpetological Review, 48*, 841.

Herodotus. (2008). *The histories* (R. Waterfield, Trans.). Oxford: Oxford University Press. (Original work published ca. 420 BC).

Herrera-Flores, J. A., Stubbs, T. L., & Benton, M. J. (2017). Macroevolutionary patterns in Rhynchocephalia: Is the tuatara (*Sphenodon punctatus*) a living fossil? *Palaeontology, 60*, 319–328.

Herzog, H. A. (1974). The vocal communication system and related behaviors of the American alligator (*Alligator mississippiensis*) and other crocodilians [Doctoral dissertation]. University of Tennessee.

Herzog, H. A., Jr. (1990). Experiential modification of defensive behaviors in garter snakes, *Thamnophis sirtalis*. *Journal of Comparative Psychology, 104*, 334–339.

Herzog, H. A., Jr., Bowers, B. B., & Burghardt, G. M. (1989). Development of antipredator responses in snakes: IV. Interspecific and intraspecific differences in habituation of defensive behavior. *Developmental Psychobiology, 22*, 489–508.

Herzog, H. A., & Burghardt, G. M. (1977). Vocalization in juvenile crocodilians. *Zeitschrift für Tierpsychologie, 44*, 294–304.

Hewlett, J. B., & Schuett, G. W. (2019). *Crotalus horridus* (timber rattlesnake). Male defense of mother and offspring. *Herpetological Review, 50*, 389–390.

Hibbard, C. W. (1964). A brooding colony of the blind snake *Leptotyphlops dulcis dissecta* Cope. *Copeia, 1964*, 222.

Hicks, J. (2002). The physiological and evolutionary significance of cardiovascular shunting patterns in reptiles. *News in Physiological Sciences, 17*, 241–245.

Hidalgo, H. (1982). Courtship and mating behavior in *Rhinoclemmys pulcherrima incisa* (Testudines: Emydidae: Batagurinae). *Transactions of the Kansas Academy of Science, 85*, 82–95.

Hildebrand, H. (1963). Hallazgo del área de anidación de la tortuga marina "lora", *Lepidochelys kempi* (Garman), en la costa Occidental del Golfo de México. *Cieneia (Mexico), 22*, 101–104.

Hill, J. G. I., Chanhome, L., Artchwakom, T., Thirakhupt, K., & Voris, H. K. (2006). Nest attendance by a female Malayan pit viper (*Calloselasma rhodostoma*) in Northeast Thailand. *The Natural History Journal of Chulalongkorn University, 6*, 57–66.

Hiltpold, I., & Shriver, W. G. (2018). Birds bug on indirect plant defenses to locate insect prey. *Journal of Chemical Ecology, 44*, 576–579.

Hirth, H. F. (1980). Some aspects of the nesting behavior and reproductive biology of sea turtles. *American Zoologist, 20*, 507–523.

Hoare, J. M., & Nelson, N. J. (2006). *Hoplodactylus maculatus* (common gecko) social assistance. *Herpetological Review, 37*, 222–223.

Hodsdon, L. A., & Pearson, J. F. W. (1943). Notes on the discovery and biology of two Bahaman freshwater turtles of the genus *Pseudemys*. *Proceedings of the Florida Academy of Sciences*, 6, 17–23.

Holcomb, S., & Carr, J. (2011). Hatchling emergence from naturally incubated alligator snapping turtle (*Macrochelys temminckii*) nests in northern Louisiana. *Chelonian Conservation and Biology*, 10, 222–227.

Holleley, C. E., O'Meally, D., Sarre, S. D., Graves, J. A. M., Ezaz, T., Matsubara, K., & Georges, A. (2015). Sex reversal triggers the rapid transition from genetic to temperature-dependent sex. *Nature*, 523, 79–82.

Honarvar, S., O'Connor, M. P., & Spotila, J. R. (2008). Density-dependent effects on hatching success of the olive ridley turtle, *Lepidochelys olivacea*. *Oecologia*, 157, 221–230.

Hone, D. W. E., & Naish, D. (2013). The "species recognition hypothesis" does not explain the presence and evolution of exaggerated structures in non-avialan dinosaurs. *Journal of Zoology*, 290, 172–180.

Hone, D. W. E., Naish, D., & Cuthill, I. C. (2011). Does mutual sexual selection explain the evolution of head crests in pterosaurs and dinosaurs? *Lethaia*, 45, 139–156.

Höner, O. P., Wachter, B., East, M. L., Streich, W. J., Wilhelm, K., Burke, T., & Hofer, H. (2007). Female mate-choice drives the evolution of male-biased dispersal in a social mammal. *Nature*, 448, 798.

Hopson, J. A. (1977). Relative brain size and behavior in archosaurian reptiles. *Annual Review of Ecology and Systematics*, 8, 429–448.

Horn, H. G. (1999). Evolutionary efficiency and success in monitors: A survey of behavior and behavioral strategies. *Mertensiella*, 11, 167–180.

Horn, H. G., & Visser, G. J. (1989). Review of reproduction of monitor lizards *Varanus spp.* in captivity. *International Zoo Yearbook*, 28, 140–150.

Horn, H. G., & Visser, G. J. (1997). Review of reproduction of monitor lizards *Varanus* spp. in captivity II. *International Zoo Yearbook*, 35, 227–246.

Horne, B. D., Brauman, R. J., Moore, M. J., & Seigel, R. A. (2003). Reproductive and nesting ecology of the yellow-blotched map turtle, *Graptemys flavimaculata*: Implications for conservation and management. *Copeia*, 2003, 729–738.

Horner, J. R. (1982). Evidence of colonial nesting and "site fidelity" among ornithischian dinosaurs. *Nature*, 297, 675–676.

Horner, J. R. (2000). Dinosaur reproduction and parenting. *Annual Review of Earth and Planetary Sciences*, 28, 19–45

Horner, J. R., & Makela, R. (1979). Nest of juveniles provides evidence of family structure among dinosaurs. *Nature*, 282, 256–257.

Hoss, S., Deutschman, D., Booth, W., & Clark, R. (2015). Post-birth separation affects the affiliative behaviour of kin in a pitviper with maternal attendance. *Biological Journal of the Linnean Society*, 116, 637–648.

Hotton, N., III. (1986). Dicynodonts and their role as primary consumers. In N. Hotton, III, P. D. MacLean, J. J. Roth & E. C. Roth (Eds.), *The ecology and biology of mammal-like reptiles* (pp. 71–82). Washington, DC: Smithsonian Institution Press, Washington.

Hotton, N., III, Maclean, P. D., Roth, J. J., & Roth, E. C. (Eds.). (1986). *The ecology and biology of mammal-like reptiles*. Washington, DC: Smithsonian Institution Press.

Houghton, J., & Hays, G. (2001). Asynchronous emergence by loggerhead turtle (*Caretta caretta*) hatchlings. *Naturwissenschaften, 88*, 133–136.

Houston, D. C. (1986). Scavenging efficiency of turkey vultures in tropical forest. *The Condor, 88*, 318–323.

Howard R. D. (1980). Mating behaviour and mating success in woodfrogs, *Rana sylvatica*. *Animal Behaviour, 28*, 705–716.

Hsiang, A. Y., Field, D. J., Webster, T. H., Behlke, A. D., Davis, M. B., Racicot, R. A., & Gauthier, J. A. (2015). The origin of snakes: Revealing the ecology, behavior, and evolutionary history of early snakes using genomics, phenomics, and the fossil record. *BMC Evolutionary Biology, 15*, 87.

Hsieh, S. T., & Lauder, G. V. (2004). Running on water: Three-dimensional force generation by basilisk lizards. *Proceedings of the National Academy of Sciences, 101*, 16784–16788.

Hu, Y., & Wu, X. B. (2010). Multiple paternity in Chinese alligator (*Alligator sinensis*) clutches during a reproductive season at Xuanzhou Nature Reserve. *Amphibia-Reptilia, 31*, 419–424.

Huang, M. (1984). An estimate of the population of *Agkistrodon shedaoensis* on Shedao Island. (Chinese, with English Abstract.) *Acta Herpetologica Sinica, 3*, 17–22.

Huang, M. (1989). Studies on *Agkistrodon shedaoensis* ecology. *Current Herpetology in East Asia*. Kyoto: Herpetological Society of Japan.

Huang, W.-S. (2008). Predation risk of whole-clutch filial cannibalism in a tropical skink with maternal care. *Behavioral Ecology, 19*, 1069–1074.

Huang, W.-S., Greene, H. W., Chang, T.-J., & Shine, R. (2011). Territorial behavior in Taiwanese kukrisnakes (*Oligodon formosanus*). *Proceedings of the National Academy of Sciences, 108*, 7455–7459.

Huang, W.-S., Lin, S. M., Dubey, S., & Pike, D. A. (2012). Predation drives interpopulation differences in parental care expression. *Journal of Animal Ecology, 82*, 429–437.

Huang, W.-S., & Wang, H.-Y. (2009). Predation risks and anti-predation parental care behavior: An experimental study in a tropical skink. *Ethology, 115*, 273–279.

Huey, R. B. (1982). Temperature, physiology, and the ecology of reptiles. In C. Gans & F. H. Pough (Eds.), *Biology of the Reptilia* (Vol. 12, Physiology (C), pp. 25–91). London: Academic Press.

Huey, R. B., Peterson, C.R., Arnold, S. J., & Porter, W. P. (1989). Hot rocks and not-so-hot rocks: Retreat-site selection by garter snakes and its thermal consequences. *Ecology, 70*, 931–944.

Hughes, D. A., & Richard, J. D. (1974). The nesting of the Pacific ridley turtle *Lepidochelys olivacea* on Playa Nancite, Costa Rica. *Marine Biology, 24*, 97–107.

Hughes, D. F., & Blackburn, D. G. (2020). Evolutionary origins of viviparity in Chamaeleonidae. *Journal of Zoological Systematics and Evolutionary Research, 58*, 284–302.

Hutchison, V. H., Dowling, H. G., & Vinegar, A. (1966). Thermoregulation in a brooding female Indian python, *Python molurus bivittatus*. *Science, 151*, 694–695.

Hutchison, V. H., & Kosh, R. J. (1974). Thermoregulatory function of the parietal eye in the lizard *Anolis carolinensis*. *Oecologia, 16*, 173–177.

Hutchinson, D. A., Mori, A., Savitzky, A. H., Burghardt, G. M., Wu, X., Meinwald, J., & Schroeder, F. C. (2007). Dietary sequestration of defensive steroids in nuchal glands of the Asian snake *Rhabdophis tigrinus*. *Proceedings of the National Academy of Sciences, 104*, 2265–2270.

Hutchinson, D. A., Savitzky, A. H., Burghardt, G. M., Nguyen, C., Meinwald, J., Schroeder, F. C., & Mori, A. (2013). Chemical defense of an Asian snake reflects local availability of toxic prey and hatchling diet. *Journal of Zoology, 289*, 270–279.

Huttenlocker, A. K., Mazierski, D., & Reisz, R. R. (2011). Comparative osteo-histology of hyperelongate neural spines in the Edaphosauridae (Amniota: Synapsida). *Palaeontology, 54*, 573–590.

Ims, R. A. (1990). On the adaptive value of reproductive synchrony as a predator-swamping strategy. *The American Naturalist, 136*, 485–498.

Isbell, L. A. (2009). *The fruit, the tree, and the serpent: Why we see so well*. Cambridge, MA: Harvard University Press.

Ito, R., & Mori, A. (2010). Vigilance against predators induced by eavesdropping on heterospecific alarm calls in a non-vocal lizard *Oplurus cuvieri cuvieri* (Reptilia: Iguania). *Proceedings of the Royal Society B: Biological Sciences, 277*, 1275–1280.

IUCN. (2020). *The IUCN Red List of threatened species*. Version 2019.3. Retrieved March 10, 2020, from http://iucnredlist.org.

Iverson, J. B. (1985). Geographic variation in sexual dimorphism in the mud turtle *Kinosternon hirtipes*. *Copeia, 1985*, 388–393.

Iverson, J. B. (1990). Nesting and parental care in the mud turtle, *Kinosternon flavescens*. *Canadian Journal of Zoology, 68*, 230–233.

Iverson, J. B., Francois, K., Jollay, J., Buckner, S. D., & Knapp, C. R. (2017). *Cyclura cychlura figginsi* (Northern Bahamian rock iguana). Allogrooming. *Herpetological Review, 48*, 188.

Iverson, J. B., Grant, T. D., Knapp, C. R., & Pasachnik, S. A. (Eds.). (2016). *Iguanas: Biology, systematics, and conservation. Herpetological Conservation and Biology, 11*, Monograph 6.

Iverson, J. B., Hines, K. N., & Valiulis, J. M. (2004). The nesting ecology of the Allen Cays rock iguana, *Cyclura cychlura inornata* in the Bahamas. *Herpetological Monographs, 18*, 1–36.

Iwaniuk, A. N., & Whishaw, I. Q. (2000). On the origin of skilled forelimb movements. *Trends in Neurosciences, 23*, 372–376.

Jackson, C. J., Jr., & Davis, J. D. (1972). A quantitative study of the courtship display of the red-eared turtle, *Chrysemys scripta elegans*. *Herpetologica, 28*, 58–64.

James, G. W. (1913). Hopi snake dance. *Outing, 36*, 302–310.

James, M., & Fox, T. (2007). The largest of lizards. *Newsletter of the Gippsland Plains Conservation Management Network, 1*, 9.

James, W. (1890). *Principles of psychology*. New York: Holt.

Janik, V. M. (2014). Cetacean vocal learning and communication. *Current Opinion in Neurobiology, 28*, 60–65.

Janse van Rensburg, D. A., Mouton, P. L. F. N., & Van Niekerk, A. (2009). Why cordylid lizards are black at the south-western tip of Africa. *Journal of Zoology*, *278*, 333–341.

Janzen, F. J. (1994). Climate change and temperature-dependent sex determination in reptiles. *Proceedings of the National Academy of Sciences*, *91*, 7487–7490.

Jarvie, S., Recio, M. R., Adolph, S. C., Seddon, P. J., & Cree, A. (2016). Resource selection by tuatara following translocation: A comparison of wild-caught and captive-reared juveniles. *New Zealand Journal of Ecology*, *40*, 334–341.

Jarvis, E. D., Güntürkün, O., Bruce, L., Csillag, A., Karten, H., Kuenzel, W., & Butler, A. B. (2005). Avian brains and a new understanding of vertebrate brain evolution. *Nature Reviews Neuroscience*, *6*, 151–159.

Jayne, B. C., & Bennett, A. F. (1990). Selection on locomotor performance capacity in a natural population of garter snakes. *Evolution*, *44*, 1204–1229.

Jellen, B. C., & Aldridge, R. D. (2014). It takes two to tango: Female movement facilitates male mate location in wild northern watersnakes (*Nerodia sipedon*). *Behaviour*, *151*, 421–434.

Jenkins, J. D. (1979). Notes on the courtship of the map turtle *Graptemys pseudogeographica* (Gray) (Reptilia, Testudines, Emydidae). *Journal of Herpetology*, *13*, 129–131.

Jerison, H. J. (1973). *Evolution of brain and intelligence*. Orlando, FL: Academic Press.

Jerison, H. J. (1982). Allometry, brain size, cortical surface, and convolutedness. In E. Armstrong & D. Falk (Eds.), *Primate brain evolution* (pp. 77–84). Boston: Springer.

Ji, Q., Ji, S. A., Cheng, Y. N., You, H. L., Lü, J. C., Liu, Y. Q., & Yuan, C. X. (2004). Palaeontology: Pterosaur egg with a leathery shell. *Nature*, *432*, 572–572.

Ji, Q., Ji, S. A., Lü, J. C., You, H. L., & Yuan, C. X. (2006). Embryos of Early Cretaceous choristodera (Reptilia) from the Jehol biota in Western Liaoning, China. *Journal of Paleontological Society of Korea*, *22*, 111–118.

Ji, Q., Wu, X. C., & Cheng, Y. N. (2010). Cretaceous choristoderan reptiles gave birth to live young. *Naturwissenschaften*, *97*, 423–428.

Joanen, T. (1969). Nesting ecology of alligators in Louisiana. In *Proceedings of the Annual Conference of Southeastern Association of Game and Fish Commissioners* (Vol. 23, pp. 141–151).

Jones, M. E., Worthy, J. P., Evans, S. E., & Worthy, T. H. (2009). A sphenodontine (Rhynchocephalia) from the Miocene of New Zealand and palaeobiogeography of the tuatara (*Sphenodon*). *Proceedings of the Royal Society. Biological Sciences*, *276*, 1385–1390.

Joy, J., & Crews, D. (1985). Social dynamics of group courtship behavior in male red-sided garter snakes (*Thamnophis sirtalis parietalis*). *Journal of Comparative Psychology*, *99*, 145.

Kabelik, D., & Crews, D. (2016). Hormones, brain, and behavior in reptiles. In D. W. Pfaff, A. Arnold, J. Balthazart & R. Rubin (Eds.), *Hormones, brain and behavior* (3rd ed., Vol. II, pp. 171–213). New York: Academic Press.

Kaiser, H., Lim, J., & O'Shea, M. (2012). Courtship entanglements: A first report of mating behavior and sexual dichromatism in the southeast Asian keel-bellied whipsnake, *Dryophiops rubescens* (Gray, 1835). *Herpetology Notes, 5,* 365–368.

Kalb, H., & Owens, D. (1994). Differences between solitary and arribada nesting olive ridley females during the internesting period. In *Proceedings of the 14th Annual Symposium on Sea Turtle Biology and Conservation. NOAA Technical Memorandum NMFS-SEFSC-351* (p. 68).

Kalb, H. J. (1999). Behavior and physiology of solitary and arribada nesting olive ridley sea turtles (*Lepidochelys olivacea*) during the internesting period [Doctoral dissertation]. Texas A & M University.

Kalmus, H. (1965). Origins and general features. *Symposium of the Zoological Society of London, 14,* 1–12.

Kamath, A., & Losos, J. (2017). The erratic and contingent progression of research on territoriality: A case study. *Behavioral Ecology and Sociobiology, 71,* 89. https://doi.org/10.1007/s00265-017-2319-z.

Kamath, A., & Losos, J. (2018). Reconsidering territoriality is necessary for understanding *Anolis* mating systems. *Behavioral Ecology and Sociobiology, 72,* 106. https://doi.org/10.1007/s00265-018-2524-4.

Kane, D., Davis, A. C., & Michael, C. J. (2019). Play behaviour by captive tree monitors, *Varanus macraei* and *Varanus prasinus. Herpetological Bulletin, 149,* 28–31. https://doi.org/10.33256/hb149.2831.

Kappeler, P. M. (2019). A framework for studying social complexity. *Behavioral Ecology and Sociobiology, 73,* 13.

Karunarathna, D., & Amarasinghe, A. (2011). Natural history and conservation status of *Calodactylodes illingworthorum* Deraniyagala, 1953 (Sauria: Gekkonidae) in south-eastern Sri Lanka. *Herpetotropicos, 6,* 5–10.

Kaufmann, J. (1992). Habitat use by wood turtles in central Pennsylvania. *Journal of Herpetology, 26,* 315–321.

Kaufmann, J. (1995). Home ranges and movements of wood turtles, *Clemmys insculpta*, in central Pennsylvania. *Copeia, 1,* 22–27.

Kawazu, I., Okabe, H., & Kobayashi, N. (2017). Direct observation of mating behavior involving one female and two male loggerhead turtles in the wild. *Current Herpetology, 36,* 69–72.

Kearney, M., Shine, R., Comber, S., & Pearson, D. (2001). Why do geckos group? An analysis of "social" aggregations in two species of Australian lizards. *Herpetologica, 57,* 411–422.

Keller, W. L., & Heske, E. J. (2001). An observation of parasitism of black rat snake (*Elaphe obsoleta*) eggs by a beetle (*Nicrophorus pustulatus*) in Illinois. *Transactions of the Illinois State Academy of Science, 94,* 167–169.

Kellert, S. R. (1993). *The biological basis for human values of nature.* Washington, DC: Island Press.

Kelly, D. (2002). The functional morphology of penile erection: Tissue designs for increasing and maintaining stiffness. *Integrative and Comparative Biology, 42,* 216–221.

Kelly, D. (2004). Turtle and mammal penis designs are anatomically convergent. *Proceedings of the Royal Society of London B, 271,* S293–S295.

Kennedy, M. L., Payne, J. F., & Heidt, G. (1990). *Laboratory studies in zoology* (4th ed.). Winston Salem, NC: Hunter Textbooks.

Kent, G. C. (1987). *Comparative anatomy of the vertebrates* (6th ed.). St. Louis: Times Mirror / Mosby.

Keren-Rotem, T., Bouskila, A., & Geffen, E. (2006). Ontogenetic habitat shift and risk of cannibalism in the common chameleon (*Chamaeleo chamaeleon*). *Behavioral Ecology and Sociobiology, 59*, 723–731.

Kermack, D. M., & Kermack, K. A. (1984). *The evolution of mammalian characters.* London: Croom Helm.

Kettler, L., & Carr, C. E. (2019). Neural maps of interaural time difference in the American alligator: A stable feature in modern archosaurs. *Journal of Neuroscience, 39*, 3882–3896.

Khan, J. J., Richardson, J. M., & Tattersall, G. J. (2010). Thermoregulation and aggregation in neonatal bearded dragons (*Pogona vitticeps*). *Physiology & Behavior, 100*, 180–186.

King, D. R., Pianka, E. R., & Green, B. (2002). Biology, ecology, and evolution. In J. B. Murphy, C. Ciofi, C. de La Panouse, & T. Walsh (Eds.), *Komodo dragons: Biology and conservation* (pp. 23–41). Washington, DC: Smithsonian Institution Press.

King, F. W., Thorbjarnarson, J. B., & Yamashita, C. (1998). Cooperative feeding, a misinterpreted and under-reported behavior of crocodilians. Retrieved from Florida Museum of Natural History website, http://www.flmnh.ufl.edu /herpetology/herpbiology/bartram.htm.

Kipling, R. (1895). The Undertakers. In *The second jungle book* (pp. 134–148). London: Macmillan & Co.

Kis, A., Huber, L., & Wilkinson, A. (2015). Social learning by imitation in a reptile (*Pogona vitticeps*). *Animal Cognition, 18*, 325–331.

Klauber, L. (1956). *Rattlesnakes.* Berkley: University of California Press.

Klauber, L. (1972). *Rattlesnakes: Their habits, life histories, and influence on mankind.* Berkley: University of California Press.

Klein, W., Abe, A., Andrade, D., & Perry, S. (2003). Structure of the posthepatic septum and its influence on visceral topology in the tegu lizard, *Tupinambis merianae* (Teidae: Reptilia). *Journal of Morphology, 258*, 151–157.

Knapp, C. R., Prince, L., & James, A. (2016). Movements and nesting of the Lesser Antillean iguana (*Iguana delicatissima*) from Dominica, West Indies: Implications for conservation. *Herpetological Conservation and Biology, 11*, 154–167.

Knell, R. J., Naish D., Tomkins, J. L., & Hone, D. W. E. (2012). Sexual selection in prehistoric animals: Detection and implications. *Trends in Ecology and Evolution, 28*, 38–47.

Kobayashi, T., & Watanabe, M. (1986). An analysis of snake-scent application behaviour in Siberian chipmunks (*Eutamias sibiricus asiaticus*). *Ethology, 72*, 40–52.

Koch, A., Guinea, M., & Whiting, S. (2008). Asynchronous emergence of flatback sea turtles, *Natator depressus*, from a beach hatchery in Northern Australia. *Journal of Herpetology, 42*, 1–9.

Koenig, W. D., & Dickinson, J. L. (2004). *Ecology and evolution of cooperative breeding in birds.* Cambridge, UK: Cambridge University Press.

Koenig, W. D., & Stacey, P. B. (1990). Acorn woodpeckers: Group-living and food storage under contrasting ecological conditions. In P. B. Stacey & W. D. Koenig (Eds.), *Cooperative breeding in birds: Long-term studies of ecology and behavior* (pp. 415-453). Cambridge, UK: Cambridge University Press.

Koford, R. R., Bowen B. S., & Vehrencamp S. L. (1990). Groove-billed anis: Joint nesting in a tropical cuckoo. In P. B. Stacey & W. D. Koenig (Eds.), *Cooperative breeding in birds: Long-term studies of ecology and behavior* (pp. 335-355). Cambridge, UK: Cambridge University Press.

Kofron, C. P., & Farris, P. A. (2015). Infrasound production by a yacare caiman *Caiman yacare* in the Pantanal, Brazil. *Herpetology Notes, 8,* 385-387.

Kojima, Y., & Mori, A. (2015). Active foraging for toxic prey during gestation in a snake with maternal provisioning of sequestered chemical defences. *Proceedings of the Royal Society B, 282,* 20142137. https://doi.org/10.1098/rspb.2014.2137.

Kovar, R., Brabec, M., Vita, R., Vodica, R., & Bogdan V. (2016). Nesting and over-wintering sites of Aesculapian snake, *Zamenis longissimus,* in an anthropogenic landscape in the northern extreme of its range. *Herpetological Bulletin, 136,* 35-36.

Kramer, M., & Burghardt, G. M. (1998). Precocious courtship and play in emydid turtles. *Ethology, 104,* 38-56.

Kramer, M., & Fritz, U. (1989). Courtship of the turtle, *Pseudemys nelsoni. Journal of Herpetology, 23,* 84-86.

Krause, J., & Ruxton, G. D. (2002). *Living in Groups.* Oxford: Oxford University Press.

Krause, J., Ruxton, G. D., & Krause, S. (2010). Swarm intelligence in animals and humans. *Trends in Ecology and Evolution, 25,* 28-34.

Kropach, C. (1971). Sea snake (*Pelamis platurus*) aggregations on slicks in Panama. *Herpetologica, 27,* 131-135.

Kropach, C. (1975). The yellow-bellied sea snakes, *Pelamis,* in the eastern Pacific. *The Biology of Sea Snakes.* Baltimore: University Park Press.

Krull, C. R., Parsons, S., & Hauber, M. (2009). The presence of ultrasonic harmonics in the calls of the rifleman (*Acanthisitta chloris*). *Notornis, 56,* 158-161.

Krysko, K. L., Hooper, A. N., & Sheehy, C. M., III. (2003). The Madagascar giant day gecko, *Phelsuma madagascariensis grandis* Gray 1870 (Sauria: Gekkonidae): A new established species in Florida. *Florida Scientist, 66,* 222-225.

Kunz, T. H., & Hosken, D. J. (2009). Male lactation: Why, why not and is it care? *Trends in Ecology & Evolution, 24,* 80-85.

Kurbanov, I. R. (1985). *Змеи Туркмении* [Snakes of Turkmenistan] (in Turkmen with Russian summary). Ashgabat: Turkmenian Book Publishing House.

Kürten, L., & Schmidt, U. (1982). Thermoperception in the common vampire bat (*Desmodus rotundus*). *Journal of Comparative Physiology, 146,* 223-228.

Kushlan, J. A., & Kushlan, M. S. (1980). Everglades alligator nests: Nesting sites for marsh reptiles. *Copeia, 1980,* 930-932.

Labra, A., & Niemeyer, H.M. (1999). Intraspecific chemical recognition in the lizard *Liolaemus tenuis*. *Journal of Chemical Ecology, 25*, 799–1811. https://doi.org/10.1023/A:1020925631314.

Labra, A., Reyes-Olivares, C., & Weymann, M. (2016). Asymmetric response to heterotypic distress calls in the lizard *Liolaemus chiliensis*. *Ethology, 122*, 758–768.

Lack, D. (1954). *The natural regulation of animal numbers*. Oxford: The Clarendon Press.

Lack, D. (1966). *Population studies of birds*. Oxford: Clarendon Press.

Lack, D. (1968a). Bird migration and natural selection. *Oikos, 19*, 1–9.

Lack, D. (1968b). *Ecological adaptations for breeding in birds*. London: Methuen.

LaDuc, T. J. (2002). Does a quick offense equal a quick defense? Kinematic comparisons of predatory and defensive strikes in the western diamond-backed rattlesnake (*Crotalus atrox*). In G. W. Schuett, M. Höggren, M. E. Douglas & H. W. Greene (Eds.), *Biology of the vipers* (pp. 267–278). Eagle Mountain, UT: Eagle Mountain Publishing.

Lafferriere, N., Antelo, R., Alda, F., Mårtensson, D., Hailer, F., Castroviejo-Fisher, S., & Vilá, C. (2016). Multiple paternity in a reintroduced population of the Orinoco crocodile (*Crocodylus intermedius*) at the El Frío Biological Station, Venezuela. *PLoS ONE, 11*, e0150245.

Laird, M. K., Thompson, M. B., & Whittington, C. M. (2019). Facultative oviparity in a viviparous skink (*Saiphos equalis*). *Biology Letters, 15*, 20180827.

Laland K. N., & Janik, V. (2006). The animal cultures debate. *Trends in Ecology & Evolution, 21*, 542–547.

Laland, K. N., Uller, T., Feldman, M. W., Sterelny, K., Müller, G. B., Moczek, A., Jablonka, E., & Odling-Smee, J. (2015). The extended evolutionary synthesis: Its structure, assumptions and predictions. *Proceedings of the Royal Society B: Biological Sciences, 282*, 20151019. https://doi.org/10.1098/rspb.2015.1019.

Lalonde, R. G., & Mangel, M. (1994). Seasonal effects on superparasitism by *Rhagoletis completa*. *Journal of Animal Ecology, 63*, 583–588.

Lambert, H., Carder, G., and D'Cruze, N. (2019). Given the cold shoulder: A review of the scientific literature for evidence of reptile sentience. *Animals, 9*, 821. https://doi.org/10.3390/ani9100821.

Lancaster, J. R., Wilson, P., & Espinoza, R. E. (2006). Physiological benefits as precursors of sociality: Why banded geckos band. *Animal Behaviour, 72*, 199–207.

Lance, V., Rostal, D., Elsey, R., & Trosclair, P., III. (2009). Ultrasonography of reproductive structures and hormonal correlates of follicular development in female American alligators, *Alligator mississippiensis*, in southwest Louisiana. *General and Comparative Endocrinology, 162*, 251–256.

Landberg, T., Mailhot, J., & Brainerd, E. (2003). Lung ventilation during treadmill locomotion in a terrestrial turtle, *Terrapene carolina*. *Journal of Experimental Biology, 206*, 3391–3404.

Lang, J. W. (2015). Behavioral ecology of gharial (*Gavialis gangeticus*). Paper presented at the Society for Integrative and Comparative Biology Annual Meeting, in the symposium "Integrative Biology of the Crocodilia," West Palm Beach, Florida.

Lang, J. W., & Kumar, P. (2016). Chambal gharial ecology project—2016 update. In *World Crocodile Conference: Proceedings of the 24th Working Meeting of the IUCN/SSC Specialist Group* (pp. 136-148). Gland, Switzerland: IUCN.

Langkilde, T., O'Connor, D., & Shine, R. (2007). Benefits of parental care: Do juvenile lizards obtain better-quality habitat by remaining with their parents? *Austral Ecology, 32*, 950-954.

Langner, C. (2017). Hidden in the heart of Borneo—shedding light on some mysteries of an enigmatic lizard: First records of habitat use, behavior, and food items of *Lanthanotus borneensis* Steindachner, 1878 in its natural habitat. *Russian Journal of Herpetology, 24*, 1-10.

Lanham, E., & Bull, C. (2000). Maternal care and infanticide in the Australian skink, *Egernia stokesii. Herpetological Review, 31*, 151.

Lanham, E., & Bull, C. (2004). Enhanced vigilance in groups in *Egernia stokesii*, a lizard with stable social aggregations. *Journal of Zoology, 263*, 95-99.

Lardie, R. L. (1975). Courtship and mating behavior in the yellow mud turtle, *Kinosternon flavescens flavescens. Journal of Herpetology, 9*, 223-227.

Larsen, K. W., Gregory, P. T., & Antoniak, R. (1993). Reproductive ecology of the common garter snake *Thamnophis sirtalis* at the northern limit of its range. *American Midland Naturalist, 129*, 336-345.

Latreille, M. (1825). *Families naturelles du règne animal.* Paris: Bailliere.

Laurin, M. R., & Reisz, R. (1995). A reevaluation of early amniote phylogeny. *Zoological Journal of the Linnean Society, 113*, 165-223.

Lawver, D. R., & Jackson, F. D. (2014). A review of the fossil record of turtle reproduction: Eggs, embryos, nests and copulating pairs. *Bulletin of the Peabody Museum of Natural History, 55*, 215-237.

Leache, A. D., Helmer, D. S., & Moritz, C. (2010). Phenotypic evolution in high-elevation populations of western fence lizards (*Sceloporus occidentalis*) in the Sierra Nevada Mountains. *Biological Journal of the Linnean Society, 100*, 630-641.

Lee, C., Pike, D. A., Tseng, H., Hsu, J., Huang, S., Shaner, P. L., Liao, C., Manica, A., & Huang, W. 2019. When males live longer: Resource-driven territorial behavior drives sex-specific survival in snakes. *Science Advances, 5*, eaar5478.

Lee, D. (1968). Possible communication between eggs of the American alligator. *Herpetologica, 24*, 88.

Lee, M. S. Y. (2013). Turtle origins: Insights from phylogenetic retrofitting and molecular scaffolds. *Journal of Evolutionary Biology, 26*, 2729-2738.

Lee, M. S. Y., Hugall, A. F., Lawson, R., & Scanlon, J. D. (2007). Phylogeny of snakes (Serpentes): Combining morphological and molecular data in likelihood, Bayesian and parsimony analyses. *Systematics and Biodiversity, 5*, 371-389.

Lee, M. S. Y., & Shine, R. (1998). Reptilian viviparity and Dollo's Law. *Evolution, 52*, 1441-1450.

Lee, P. (2008). Molecular ecology of marine turtles: New approaches and future directions. *Journal of Experimental Marine Biology and Ecology, 356*, 25-42.

Lee, S. A. (2019). Trends in embryonic and ontogenetic growth metabolisms in nonavian dinosaurs and extant birds, mammals, and crocodylians with implications for dinosaur egg incubation. *Physical Review E, 99*, 052405.

Legler, J. M. (1960). Natural history of the ornate box turtle, *Terrapene ornata ornata* Agassiz. *University of Kansas Publications of the Museum of Natural History*, *11*, 527–669.

Legler, J. M., & Vogt, R. C. (2013). *The turtles of Mexico: Land and freshwater forms*. Berkeley: University of California Press.

Lehrman, D. S. (1953). A critique of Konrad Loernz's theory of instinctive behavior. *Quarterly Review of Biology*, *28*, 337–363.

Lehtonen, J. (2020). The Price equation and the unity of social evolution theory. *Philosophical Transaction of the Royal Society B*, *375*, 20190362. https://doi.org/10.1098/rstb.2019.0362.

Leighton, P. A., Horrocks, J. A., Krueger, B. H., Beggs, J. A., & Kramer, D. L. (2008). Predicting species interactions from edge responses: Mongoose predation on hawksbill sea turtle nests in fragmented beach habitat. *Proceedings of the Royal Society, B: Biological Sciences*, *275*, 2465–2472.

LeMaster, M. P., & Mason, R. T. (2002). Variation in a female sexual attractiveness pheromone controls male mate choice in garter snakes. *Journal of Chemical Ecology*, *28*, 1269–1285.

Lemos-Espinal, J., Ballinger, R., Sarabia, S., & Smith, G. (1997). Thermal Ecology of the Lizard *Sceloporus muscronatus muscronatui* in Sierra Del Ajusco, Mexico. *Southwestern Naturalist*, *42*, 344–347.

Leonardi, G. (1999). Cooperative hunting of the jackdaws by the lanner falcon (*Falco biarmicus*). *Journal of Raptor Research*, *33*, 123–137.

Leu, S., Burzacott, D., Whiting, M., & Bull, C. (2015). Mate familiarity affects pairing behaviour in a long-term monogamous lizard: Evidence from detailed bio-logging and a 31-year field study. *Ethology*, *121*, 760–768.

Levine, B. A., Smith, C. F., Schuett, G. W., Douglas, M. R., Davis, M. A., & Douglas, M. E. (2015). Bateman-Trivers in the 21st Century: Sexual selection in a North American pitviper. *Biological Journal of the Linnean Society*, *114*, 436–445.

Leyhausen, P. (1965). The communal organization of solitary mammals. *Symposia of the Zoological Society of London*, *14*, 249–263.

Li, J. (1995). *China Snake Island*. Dalian: Liaoning Science and Technology Press.

Li, Q., Gao K.-Q., Meng, Q., Clarke, J. A., Shawkey, M. D., D'Alba, L., Pei, R., Ellison, M., Norell, M. A., & Vinther, J. (2012). Reconstruction of *Microraptor* and the evolution of iridescent plumage. *Science*, *335*, 1215–1219.

Ligon, J. D. (1999). *The evolution of avian breeding systems*. Oxford: Oxford University Press.

Ligon, R. A., & McGraw, K. J. (2016). Social costs enforce honesty of a dynamic signal of motivation. *Proceedings of the Royal Society B: Biological Sciences*, *283*, 20161873.

Lillywhite, H. B. (2014). *How snakes work: Structure, function and behavior of the world's snakes*. Oxford: Oxford University Press.

Lillywhite, H. B., Babonis, L., Sheehy, C., III, & Tu, M., III. (2008). Sea snakes (*Laticauda* spp.) require fresh drinking water: Implication for the distribution and persistence of populations. *Physiological and Biochemical Zoology*, *81*, 785–796.

Lillywhite, H. B., Brischoux, F., Sheehy, C., III, & Pfaller, J. (2012). Dehydration and drinking responses in a pelagic sea snake. *Integrative and Comparative Biology, 52*, 227–234

Lillywhite, H. B., Heatwole, H., & Sheehy, C. (2015). Dehydration and drinking behavior in true sea snakes (Elapidae: Hydrophiinae: Hydrophiini). *Journal of Zoology, 296*, 261–269.

Lillywhite, H. B., Iii, C., Brischoux, F., & Pfaller, J. (2015). On the abundance of a pelagic sea snake. *Journal of Herpetology, 49*, 184–189.

Lillywhite, H. B., Solórzano, A., Sheehy, C., III, Ingley, S., & Sasa, M. (2010). New perspectives on the ecology and natural history of the yellow-bellied sea snake (*Pelamis platurus*) in Costa Rica: Does precipitation influence distribution. *IRCF Reptiles and Amphibians, 17*, 69–72.

Lima, D. C., Passos, D. C., & Borges-Nojosa, D. M. (2011). Communal nests of *Phyllopezus periosus*, an endemic gecko of the Caatinga of northeastern Brazil. *Salamandra, 47*, 227–228.

Limpus, C. J., Miller, J. D., Parmenter, C. J., & Limpus, D. J. (2003). The green turtle, *Chelonia mydas*, population of Raine Island and the northern Great Barrier Reef: 1843–2001. *Memoirs of the Queensland Museum, 49*, 349–440.

Lindeman, P. V. (1991). Survivorship of overwintering hatchling painted turtles, *Chrysemys picta*, in northern Idaho. *Canadian Field Naturalist, 105*, 263–266.

Lindeman, P. V. (1999). Aggressive interactions during basking among four species of emydid turtles. *Journal of Herpetology, 33*, 214–219.

Lindzey, J., & Crews, D. (1986). Hormonal control of courtship and copulatory behavior in male *Cnemidophorus inornatus*, a direct sexual ancestor of a unisexual, parthenogenetic lizard. *General and Comparative Endocrinology, 64*, 411–418.

Liu, Y., He, B., Shi, H., Murphy, R. W., Fong, J. J., Wang J., Fu L., & Ma, Y. (2008). An analysis of courtship behaviour in the four-eyed spotted turtle, *Sacalia quadriocellata* (Reptilia: Testudines: Geoemydidae). *Amphibia-Reptilia, 29*, 185–195.

Lockley, M. G., McCrea, R. T., Buckley, L. G., Lim, J. D., Matthews, N. A., Breithaupt, B. H., Houck, K. J., Gierliński, G. D., Surmik, D., Kim, K. S., & Xing, L. (2016). Theropod courtship: Large scale physical evidence of display arenas and avian-like scrape ceremony behaviour by Cretaceous dinosaurs. *Scientific Reports, 6*, 18952.

Lohmann, K. J., Lohmann, C. M., Brothers, J. R., & Putman, N. F. (2013). Natal homing and imprinting in sea turtles. In J. Wyneken, K. J. Lohmann & J. A. Musick (Eds.), *The biology of sea turtles* (pp. 59–77). New York: CRC Press.

Londoño, C., Bartolomé, A., Carazo, P., & Font, E. (2018). Chemosensory enrichment as a simple and effective way to improve the welfare of captive lizards. *Ethology, 124*, 674–683.

López, P., Aragón, P., & Martín, J. (1998). Iberian rock lizards (*Lacerta monticola cyreni*) assess conspecific information using composite signals from faecal pellets. *Ethology, 104*, 809–820.

López, P., & Martín, J. (2001). Pheromonal recognition of females takes precedence over the chromatic cue in male Iberian wall lizards *Podarcis hispanica*. *Ethology, 107*, 901–912.

López, P., & Martín, J. (2009). Potential chemosignals associated with male identity in the amphisbaenian *Blanus cinereus*. *Chemical Senses, 34*, 479–486.

López-Ortiz, R., & Lewis, A. R. (2002). Seasonal abundance of hatchlings and gravid females of *Sphaerodactylus nicholsi* in Cabo Rojo, Puerto Rico. *Journal of Herpetology, 36*, 276–280.

Lorenz, K. (1937). Über den Begriff der Instinkhandlung. *Folia Biotheoretica, 2*, 17–50.

Lorenz, K. (1950). The comparative method in studying innate behaviour patterns. *Symposia of the Society for Experimental Biology, 4*, 221–68.

Lorenz, K. (1966). *On aggression*. New York: Harcourt, Brace & World, Inc.

Lorenz, K. (1970). *Studies in animal and human behavior* (Vol. 1). London: Methuen.

Lorioux, S., Lisse, H., & Lourdais, O. (2013). Dedicated mothers: Predation risk and physical burden do not alter thermoregulatory behaviour of pregnant vipers. *Animal Behaviour, 86*, 401–408.

Losos, J. B. (1985). An experimental demonstration of the species-recognition role of *Anolis* dewlap color. *Copeia, 1985*, 905–910.

Losos, J. B. (2009). *Lizards in an evolutionary tree: Ecology and adaptive radiation of anoles*. Berkeley, Los Angeles & London: University of California Press.

Losos, J. B., & Chu, L. R. (1998). Examination of factors potentially affecting dewlap size in Caribbean anoles. *Copeia, 1998*, 430–438.

Losos, J. B., Warheit, K. I., & Schoener, T. W. (1997). Adaptive differentiation following experimental island colonization on *Anolis* lizards. *Nature, 387*, 70–73.

Lott, D. F. (1991). *Intraspecific variation in the social systems of wild vertebrates* (Vol. 2). Cambridge, UK: Cambridge University Press.

Lougheed, S. C., Gibbs, H. L., Prior, K. A., & Weatherhead, P. J. (1999). Hierarchical patterns of genetic population structure in black rat snakes (*Elaphe obsolete obsolete*) as revealed by microsatellite DNA analysis. *Evolution, 53*, 1995–2001.

Lovern, M. B., Holmes, M. M., & Wade, J. (2004). The green anole (*Anolis carolinensis*): A reptilian model for laboratory studies of reproductive morphology and behavior. *ILAR Journal, 45*, 54–64.

Lovich, J. E., Garstka, W. R., & Cooper, W. E., Jr. (1990). Female participation in courtship behavior of the turtle *Trachemys s. scripta. Journal of Herpetology, 24*, 422–424.

Lowe, C. (1948). Territorial behavior in snakes and the so-called courtship dance. *Herpetologica, 4*, 129–135.

Lowe, W. (1932). *The trail that is always new*. London: Gurney and Jackson.

Löwenborg, K., Gotthard, K., & Hagman, M. (2012). How a thermal dichotomy in nesting environments influences offspring of the world's most northerly oviparous snake, *Natrix natrix* (Colubridae). *Biological Journal of the Linnean Society, 107*, 833–844.

Lü, J., Kobayashi, Y., Deeming, D. C., & Liu, Y. (2014). Post-natal parental care in a Cretaceous diapsid from northeastern China. *Geosciences Journal, 19*, 273–280.

Lucas, J. R., Gentry, K. E., Sieving, K. E., & Freeberg, T. M. (2018). Communication as a fundamental part of Machiavellian intelligence. *Journal of Comparative Psychology, 132*, 442–454.

Luiselli, L., Madsen, T., Capizzi, D., Rugiero, L., Pacini, N., & Capula, M. (2011). Long-term population dynamics in a Mediterranean aquatic snake. *Ecological Research, 26,* 745-753.

Lutz, D. (2005). *Tuatara: A living fossil.* Salem, OR: Dimi Press.

Lyman-Henley, L. P., & Burghardt, G. M. (1994). Opposites attract: Effects of social and dietary experience on snake aggregation behaviour. *Animal Behaviour, 47,* 980-982.

Machado, L. L., Galdino, C. A., & Sousa, B. M. (2007). Defensive behavior of the lizard *Tropidurus montanus* (Tropiduridae): Effects of sex, body size and social context. *South American Journal of Herpetology, 2,* 136-141.

Mack, A. L., & Jones, J. (2003). Low-frequency vocalizations by cassowaries (*Casuarius* spp.). *Auk, 120,* 1062-1068.

MacLean, P. D. (1985). Brain evolution relating to family, play, and the separation call. *Archives of General Psychiatry, 42,* 405-417.

MacLean, P. D. (1990). *The triune brain in evolution.* New York: Plenum.

Maddin, H. C., Mann, A., & Hebert, B. (2020). Varanopid from the Carboniferous of Nova Scotia reveals evidence of parental care in amniotes. *Nature Ecology & Evolution, 4,* 50-56.

Maddock, L., Bone, Q., & Rayner, J. M. (1994). *The mechanics and physiology of animal swimming.* Cambridge, UK: Cambridge University Press.

Madsen, T. (1984). Movements, home range size and habitat use of radio-tracked grass snakes (*Natrix natrix*) in southern Sweden. *Copeia, 1984,* 707-713.

Magnusson, W. E. (1980). Hatching and creche formation by *Crocodylus porosus. Copeia, 1980,* 359-362.

Magnusson, W. E. (1982). Mortality of eggs of the crocodile *Crocodylus porosus* in Northern Australia. *Journal of Herpetology, 16,* 121-130.

Magnusson, W. E. (1989). Termite mounds as nest sites. In C. Ross (Ed.), *Crocodiles and alligators* (pp. 122). New York: Facts on File Publications.

Magnusson, W. E., & Lima, A. P. (1984). Perennial communal nesting by *Kentropyx calcaratus. Journal of Herpetology, 18,* 73-75.

Magnusson, W. E., Lima, A. P., & Sampaio, R. M. (1985). Sources of heat for nests of *Paleosuchus trigonatus* and a review of crocodilian nest temperatures. *Journal of Herpetology, 19,* 199-207.

Mahmoud, I. Y. (1967). Courtship behavior and sexual maturity in four species of kinosternid turtles. *Copeia, 1967,* 314-319.

Maina, J. N. (2006). Development, structure, and function of a novel respiratory organ, the lung-air sac system of birds: To go where no other vertebrate has gone. *Biological Reviews, 81,* 545-579.

Mallow, D., Ludwig, D., & Nilson, G. (2003). *True vipers: Natural history and toxinology of Old World vipers.* Malabar: Krieger Publishing Company.

Maloof, A. C., Porter, S. M., Moore, J. L., Dudás, F. Ö., Bowring, S. A., Higgins, J. A., & Eddy, M. P. (2010). The earliest Cambrian record of animals and ocean geochemical change. *Bulletin, 122,* 1731-1774.

Mangiacotti, M., Gaggiani, S., Coladonato, A. J., Scali, S., Zuffi, M. A. L., & Sacchi, R. (2019). First experimental evidence that proteins from femoral glands convey identity-related information in a lizard. *Acta Ethologica, 22,* 57-65.

Mangiacotti, M., Martín, J., López, P., Reyes-Olivares, C. V., Rodríguez-Ruiz, G., Coladonato, A. J., . . . Sacchi, R. (2020). Proteins from femoral gland secretions of male rock lizards Iberolacerta cyreni allow self—but not individual—recognition of unfamiliar males. *Behavioral Ecology and Sociobiology, 74*, 68. https://doi.org/10.1007/s00265-020-02847-8.

Mann, A., McDaniel, E. J., McColville, E. R., & Maddin, H. C. (2019). *Carbonodraco lundi* gen et sp. nov., the oldest parareptile, from Linton, Ohio, and new insights into the early radiation of reptiles. *Royal Society Open Science, 6*, 191191.

Manning, C. J., Dewsbury, D. A., Wakeland, E. K., & Potts, W. K. (1995). Communal nesting and communal nursing in house mice, *Mus musculus domesticus. Animal Behaviour, 50*, 741–751.

Mannion, P. D., Benson, R. B., Carrano, M. T., Tennant, J. P., Judd, J., & Butler, R. J. (2015). Climate constrains the evolutionary history and biodiversity of crocodylians. *Nature Communications, 6*, 8438.

Manrod, J. D., Hartdegen, R., & Burghardt, G. M. (2008). Rapid solving of a problem apparatus by juvenile black-throated monitor lizards (*Varanus albigularis albigularis*). *Animal Cognition, 11*, 267–273.

Manzig, P. C., Kellner, A. W., Weinschütz, L. C., Fragoso, C. E., Vega, C. S., Guimarães, G. B., Godoy, L. C., Liccardo, A., Ricetti, J. H., & Moura, C. C. (2014). Discovery of a rare pterosaur bone bed in a Cretaceous desert with insights on ontogeny and behavior of flying reptiles. *PLoS ONE, 9*, 100005.

Marcellini, D. (1977). Acoustic and visual display behavior of gekkonid lizards. *American Zoologist, 17*, 251–260.

Marchand, M. N., & Litvaitis, J. A. (2004). Effects of habitat features and landscape composition on the population structure of a common aquatic turtle in a region undergoing rapid development. *Conservation Biology, 18*, 758–767.

Marchand, M. N., Litvaitis, J. A., Maier, T. J., & DeGraaf, R. M. (2002). Use of artificial nests to investigate predation on freshwater turtle nests. *Wildlife Society Bulletin, 30*, 1092–1098.

Marco, A., Díaz-Paniagua, C., & Hidalgo-Vila, J. (2004). Influence of egg aggregation and soil moisture on incubation of flexible-shelled lacertid lizard eggs. *Canadian Journal of Zoology, London, 82*, 60–65.

Maritz, B., Alexander, G. J., & Maritz, R. A. (2019). The underappreciated extent of cannibalism and ophiophagy in African cobras. *Ecology, 100*, e02522.

Marquez, M. (1994). Synopsis of biological data on the Kemp's ridley turtle, *Lepidochelys kempi* (Garman, 1880). Miami, FL: National Oceanic and Atmospheric Administration.

Marsola, J. C. A., Batezelli, A., Montefeltro, F. C., Grellet-Tinner, G., & Langer, M. C. (2016). Palaeoenvironmental characterization of a crocodilian nesting site from the Late Cretaceous of Brazil and the evolution of crocodyliform nesting strategies. *Palaeogeography, Palaeoclimatology, Palaeoecology, 457*, 221–232.

Martill, D. M., & Naish, D. (2006). Cranial crest development in the azhdarchoid pterosaur *Tupuxuara*, with a review of the genus and tapejarid monophyly. *Palaeontology, 49*, 925–941.

Martill, D. M., Tischlinger, H., & Longrich, N. R. (2015). A four-legged snake from the Early Cretaceous of Gondwana. *Science*, *349*, 416-419.

Martin, G. R. (1982). An owl's eye: Schematic optics and visual performance in *Strix aluco*. *Journal of Comparative Physiology*, *145*, 341-349.

Martín, J., & López, P. (2011). Pheromones and reproduction in reptiles. In D. Norris & K. Lopez (Eds.), *Hormones and reproduction of vertebrates* (Vol. 3, pp. 141-167). Cambridge, MA: Academic Press.

Martín, J., & López, P. (2014). Pheromones and chemical communication in lizards. In J. L. Rheubert, D. S. Siegel & S. E. Trauth (Eds.), *Reproductive biology and phylogeny of lizards and tuatara* (pp. 43-77). Boca Raton, FL: CRC Press.

Martín, J., Polo-Cavia, N., Gonzalo, A., López, P., & Civantos, E. (2011). Social aggregation behaviour in the North African amphisbaenian *Trogonophis wiegmanni*. *African Journal of Herpetology*, *60*, 171-176.

Martin, J. E. (2013). Surviving a potentially lethal injury? Bite mark and associated trauma in the vertebra of a dyrosaurid crocodilian. *Palaios*, *28*, 6-8.

Martin, J. H., & Bagby, R.M. (1973). Properties of rattlesnake shaker muscle. *Journal of Experimental Zoology*, *185*, 293-300.

Martin, K. (1999). Ready and waiting: Delayed hatching and extended incubation of anamniotic vertebrate terrestrial eggs. *American Zoologist*, *39*, 279-288.

Martin, L. D., & Bennett, D. K. (1977). The burrows of the Miocene beaver *Palaeocastor*, western Nebraska, USA. *Palaeogeography, Palaeoclimatology and Palaeoecology*, *22*, 173-193.

Martin, T., Marugán-Lobón, J., Vullo, R., Martín-Abad, H., Luo, Z.-X., & Buscalioni, A. D. (2015). A Cretaceous eutriconodont and integument evolution in early mammals. *Nature*, *526*, 380.

Martins, E. P. (1993). A comparative study of the evolution of *Sceloporus* push-up displays. *The American Naturalist*, *142*, 994-1018.

Martins, E. P. (1994). Structural complexity in a lizard communication system: The *Sceloporus graciosus* "push-up" display. *Copeia*, *1994*, 944-955.

Mason, P., & Adkins, E. K. (1976). Hormones and social behavior in the lizard, *Anolis carolinensis*. *Hormones and Behavior*, *7*, 75-86.

Mason, R. T., & Crews, D. (1985). Female mimicry in garter snakes. *Nature*, *316*, 59-60.

Mason, R. T., & Gutzke, W. H. N. (1990). Sex recognition in the leopard gecko, *Eublepharis macularius* (Sauria: Gekkonidae): Possible mediation by skin-derived semiochemicals. *Journal of Chemical Ecology*, *16*, 27-36.

Mason, R. T., Jones, T., Fales, H., Pannell, L., & Crews, D. (1990). Characterization, synthesis, and behavioral responses to sex attractiveness pheromones of red-sided garter snakes (*Thamnophis sirtalis parietalis*). *Journal of Chemical Ecology*, *16*, 2353-2369.

Mason, R. T., & Parker, M. R. (2010). Social behavior and pheromonal communication in reptiles. *Journal of Comparative Physiology A*, *196*, 729-749.

Mateo, J. A., & Cuadrado, M. (2012). Communal nesting and parental care in Oudri's Fan-Footed Gecko (*Ptyodactylus oudrii*): Field and experimental evidence of an adaptive behavior. *Journal of Herpetology*, *46*, 209-213.

Mateo, J. A., & Pleguezuelos, J. M. (2015). Cannibalism of an endemic island lizard (genus *Gallotia*). *Zoologischer Anzeiger, 259*, 131–134.

Mathevon, N., Vergne, A., & Aubin, T. (2013). Acoustic communication in crocodiles: How do juvenile calls code information? *Proceedings of Meetings on Acoustics, 19*, 1–5.

Matsubara, S., Deeming, D., & Wilkinson, A. (2017) Cold-blooded cognition: New directions in reptile cognition. *Current Opinion in Behavioral Science, 16*, 126–130.

Matuschka, F. R., & Bannert, B. (1989). Recognition of cyclic transmission of *Sarcocystis stehlinii* n. sp. in the Gran Canarian giant lizard. *The Journal of Parasitology, 75*, 383–387.

Maxwell, W. D., & Ostrom, J. H. (1995). Taphonomy and paleobiological implications of *Tenontosaurus-Deinonychus* associations. *Journal of Vertebrate Paleontology, 15*, 707–712.

Mayer, M., Shine, R., & Brown, G. P. (2016). Bigger babies are bolder: Effects of body size on personality of hatchling snakes. *Behaviour, 153*, 313–323.

McAlpin, S., Duckett, P., & Stow, A. (2011). Lizards cooperatively tunnel to construct a long-term home for family members. *PLoS ONE, 6*, e19041.

McCall, P. J. (1995). Oviposition aggregation pheromone in the *Simulium damnosum* complex. *Medical and Veterinary Entomology, 9*, 101–108.

McCall, P. J., Heath, R. R., Dueben, B. D., & Wilson, M. D. (1997). Oviposition pheromone in the *Simulium damnosum* complex: Biological activity of chemical fractions from gravid ovaries. *Physiological Entomology, 22*, 224–230.

McCallum, M. L., & McCallum, J. L. (2006). Publication trends of natural history and field studies in herpetology. *Herpetological Conservation and Biology, 1*, 63–68.

McClain, C. R., Balk, M. A., Benfield, M. C., Branch, T. A., Chen, C., Cosgrove, J., ... Thaler, A. D. (2015). Sizing ocean giants: Patterns of intraspecific size variation in marine megafauna. *PeerJ, 3*, e715.

McCoy, J. K., Baird, T. A., & Fox, S. F. (2003). Sexual selection, social behavior, and the environmental potential for polygyny. In S. F. Fox, J. K. McCoy & T. A. Baird (Eds.), *Lizard social behavior* (pp. 149–171). Baltimore: Johns Hopkins University Press.

McCracken, G. F. (1984). Communal nursing in Mexican free-tailed bats. *Science, 223*, 1090–1091.

McCrea, R. T., & Buckley, L. G. (2011). Preliminary observations and interpretations on a hadrosaur (Lambeosaurinae) associated with an abundance of juvenile *Tyrannosaur* (Albertosaurinae) teeth from the upper Cretaceous (Campanian/Maastrichtian) Wapiti Formation of northeastern British Columbia. Paper presented at the International Hadrosaur Symposium at the Royal Tyrrell Museum of Palaeontology, Drumheller, Alberta, Canada.

McCrea, R. T, Buckley, L G., Farlow, J. O., Lockley, M. G., Currie, P. J., Matthews, N. A., & Pemberton, S. G. (2014) A "terror of tyrannosaurs": The first trackways of tyrannosaurids and evidence of gregariousness and pathology in Tyrannosauridae. *PLoS ONE, 9*, e103613. https://doi.org/10.1371/journal.pone.0103613.

McCue, M. D. (2010). Starvation physiology: Reviewing the different strategies animals use to survive a common challenge. *Comparative Biochemistry and Physiology Part A: Molecular & Integrative Physiology, 156*, 1–18.

McGlashan, J., Loudon, F., Thompson, M., & Spencer, R. (2015). Hatching behavior of eastern long-necked turtles (*Chelodina longicollis*): The influence of asynchronous environments on embryonic heart rate and phenotype. *Comparative Biochemistry and Physiology Part A: Molecular & Integrative Physiology, 188*, 58–64.

McGlashan, J., Spencer, R., & Old, J. (2011). Embryonic communication in the nest: Metabolic responses of reptilian embryos to developmental rates of siblings. *Proceedings of the Royal Society B: Biological Sciences, 279*, 1709–1715.

McGlashan, J., Thompson, M., Janzen, F., & Spencer, R. (2018). Environmentally induced phenotypic plasticity explains hatching synchrony in the freshwater turtle *Chrysemys picta*. *Journal of Experimental Zoology Part A: Ecological and Integrative Physiology, 329*, 362–372.

McGowan, A., Broderick, A., Deeming, J., Godley, B., & Hancock, E. (2001). Dipteran infestation of loggerhead (*Caretta caretta*) and green (*Chelonia mydas*) sea turtle nests in northern Cyprus. *Journal of Natural History, 35*, 573–581.

McGowan, A., Rowe, L., Broderick, A., & Godley, B. (2001). Nest factors predisposing loggerhead sea turtle (*Caretta caretta*) clutches to infestation by dipteran larvae on northern Cyprus. *Copeia, 2001*, 808–812.

McGowan, K. J., & Woolfenden, G. E. (1989). A sentinel system in the Florida scrub jay. *Animal Behaviour, 37*, 1000–1006.

McGuire, J. A., & Dudley, R. (2005). The cost of living large: Comparative gliding performance in flying lizards (Agamidae: *Draco*). *The American Naturalist, 166*, 93–106.

McGuire, J. A., & Dudley, R. (2011). The biology of gliding in flying lizards (genus *Draco*) and their fossil and extant analogs. *Integrative and Comparative Biology, 51*, 983–990.

McIlhenny, E. A. (1935). *The alligator's life history*. Boston: Christopher Publishing House.

McKenna, L. (2016). Vocalizations of sea turtle hatchlings and embryos [Unpublished Master's thesis]. Indiana University–Purdue University.

McKenna, L., Paladino, F., Tomillo, P., & Robinson, N. (2019). Do sea turtles vocalize to synchronize hatching or nest emergence? *Copeia, 107*, 120–123.

McKeown, S. (1996). *A field guide to reptiles and amphibians in the Hawaiian Islands*. Honolulu: Diamond Head Publishing.

McLain, M. A., Nelsen, D., Snyder, K., Griffin, C. T., Siviero, B., Brand, L. R., & Chadwick, A. V. (2018). Tyrannosaur cannibalism: A case of a tooth-traced tyrannosaurid bone in the Lance Formation (Maastrichtian), Wyoming. *Palaios, 33*, 164–173.

McMillan, V. E. (2000). Aggregating behavior during oviposition in the dragonfly *Sympetrum vicinum* (Hagen) (Odonata: Libellulidae). *American Midland Naturalist, 144*, 11–19.

McVay, J., Rodriguez, D., Rainwater, T., Dever, J., Platt, S., McMurry, S., & Densmore, L., III. (2008). Evidence of multiple paternity in Morelet's crocodile (*Crocodylus moreletii*) in Belize, CA, inferred from microsatellite markers. *Journal of Experimental Zoology Part A: Ecological Genetics and Physiology, 309*, 643–648.

Mead, J. I., Steadman, D. W., Bedford, S. H., Bell. C. J., & Spriggs, M. (2002). New extinct mekosuchine crocodile from Vanuatu, South Pacific. *Copeia, 2002*, 632–641.

Medica, P., Bury, R., & Luckenbach, R. (1980). Drinking and construction of water catchments by the desert tortoise, *Gopherus agassizii*, in the Mojave Desert. *Herpetologica, 36*, 301–304.

Medina-Rangel, G. F., & Lopez-Parilla, Y. R. (2014). *Ptychoglossus festae* (Per-acca's largescale lizard). Nesting. *Herpetological Review, 45*, 504.

Meek, R. (2017). Repeated use of roadside tunnels of the European mole (*Talpa europea*) as a communal nesting area by grass snakes, *Natrix natrix*: Are there thermal benefits? *Herpetological Bulletin, 139*, 16–19.

Melstrom, K. M., & Irmis, R. B. (2019). Repeated evolution of herbivorous Crocodyliforms during the age of dinosaurs. *Current Biology, 29*, 2389–2395.e3.

Mendelson, J. R., Schuett, G. W., & Lawson, D. (2019). Krogh's principle and why the modern zoo is important to academic research. In A. B. Kaufman, M. J. Bashaw & T. L. Maple (Eds.), *Scientific foundations of zoo and aquariums: Their role in conservation and research* (pp. 586–617). Cambridge, UK: Cambridge University Press.

Mendl, M., Burman, O. H. P., & Paul, S. P. (2010). An integrative and functional framework for the study of animal emotion and mood. *Proceedings of the Royal Society B: Biological Sciences, 277*, 2895–2904.

Mendyk, R. W., & Horn, H. G. (2011). Skilled forelimb movements and extractive foraging in the arboreal monitor lizard *Varanus beccarii* (Doria, 1874). *Herpetological Review, 42*, 343–349.

Meng, Q., Liu, J., Varricchio, D. J., Huang, T., & Gao, C. (2004). Parental care in an ornithischian dinosaur. *Nature, 431*, 145–146.

Meredith, M., & Burghardt, G. M. (1978). Electrophysiological studies of the tongue and accessory olfactory bulb in garter snakes. *Physiology and Behavior, 21*, 1001–1008.

Merlen, G., & Thomas, R. (2013). A Galapagos ectothermic terrestrial snake gambles a potential chilly bath for a protein-rich dish of fish. *Herpetological Review, 44*, 415–417.

Meyer-Rochow, V. B. (2014). Polarization sensitivity in reptiles. In G. Horváth (Ed.), *Polarized light and polarization vision in animal sciences* (Vol. 2, pp. 265–274). Berlin: Springer.

Meylan, A. B., Bowen, B. W., & Avise, J. C. (1990). A genetic test of the natal homing versus social facilitation models for green turtle migration. *Science, 248*, 724–727.

Michael, D. R., Cunningham, R. B., & Lindenmayer, D. B. (2010). The social elite: Habitat heterogeneity, complexity and quality in granite inselbergs influence patterns of aggregation in *Egernia striolata* (Lygosominae: Scinci-dae). *Austral Ecology, 35*, 862–870.

Michener, C. D. (1969). Comparative social behavior of bees. *Annual Review of Entomology, 14,* 299-342.

Minta, S. C., K. A. Minta, & D. F. Lott. (1992). Hunting associations between badgers (*Taxidea taxus*) and coyotes (*Canis latrans*). *Journal of Mammalogy, 73,* 814-820.

Mitchell, J. C. (1986). *Cannibalism in reptiles: A worldwide review.* Herpetological Circular No 15. Lawrence, KS: Society for the Study of Amphibians and Reptiles.

Mitchell, T. S., Janzen, F. J., & Warner, D. A. (2018). Quantifying the effects of embryonic phenotypic plasticity on adult phenotypes in reptiles: A review of current knowledge and major gaps. *Journal of Experimental Zoology, 329,* 203-214.

Mitchell, T. S., Maciel, J. A., & Janzen, F. J. (2015). Maternal effects influence phenotypes and survival during early life stages in an aquatic turtle. *Functional Ecology, 29,* 268-276.

Mociño-Deloya, E., Setser, K., Pleguezuelos, J. M., Kardon, A., & Lazcano, D. (2009). Cannibalism of nonviable offspring by postparturient Mexican lance-headed rattlesnakes, *Crotalus polystictus. Animal Behaviour, 77,* 145-150.

Mock, D. W., Drummond, H., & Stinson, C. H. (1990). Avian siblicide. *American Scientist, 78,* 438-449.

Moldowan, P. D. (2014). Sexual dimorphism and alternative reproductive tactics in the Midland Painted Turtle (*Chrysemys picta marginata*) [Unpublished Master's thesis]. Laurentian University.

Moll, D., & Moll, E. O. (2004). *The ecology, exploitation and conservation of river turtles.* Oxford: Oxford University Press.

Moll, E. O. (1979). Reproductive cycles and adaptations. In M. Harless and H. Morelock (Eds.), *Turtles: Perspectives and research* (pp. 305-331). New York: Wiley Interscience.

Moll, E. O., & Legler, J. M. (1971). The life history of a neotropical slider turtle, *Pseudemys scripta* (Schoepff), in Panama. *Bulletin of the Los Angeles County Museum of Natural History, 11,* 1-98.

Møller, A. P., & Laursen, K. (2019). The ecological significance of extremely large flocks of birds. *Ecology and Evolution, 9,* 6559-6567.

Molnar, R. E. (1982). A longirostrine crocodilian from Murua (Woodlark), Solomon Sea. *Memoirs of the Queensland Museum, 20,* 675-685.

Molnar, R. E. (2004). *Dragons in the dust: The paleobiology of the giant monitor lizard Megalania.* Bloomington: Indiana University Press.

Monk, O. (2000). Histology of the fusion area between the parasitic male and the female in the deep-sea anglerfish *Neoceratias spinifer* Pappenheim, 1914 (Teleostei, Ceratioidei). *Acta Zoologica, 81,* 315-324.

Montgomery, C. E., Griffith Rodriquez, E. J., Ross, H. L., & Lips, K. R. (2011). Communal nesting in the anoline lizard *Norops lionotus* (Polychrotidae) in Central Panama. *Southwestern Naturalist, 56,* 83-88.

Moore, J. A, Daugherty, C. H., Godfrey, S., & Nelson, N. (2009). Seasonal monogamy and multiple paternity in a wild population of a territorial reptile (tuatara). *Biological Journal of the Linnean Society, 98,* 161-170.

Moore, J. A., Daugherty, C. H., & Nelson, N. (2009). Large male advantage: Phenotypic and genetic correlates of territoriality in tuatara. *Journal of Herpetology, 43,* 570–579.

Moore, J. A., Hoare, J. M., Daugherty, C. H., & Nelson, N. J. (2007). Waiting reveals waning weight: Monitoring over 54 years shows a decline in body condition of a long-lived reptile (tuatara, *Sphenodon punctatus*). *Biological Conservation, 135,* 181–188.

Moore, J. R., & Varricchio, D. J. (2016). The evolution of diapsid reproductive strategy with inferences about extinct taxa. *PLoS ONE, 11,* e0158496.

Moore, M. J. C., & Seigel, R.A. (2006). No place to nest or bask: Effects of human disturbance on the nesting and basking habits of yellow-blotched map turtles (*Graptemys flavimaculata*). *Biological Conservation, 130,* 386–393.

Mora, J. M. (1989). Eco-behavioral aspects of two communally nesting iguanines and the structure of their shared nesting burrows. *Herpetologica, 45,* 293–298.

Morand-Ferron, J., Sol, D., & Lefebvre, L. (2007). Food stealing in birds: Brain or brawn? *Animal Behaviour, 74*(6), 1725–1734.

Moreira, P. L., López, P., & Martín, J. (2006). Femoral secretions and copulatory plugs convey chemical information about male identity and dominance status in Iberian rock lizards (*Lacerta monticola*). *Behavioral Ecology and Sociobiology, 60,* 166–174.

Mori, A., & Burghardt, G. M. (2000). Does prey matter? Geographic variation in antipredator responses of hatchlings of a Japanese natricine snake, *Rhabdophis tigrinus. Journal of Comparative Psychology, 114,* 408–413.

Mori, A., & Burghardt, G. M. (2008). Comparative experimental tests of natricine antipredator displays, with special reference to the apparently unique displays in the Asian genus, *Rhadophis. Journal of Ethology, 26,* 61–68.

Mori, A., & Burghardt, G. M. (2017). Do tiger keelback snakes (*Rhabdophis tigrinus*) recognize how toxic they are? *Journal of Comparative Psychology, 131,* 257–265.

Mori, A., Burghardt, G. M., Savitzky, A. H., Roberts, K. A., Hutchinson, D. A., & Goris, R. C. (2012). Nuchal glands: A novel defensive system in snakes. *Chemoecology, 22,* 187–198.

Mori, A., Toda, M., & Ota, H. (2002). Winter activity of the hime-habu (*Ovophis okinavensis*) in the humid subtropics: Foraging on breeding anurans at low temperatures. In G. W. Schuett, M. Höggren, M. E. Douglas & H. W. Greene (Eds.), *Biology of the vipers* (pp. 329–344). Eagle Mountain, UT: Eagle Mountain Publishing.

Morrison, S., Harlow, P., & Keogh, J. (2009). Nesting ecology of the critically endangered Fijian crested iguana *Brachylophus vitiensis* in a Pacific tropical dry forest. *Pacific Conservation Biology, 15*(2), 135–147.

Mosqueira Manso, J. (1960). *Las tortugas del Orinoco* (2nd ed.). Buenos Aires: Editorial Citania.

Moss, B. (2017). Marine reptiles, birds and mammals and nutrient transfers among the seas and the land: An appraisal of current knowledge. *Journal of Experimental Marine Biology and Ecology, 492,* 63–80.

Moss, J. B., Gerber, G. P., Laaser, T., Goetz, M., Oyog, T., & Welch, M. E. (2020). Conditional female strategies influence hatching success in a communally

nesting iguana. *Ecology and Evolution, 10,* 3424–3438. https://doi.org/10.1002
/ece3.6139.

Motani, R. (2010). Warm-blooded "sea dragons"? *Science, 328,* 1361–1362.

Motani, R., Jiang, D.-Y., Tintori, A., Rieppel, O., & Chen, G.-B. (2014). Terrestrial
origin of viviparity in Mesozoic marine reptiles Indicated by Early Triassic
embryonic fossils. *PLoS ONE, 9,* e88640. https://doi.org/10.1371/journal
.pone.0088640.

Motani, R., Rothschild, B. M., & Wahl, W. (1999). Large eyeballs in diving
ichthyosaurs. *Nature, 402,* 747–747.

Mourer-Chauviré, C., & Popin, F. (1985). Le mystère des tumulus de Nouvelle-
Calédonie. *Recherche, 16,* 1094.

Mouton, P. (2011). Aggregation behaviour of lizards in the arid western regions of
South Africa. *African Journal of Herpetology, 60,* 155–170.

Mrosovsky, N. (1972). Spectrographs of the sounds of leatherback turtles.
Herpetologica, 28, 256–258.

Müller, J., Hipsley, C. A., Head, J. J., Kardjilov, N., Hilger, A., Wuttke, M., & Reisz,
R. R. (2011). Eocene lizard from Germany reveals amphisbaenian origins.
Nature, 473, 364.

Müller, J., & Tsuji, L. A. (2007). Impedance-matching hearing in Paleozoic
reptiles: Evidence of advanced sensory perception at an early stage of
amniote evolution. *PLoS ONE, 2,* e889.

Muniz, F., Da Silveira, R., Campos, Z., Magnusson, W., Hrbek, T., & Farias, I. (2011).
Multiple paternity in the black caiman (*Melanosuchus niger*) population in the
Anavilhanas National Park, Brazilian Amazonia. *Amphibia-Reptilia, 32,* 428–434.

Murphy, J. B., Ciofi, C., de La Panouse, C., & Walsh, T. (Eds.). (2002). *Komodo
dragons: Biology and conservation.* Chicago: Smithsonian Institution.

Murphy, J. B., Evans, M., Augustine, L., & Miller, K. (2016). Behaviors in the
Cuban crocodile (*Crocodylus rhombifer*). *Herpetological Review, 47,* 235–240.

Murphy, J. B., & Lamoreaux, W. E. (1978). Mating behavior in three Australian chelid turtles (Testudines: Pleurodira: Chelidae). *Herpetologica, 34,*
398–405.

Murphy, J. B., & Mitchell, L. A. (1974). Ritualized combat behavior of the pygmy
mulga monitor lizard, *Varanus gilleni* (Sauria: Varanidae). *Herpetologica, 30,*
90–97.

Murray, C. M., Crother, B. I., & Doody, J. S. (2020). The evolution of crocodilian
nesting ecology and behavior. *Ecology and Evolution, 10,* 131–149.

Murray, C. M., Easter, M., Padilla, S., Marin, M. S., & Guyer, C. (2016). Regional
warming and the thermal regimes of American crocodile nests in the
Tempisque Basin, Costa Rica. *Journal of Thermal Biology, 60,* 49–59.

Myburgh, J., & Warner, J. (2011). *Crocodylus niloticus* (Nile crocodile). Induced
hatching. *Herpetological Review, 42,* 368.

Myers, T. S., & Fiorillo, A. R. (2009). Evidence for gregarious behavior and age
segregation in sauropod dinosaurs. *Palaeogeography, Palaeoclimatology,
Palaeoecology, 274,* 96–104.

Nagle, R., Lutz, C., & Pyle, A. (2004). Overwintering in the nest by hatchling map
turtles (*Graptemys geographica*). *Canadian Journal of Zoology, 82,* 1211–1218.

Naish, D. (2012, June 20). Terrifying sex organs of male turtles. Accessed August 9, 2020, from https://io9.gizmodo.com/the-terrifying-sex-organs-of -male-turtles-5919870.

Naish, D. (2014). The fossil record of bird behavior. *Journal of Zoology, 292,* 268–280.

Naish, D., & Martill, D. M. (2003). Pterosaurs—a successful invasion of prehistoric skies. *Biologist, 50,* 213–216.

Narins, P. M., Feng, A. S., Lin, W., Schnitzler, H. U., Denzinger, A., Suthers, R. A., & Xu, C. (2004). Old World frog and bird vocalizations contain prominent ultrasonic harmonics. *Journal of the Acoustical Society of America, 115,* 910–913.

Nečos, P. (1999). *Chameleons: Nature's hidden jewels.* Frankfurt am Main: Edition Chimaira.

Nedyalkov, A. G. (1967). *Ловцы змей [Snake hunters].* Moscow: Znanie.

Neill, T. W. (1971). *The last of the ruling reptiles: Alligators, crocodiles, and their kin.* New York: Columbia University Press.

Nelson, G. L., & Graves, B. M. (2004). Anuran population monitoring: Comparison of the North American Amphibian Monitoring Program's calling index with mark-recapture estimates for *Rana clamitans. Journal of Herpetology, 38,* 355–360.

Nelson, J. T., Thompson, M. B., Pledger, S., Keall, S. N., & Daugherty, C. H. (2006). Performance of juvenile tuatara depends on age, clutch, and incubation regime. *Journal of Herpetology, 40,* 339–403.

Nelson, N. J., Keall, S. N., Brown, D., & Daugherty, C. H. (2002). Establishing a new wild population of tuatara (*Sphenodon guntheri*). *Conservation Biology, 16,* 887–894.

Nesbitt, S. J. (2011). The early evolution of archosaurs: Relationships and the origin of major clades. *Bulletin of the American Museum of Natural History, 2011,* 1–292.

Newman, E. A., & Hartline, P. H. (1982). The infrared "vision" of snakes. *Scientific American, 246,* 116–127.

Nichols, W. J., Resendiz, A., Seminoff, J. A., & Resendiz, B. (2000). Transpacific migration of a loggerhead turtle monitored by satellite telemetry. *Bulletin of Marine Science, 67,* 937–947.

Nicol, S., & Andersen, N. A. (2006). Body temperature as an indicator of egg-laying in the echidna, *Tachyglossus aculeatus. Journal of Thermal Biology, 31,* 483–490.

Nieuwoudt, C., Mouton, P., & Flemming, A. (2003). Sex ratio, group composition and male spacing in the large-scaled girdled lizard, *Cordylus macropholis. Journal of Herpetology, 37,* 577–580.

Nisa Ramiro, C., Rodríguez-Ruiz, G., López, P., da Silva Junior, P. I., Trefaut Rodrigues, M., & Martín, J. (2019). Chemosensory discrimination of male age by female *Psammodromus algirus* lizards based on femoral secretions and feces. *Ethology, 125,* 82–89.

Noble, D. W. A., Byrne, R. W., & Whiting, M. J. (2014). Age-dependent social learning in a lizard. *Biology Letters, 10,* 20140430.

Noble, D. W. A., Stenhouse, V., & Schwanz, L. E. (2018). Developmental temperatures and phenotypic plasticity in reptiles: A systematic review and meta-analysis. *Biological Reviews, 93,* 72–97.

Noble, D. W. A., Wechmann, K., Keogh, J., & Whiting, M. J. (2013). Behavioral and morphological traits interact to promote the evolution of alternative reproductive tactics in a lizard. *The American Naturalist, 182,* 726-742.

Noble, G. K., & Bradley, H. T. (1933). The mating behavior of lizards; its bearing on the theory of sexual selection. *Annals of New York Academy of Sciences, 35,* 25-100.

Noble, G. K., & Clausen, H. J. (1936). The aggregation behaviour of *Storeria dekayi* and other snakes, with especial reference to the sense organs involved. *Ecological Monographs, 6,* 269-316.

Noble, G. K., & Kumpf, K. F. (1936). The function of Jacobson's organ in lizards. *Journal of Genetic Psychology, 48,* 371-382.

Noble, G. K., & Mason E. R. (1933). Experiments on the brooding habits of the lizards *Eumeces* and *Ophisaurus. American Museum Novitates, 619,* 1-29.

Nordberg, E. J., & McKnight, D. T. (2020). Nocturnal basking behavior in a freshwater turtle. *Ecology, 101,* e03048.

Norell, M. A., Clark, J. M., Dashzeveg, D., Barsbold, T., Chiappe, L. M., Davidson, A. R., McKenna, M. C., & Novacek, M. J. (1994). A theropod dinosaur embryo, and the affinities of the Flaming Cliffs dinosaur eggs. *Science, 266,* 779-782.

Norell, M. A., Wiemann, J., Fabbri, M., Yu, C., Marsicano, C. A., Moore-Nall, A., . . . Zelenitsky, D. K. (2020). The first dinosaur egg was soft. *Nature, 583,* 406-410.

Nowak, R. M. (2018). *Walker's mammals of the world* (7th ed.). Baltimore and London: Johns Hopkins University Press.

O'Connor, D., & Shine, R. (2003). Lizards in "nuclear families": A novel reptilian social system in *Egernia saxatilis* (Scincidae). *Molecular Ecology, 12,* 743-752.

O'Connor, D., & Shine, R. (2004). Parental care protects against infanticide in the lizard *Egernia saxatilis* (Scincidae). *Animal Behaviour, 68,* 1361-1369.

O'Connor, D., & Shine, R. (2006). Kin discrimination in the social lizard *Egernia saxatilis* (Scincidae). *Behavioral Ecology, 17,* 206-211.

O'Connor, J. K., Sun, C., Xu, X., Wang, X., & Zhou, Z. (2012). A new species of *Jeholornis* with complete caudal integument. *Historical Biology, 24,* 29-41.

Oda, W. Y. (2004). Communal egg laying by *Gonatodes humeralis* (Sauria, Gekkonidae) in Manaus primary and secondary forest areas. *Acta Amazonica, 34,* 331-332.

O'Donnell, R. P., Ford, N. B., Shine, R., & Mason, R. T. (2004). Male red-sided garter snakes, *Thamnophis sirtalis parietalis,* determine female mating status from pheromone trails. *Animal Behaviour, 68,* 677-683.

Ojeda, G., Amavet, P., Rueda, E., Siroski, P., & Larriera, A. (2016). Mating system of *Caiman yacare* (Reptilia: Alligatoridae) described from microsatellite genotypes. *Journal of Heredity, 108,* 135-141.

Olave, M., Martinez, L. E., Avila, L. J., Sites, J. W., & Morando, M. (2011). Evidence of hybridization in the Argentinean lizards *Liolaemus gracilis* and *Liolaemus bibronii* (Iguania: Liolaemini): An integrative approach based on genes and morphology. *Molecular Phylogenetics and Evolution, 61,* 381-391.

Oliveira, D., Marioni, B., Farias, I., & Hrbek, T. (2014). Genetic evidence for polygamy as a mating strategy in *Caiman crocodilus. Journal of Heredity, 105,* 485-492.

Olsson, M., & Madsen, T. (1998). Sexual selection and sperm competition in reptiles. In T. R. Birkhead & A. P. Møller (Eds.). *Sperm competition and sexual selection* (pp. 503-577). San Diego: Academic Press.

Olsson, M., & Shine, R. (1998). Chemosensory mate recognition may facilitate prolonged mate guarding by male snow skinks, *Niveoscincus microlepidotus*. *Behavioral Ecology and Sociobiology, 43,* 359-363.

Olsson, M., & Shine, R. (2000). Ownership influences the outcome of male-male contests in the scincid lizard, *Niveoscincus microlepidotus. Behavioral Ecology, 11,* 587-590.

Oppenheim, R. W. (1972). Prehatching and hatching behaviour in birds: A comparative study of altricial and precocial species. *Animal Behaviour, 20,* 644-655.

Ord, T. J., Peters, R. A., Evans, C. S., & Taylor, A. J. (2002). Digital video playback and visual communication in lizards. *Animal Behaviour, 63,* 879-890.

Ota, H., Hikida, T., Kon, M., & Hidaka, T. (1989). Unusual nest site of a scincid lizard *Sphenomorphus kinabalensis* from Sabah, Malaysia. *Herpetological Review, 20,* 38-39.

Owen, R. (1866). *On the anatomy of vertebrates. Vol. 1. Fishes and reptiles.* London: Longmans, Green.

Padykula, H. A., & Taylor, J. M. (1982). Marsupial placentation and its evolutionary significance. *Journal of Reproduction and Fertility, Supplement, 31,* 95-104.

Palacín, C., Alonso, J. C., Alonso, J. A., Magaña, M., & Martín, C. A. (2011). Cultural transmission and flexibility of partial migration patterns in a long-lived bird, the great bustard *Otis tarda. Journal of Avian Biology, 42,* 301-308.

Paladino, F. V., O'Connor, M. P., & Spotila, J. R. (1990). Metabolism of leatherback turtles, gigantothermy, and thermoregulation of dinosaurs. *Nature, 344,* 858-860.

Pandav, B. N., Shanbhag, B. A., & Saidapur, S. K. (2007). Ethogram of courtship and mating behaviour of garden lizard, *Calotes versicolor. Current Science, 93,* 1164-1167.

Panov, E., & Zykova, L. (1993). Social organization and demography of Caucasian agama, *Stellio caucasius* (Squamata, Agamidae). *Zoologicheskii Zhurnal, 72,* 74-93.

Papaj, D. R., Averill, A. L., Prokopy, R. J., & Wong, T. T. Y. (1992). Host-marking parental behavior by vipers. In G. W. Schuett, M. Höggren, M. E. Douglas & H. W. Greene (Eds.), *Biology of the vipers* (pp. 179-206). Eagle Mountain, UT: Eagle Mountain Publishing.

Parker, J. M., Spear, S. F., & Oyler-McCance, S. (2013). Natural history notes. *Crotalus oreganus concolor* (midget faded rattlesnake). Nursery aggregation. *Herpetological Review, 43,* 658-659.

Parker, W. S., & Brown, W. S. (1972). Telemetric study of movements and oviposition of two female *Masticophis t. taeniatus. Copeia, 1972,* 892-895.

Passek, K. M. (2002). Extra-pair paternity within the female-defense polygyny of the lizard, *Anolis carolinensis*: Evidence of alternative mating strategies [PhD dissertation]. Virginia Polytechnic Institute and State University.

Patino-Martinez, J., Marco, A., Quiñones, L., & Hawkes, L. (2012). A potential tool to mitigate the impacts of climate change to the Caribbean leatherback sea turtle. *Global Change Biology, 18,* 401-411.

Patterson, R. (1971). Aggregation and dispersal behavior in captive *Gopherus agassizi. Journal of Herpetology, 5,* 214-216.

Paull, P. (2010). Communal nesting in the delicate skink [Unpublished Honours thesis]. Monash University.

Paullin, D. G. (1987). Cannibalism in American coots induced by severe spring weather and avian cholera. *The Condor, 89,* 442-443.

Pawar, S. (2003). Taxonomic chauvinism and the methodologically challenged. *Bioscience, 53,* 861-864.

Pearse, D. E., & Avise, J. C. (2001). Turtle mating systems: Behavior, sperm storage, and genetic paternity. *Journal of Heredity, 92,* 206-211.

Pearse, D. E., Janzen, F. J., & Avise, J. C. (2002). Multiple paternity, sperm storage, and reproductive success of female and male painted turtles (*Chrysemys picta*) in nature. *Behavioral Ecology and Sociobiology, 51,* 164-171.

Pearson, D., Shine, R., & Williams, A. (2002). Geographic variation in sexual size dimorphism within a single snake species (*Morelia spilota,* Pythonidae). *Oecologia, 131,* 418-426.

Pellitteri-Rosa, D., Sacchi, R., Galeotti, P., Marchesi, M., & Fasola, M. (2011). Courtship displays are condition-dependent signals that reliably reflect male quality in Greek tortoises, *Testudo graeca. Chelonian Conservation and Biology, 10,* 10-17.

Peñalver-Alcázar, M., Romero-Díaz, C., & Fitze, P. S. (2015). Communal egg-laying in oviparous *Zootoca vivipara louislantzi* of the Central Pyrenees. *Herpetology Notes, 8,* 4-7.

Pérez-Cembranos, A., & Pérez-Mellado, V. (2015). Local enhancement and social foraging in a non-social insular lizard. *Animal cognition, 18,* 629-637.

Pérez-Higareda, G., & Smith, H. M. (1988). Courtship behavior in *Rhinoclemmys areolata* from Western Tabasco, Mexico (Testudines: Emydidae). *The Great Basin Naturalist, 48,* 263-266.

Perry, G., & Dmi'el, R. (1994). Needles and haystacks: Searching for lizard eggs in a coastal sand dune. *Amphibia-Reptilia, 15,* 395-401.

Persons, W. S., & Currie, P. J. (2015). Bristles before down: A new perspective on the functional origin of feathers. *Evolution, 69,* 857-862.

Peterson, C. (1996a). Anhomeostasis: Seasonal water and solute relations in two populations of the desert tortoise (*Gopherus agassizii*) during chronic drought. *Physiological Zoology, 69,* 1324-1358.

Peterson, C. (1996b). Ecological energetics of the desert tortoise (*Gopherus agassizii*): Effects of rainfall and drought. *Ecology, 77,* 1831-1844.

Peterson, J. A. (1984). The locomotion of *Chamaeleo* (Reptilia: Sauria) with particular reference to the forelimb. *Journal of Zoology, 202,* 1-42.

Petranka, J. W., & Petranka, J. G. (1980). Selected aspects of the larval ecology of the marbled salamander *Ambystoma opacum* in the southern portion of its range. *American Midland Naturalist, 104,* 352-363.

Petrovsky, I. V. (Ed.). (1964). *Индийские Сказки* [Fairy Tales of India]. Leningrad: Hudozhestvennaya Literatura (in Russian).

Petzold H. G. (1971). *Blindschleiche und Scheltopusik*. Neue Brehm Bucherei, No. 448. Lutherstadi, GDR: Witenberg.

Pezaro, N., Thompson, M., & Doody, J. (2016). Seasonal sex ratios and the evolution of temperature-dependent sex determination in oviparous lizards. *Evolutionary Ecology, 30*, 551–565.

Pfennig, D. W., Reeve, H. K., & Sherman, P. W. (1993). Kin recognition and cannibalism in spadefoot toad tadpoles. *Animal Behaviour, 46*, 87–94.

Phillips, J. A., Alberts, A. C., & Pratt, N. C. (1993). Differential resource use, growth, and the ontogeny of social relationships in the green iguana. *Physiology and Behavior, 53*, 81–88.

Phillips, J. A., & Millar, R. P. (1998). Reproductive biology of the white-throated savanna monitor, *Varanus albigularis*. *Journal of Herpetology, 32*, 366–377.

Pianka, E. R., & King, D. (2004). *Varanoid lizards of the world*. Bloomington: Indiana University Press.

Pianka, E. R., & Vitt, L. J. (2003). *Lizards: Windows to the evolution of diversity*. Berkeley: University of California Press.

Pigliucci, M., & Müller, G. B. (2010). *Evolution, the extended synthesis*. Cambridge, MA: MIT Press.

Pike, D. A., & Seigel, R. A. (2006). Variation in hatchling tortoise survivorship at three geographic localities. *Herpetologica, 62*, 125–131.

Pike, D. A., Webb, J. K., & Shine, R. (2010). Nesting in a thermally challenging environment: Nest-site selection in a rock-dwelling gecko, *Oedura lesueurii* (Reptilia: Gekkonidae). *Biological Journal of the Linnean Society, 99*, 250–259.

Pilastro, A. (1992). Communal nesting between breeding females in a free-living population of fat dormouse (*Glis glis* L.). *Italian Journal of Zoology, 59*, 63–68.

Piñeiro, G., Ferigolo, J., Meneghel, M., & Laurin, M. (2012). The oldest known amniotic embryos suggest viviparity in mesosaurs. *Historical Biology, 24*, 620–630.

Pinheiro, M. S. (1996). Crescimento de filhotes de jacare-dopapo-amarelo, *Caiman latirostris* (Daudin, 1802), alimentados com fontes proteicas de origem animal. Escola Superior de Agricultura Luiz de Queiroz, Piracicaba, San Paulo, Brazil.

Pinker, S., & Jackendoff, R. (2005). The faculty of language: What's special about it? *Cognition, 95*, 201–236.

Pitman, C. R. S. (1931). *A game warden among his charges*. London: Nisbet.

Pizzatto, L., Manfio, R. H., & Almeida-Santos, S. M. (2006). Male-male ritualized combat in the Brazilian rainbow boa, *Epicrates cenchria crassus*. *Herpetological Bulletin, 95*, 16–20.

Pizzatto, L., & Marques, O. A. V. (2007). Reproductive ecology of boine snakes with emphasis on Brazilian species and a comparison to pythons. *South American Journal of Herpetology, 2*, 107–122.

Placyk, J. S., Jr., & Burghardt, G. M. (2011). Evolutionary persistence of chemically elicited ophiophagous antipredator responses in gartersnakes, *Thamnophis sirtalis*. *Journal of Comparative Psychology, 125*, 134–142

Placyk, J. S., Jr., Fitzpatrick, B. M., Casper, G. S., Small, R. L., Reynolds, R. G., Noble, D. W. A., Brooks, R. J., & Burghardt, G. M. (2012). Hybridization between two gartersnake species (*Thamnophis*) of conservation concern: A threat or an important natural interaction? *Conservation Genetics, 13*, 649-663.

Platnick, N. I. (1991). Patterns of biodiversity: Tropical vs temperate. *Journal of Natural History, 25*, 1083-1088.

Platt, J. R. (1964). Strong inference. *Science, 146*, 347-353.

Platt, S. G., & Elsey, R. M. (2017). *Alligator mississippiensis* (American Alligator). Feeding aggregation and behavior. *Herpetological Review, 48*, 628-629.

Platt, S. G., Thongsavath, O., Hallam, C. D., & Rainwater, T. R. (2020). Crocodylus siamensis (Siamese Crocodile). Nesting and nest attendance. *Herpetological Review, 51*, 588-590.

Platt, S. G., Win M. M., Platt, K., Reh, B., Haislip, N. A., & Rainwater, T. R. (2019). *Batagur trivittata* (Burmese roofed turtle) behavior. *Herpetological Review, 50*, 765-766.

Plotkin, P. T. (2007). *Biology and conservation of ridley sea turtles*. Baltimore: Johns Hopkins University Press.

Plummer, M. V. (1977). Notes on the courtship and mating behavior of the soft-shell turtle, *Trionyx muticus* (Reptilia, Testudines, Trionychidae). *Journal of Herpetology, 11*, 90-92.

Plummer M. V. (1981). Communal nesting of *Opheodrys aestivus* in the laboratory. *Copeia, 1981*, 243-246.

Plummer, M. V. (1989). Observations on the nesting ecology of green snakes. *Herpetological Review, 20*, 87-89.

Plummer, M. V. (1990). Nesting movements, nesting behavior, and nest sites of green snakes (*Opheodrys aestivus*) revealed by radiotelemetry. *Herpetologica, 46*, 190-195.

Plummer, M. V. (2007). Nest emergence of smooth softshell turtle (*Apalone mutica*) hatchlings. *Herpetological Conservation and Biology, 2*, 61-64.

Polis, G. A. (1981). The evolution and dynamics of intraspecific predation. *Annual Review of Ecology and Systematics, 12*, 225-251.

Polis, G. A., & Myers, C. A. (1985). A survey of intraspecific predation among reptiles and amphibians. *Journal of Herpetology, 19*, 99-107.

Polis, G. A., Myers, C. A., & Hess, W. R. (1984). A survey of intraspecific predation within the class Mammalia. *Mammal Review, 14*, 187-198.

Pooley, A. C. (1962). The Nile crocodile *Crocodylus niloticus*. Notes on the incubation period and growth rates of juveniles. *Lammergeyer, 2*, 1-55.

Pooley, A. C. (1982). The ecology of the Nile crocodile *Crocodylus niloticus* in Zululand [Unpublished Master's thesis]. University of Natal, Pietermaritzburg.

Pooley, A. C., & Gans, C. (1976). The Nile crocodile. *Scientific American, 234*, 114-124.

Pope, C. H. (1939). *Turtles of the United States and Canada*. New York: Alfred A. Knopf, Inc.

Port, M., & Cant, M. (2013). Longevity suppresses conflict in animal societies. *Biology Letters, 9*, 20130680.

Post, M. J. (2000). The captive husbandry and reproduction of the Hosmer's Skink *Egernia hosmeri. Herpetofauna, 30,* 2-6.

Pough, F. H. (1973). Lizard energetics and diet. *Ecology, 54,* 837-844.

Pough, F. H., Andrews, R. M., Cadle, J. E., Savitzky, A. H., Wells, K. D., & Bradley, M. C. (2015). *Herpetology* (4th ed.). Sunderland: Sinauer Associates.

Povel, D., & Van Der Kooij, J. (1996). Scale sensillae of the file snake (Serpentes: Acrochordidae) and some other aquatic and burrowing snakes. *Netherlands Journal of Zoology, 47,* 443-456.

Powers, A. S. (1990). Brain mechanisms of learning in reptiles. In R. P. Kresner & D. S. Olton (Eds.), *Neurobiology of comparative cognition* (pp. 157-177). Hillsdale, NJ: LEA. (Reprinted by Taylor and Francis, 2014).

Prater, S. H. (1933). The social life of snakes. *Journal of the Bombay Natural History Society, 36,* 469-476.

Prévost, J. (1961). *Écologie du Manchot Empereur* Aptenodytes forsteri *Gray*. Paris: Éditions Hermann.

Prieto-Marquez, A., & Guenther, M. F. (2018). Perinatal specimens of *Maiasaura* from the Upper Cretaceous of Montana (USA): Insights into the early ontogeny of saurolophine hadrosaurid dinosaurs. *PeerJ, 6,* e4734.

Pritchard, P. C. H. (1969). Endangered species: Kemp's ridley turtle. *Florida Naturalist, 49,* 15-19.

Pritchard, P. C. H. (2007). Arribadas I have known. In P. T. Plotkin (Ed.), *Biology and conservation of ridley sea turtles* (pp. 7-22). Baltimore: Johns Hopkins University Press.

Pritchard, P. C. H., & Trebbau, P. (1984). *The turtles of Venezuela.* Oxford: Society for the Study of Amphibians and Reptiles.

Prötzel, D., Heß, M., Scherz, M. D., Schwager, M., van't Padje, A., & Glaw, F. (2018). Widespread bone-based fluorescence in chameleons. *Scientific Reports, 8*(1), 698.

Pruett, J. E., Fargevieille, A., & Warner, D. A. (2020). Temporal variation in maternal nest choice and its consequences for lizard embryos. *Behavioral Ecology, 31,* 902-910.

Pruett-Jones, S. (1992). Independent versus non-independent mate choice: Do females copy each other? *The American Naturalist, 140,* 1000-1009.

Prum, R. O. (2008). Who's your daddy? *Science, 322,* 1799-1800.

Prum, R. O., Berv, J. S., Dornburg, A., Field, D. J., Townsend, J. P., Lemmon, E. M., & Lemmon, A. R. (2015). A comprehensive phylogeny of birds (Aves) using targeted next-generation DNA sequencing. *Nature, 526,* 569-573.

Purwandana, D., Imansyah, M. J., Ariefiandy, A., Rudiharto, H., & Jessop, T. S. (2020). Insights into the nesting ecology and annual hatchling production of the Komodo dragon. *Copeia, 2020,* 855-862.

Putman, B. J., & Clark, R. W. (2015). Habitat manipulation in hunting rattlesnakes (*Crotalus* species). *Southwestern Naturalist, 60,* 374-377.

Pyron, R. A., & Burbrink, F. T. (2014). Early origin of viviparity and multiple reversions to oviparity in squamate reptiles. *Ecology Letters, 17,* 13-21.

Pyron, R. A., Burbrink, F. T., & Wiens, J. J. (2013). A phylogeny and revised classification of Squamata, including 4161 species of lizards and snakes. *BMC Evolutionary Biology, 13,* 93.

Pytte, C. L., Moiseff, A., & Ficken, M. S. (2004). Ultrasonic singing by the blue-throated hummingbird: A comparison between production and perception. *Journal of Comparative Physiology A*, *190*, 665-673.

Qi, Z., Barrett, P. M., & Eberth, D. A. (2007). Social behaviour and mass mortality in the basal ceratopsian dinosaur *Psittacosaurus* (Early Cretaceous, People's Republic of China). *Palaeontology*, *50*, 1023-1029.

Rabosky, A. R. D., Corl, A., Liwanag, H. E., Surget-Groba, Y., & Sinervo, B. (2012). Direct fitness correlates and thermal consequences of facultative aggregation in a desert lizard. *PLoS ONE*, *7*, e40866.

Radder, R., & Shine, R. (2006). Thermally induced torpor in full-term lizard embryos synchronizes hatching with ambient conditions. *Biology Letters*, *2*, 415-416.

Radder, R. S., & Shine, R. (2007). Why do female lizards lay their eggs in communal nests? *Journal of Animal Ecology*, *76*, 881-887.

Radzio, T. A, Cox, J. A., & O'Connor, M. P. (2017). Behavior and conspecific interactions of nesting gopher tortoises (*Gopherus polyphemus*). *Herpetological Conservation and Biology*, *12*, 373-383.

Radzio, T. A., Cox, J. A., Spotila, J. R., O'Connor, M. P. (2016). Aggression, combat, and apparent burrow competition in hatchling and juvenile gopher tortoises (*Gopherus polyphemus*). *Chelonian Conservation and Biology*, *15*, 231-237.

Rahn, H. (1939). Structure and function of placenta and corpus luteum in viviparous snakes. *Proceedings of the Society for Experimental Biology and Medicine*, *40*, 381-382.

Ramos-Pallares, E., Meza-Joya, F. L., & Ramírez-Pinilla, M. P. (2013). A case of communal egg laying in a population of *Cercosaura ampuedai* (Squamata: Gymnophthalmidae) in the Colombian Andes. *Herpetological Review*, *44*, 226-229.

Rand, A. S. (1967). The adaptive significance of territoriality in iguanid lizards. In W. W. Milstead (Ed.), *Lizard ecology: A symposium* (pp. 106-115). Columbia: University of Missouri Press.

Rand, A. S. (1968). A nesting aggregation of iguanas. *Copeia*, *1968*, 552-561.

Rand, A. S., & Dugan, B. (1983). Structure of complex iguana nests. *Copeia*, *1983*, 705-711.

Rand, A. S., & Greene, H. W. (1982). Latitude and climate in the phenology of reproduction in the green iguana, *Iguana iguana*. In G. M. Burghardt & A. S. Rand (Eds.), *Iguanas of the world: Their behavior, ecology, and conservation* (pp. 142-149). Park Ridge, NJ: Noyes Publications.

Rand, A. S., & Rand, W. M. (1978). Display and dispute settlement in nesting iguanas. In N. Greenberg & P. D. MacLean (Eds.), *Behavior and neurology of lizards: An interdisciplinary colloquium* (pp. 245-251). Rockville: N.I.M.H. DHEW Publication No. ADM 77-491.

Rand, W. M., & Rand, A. S. (1976). Agonistic behavior in nesting iguanas: A stochastic analysis of dispute settlement dominated by the minimization of energy cost. *Zeitschrift für Tierpsychologie*, *40*, 279-299.

Randall, J. A. (2010). Drummers and stompers: Vibrational communication in mammals. In C. E. O'Connell-Rodwell (Ed.), *The use of vibrations in communication: Properties, mechanisms and function across taxa* (pp. 99–120). Trivandrum, Kerala, India: Transworld.

Rangel, G. F., & Lopez-Parilla, Y. R. (2014). *Ptychoglossus festae* (Paracca's largescale lizard). Nesting. *Herpetological Review, 45*, 504.

Rasmussen, A. R., Murphy, J. C., Ompi, M., Gibbons, J. W., & Uetz, P. (2011). Marine reptiles. *PLoS ONE, 6*, e27373.

Rauch N. (1988). Competition of marine iguana females (*Amblyrhynchus cristatus*) for egg-laying sites. *Behaviour, 107*, 91–106.

Reber, S. A., Nishimura, T., Janisch, J., Robertson, M., & Fitch, W. T. (2015). A Chinese alligator in heliox: Formant frequencies in a crocodilian. *Journal of Experimental Biology, 218*, 2442–2447.

Reed, W., & Clark, M. (2011). Beyond maternal effects in birds: Responses of the embryo to the environment. *Integrative & Comparative Biology, 51*, 73–80.

Reeder, T. W., Dessauer, H. C., & Cole, C. J. (2002). Phylogenetic relationships of whiptail lizards of the genus *Cnemidophorus* (Squamata, Teiidae): A test of monophyly, reevaluation of karyotypic evolution, and review of hybrid origins. *American Museum Novitates, 3365*, 1–61.

Reeder, T. W., Townsend, T. M., Mulcahy, D. G., Noonan, B. P., Wood, P. L., Jr., Sites, J. W., Jr., & Wiens, J. J. (2015). Integrated analyses resolve conflicts over squamate reptile phylogeny and reveal unexpected placements for fossil taxa. *PLoS ONE, 10*, e0118199.

Reese, A. (1915). *The alligator and its allies.* London: GP Putnam's Sons.

Refsnider, J. M., Daugherty, C. H., Godfrey, S. S., Keall, S. N., Moore, J. A., & Nelson, N. J. (2013). Patterns of nesting migrations in the tuatara (*Sphenodon punctatus*), a colonially nesting island reptile. *Herpetologica, 69*, 282–290.

Refsnider, J. M., Keall, S. N., Daugherty, C. H., & Nelson, N. J. (2009). Does nest-guarding in female tuatara (*Sphenodon punctatus*) reduce nest destruction by conspecific females? *Journal of Herpetology, 43*, 294–299.

Refsnider, J. M., Nelson, N. J., & Keall, S. N. (2007). Manipulative females: Intra-sexual interactions in nest site choice, nest guarding, and nest defense in tuatara. *New Zealand Journal of Zoology, 34*, 270–271.

Regalado, R. (2003). Social behavior and sex recognition in the Puerto Rican dwarf gecko *Sphaerodactylus nicholsi. Caribbean Journal of Science, 39*, 77–93.

Reichenbach, N. G. (1983). An aggregation of female garter snakes under corrugated metal sheets. *Journal of Herpetology, 17*, 412–413

Reilly, S. M., & Delancey, M. J. (1997). Sprawling locomotion in the lizard *Sceloporus clarkii*: The effects of speed on gait, hindlimb kinematics, and axial bending during walking. *Journal of Zoology, 243*, 417–433.

Reiserer, R., Schuett, G., & Earley, R. (2008). Dynamic aggregations of newborn sibling rattlesnakes exhibit stable thermoregulatory properties. *Journal of Zoology, 274*, 277–283.

Reisz, R. R., Evans, D. C., Roberts, E. M., Sues, H. D., & Yates, A. M. (2012). Oldest known dinosaurian nesting site and reproductive biology of the Early Jurassic sauropodomorph *Massospondylus. Proceedings of the National Academy of Sciences, 109*, 2428–2433.

Reisz, R. R., Scott, D., Sues, H. D., Evans, D. C., & Raath, M. A. (2005). Embryos of an Early Jurassic prosauropod dinosaur and their evolutionary significance. *Science, 309*, 761-764.

Reiter, S., Liaw, H. P., Yamawaki, T. M., Naumann, R. K., Laurent, G. (2017). On the value of reptilian brains to map the evolution of the hippocampal formation. *Brain, Behavior and Evolution, 90*, 41-52.

Repp, R. A. (1998). Wintertime observations of five species of reptiles in the Tucson area: Sheltersite selections/fidelity to sheltersites/notes on behavior. *Bulletin of the Chicago Herpetological Society, 33*, 49-56.

Repp, R. A., & Schuett, G. W. (2009). *Crotalus atrox* (western diamond-backed rattlesnake). Adult predation on lizards. *Herpetological Review, 40*, 353-354.

Reyes-Arriagada, R., Campos-Ellwanger, P., Schlatter, R. P., & Baduini, C. (2007). Sooty shearwater (*Puffinus griseus*) on Isla Guafo: The largest seabird colony in the world? *Biodiversity and Conservation, 16*, 913-930.

Reynolds, R. G., Booth, W., Schuett, G. W., Fitzpatrick, B. M., & Burghardt, G. M. (2012). Successive virgin births of viable male progeny in the checkered gartersnake, *Thamnophis marcianus. Biological Journal of the Linnean Society, 107*, 566-572.

Rhen, T., & Crews, D. (1999). Embryonic temperature and gonadal sex organize male-typical sexual and aggressive behavior in a lizard with temperature-dependent sex determination. *Endocrinology, 140*, 4501-4508.

Richmond, J. Q., Jockusch, E. L., & Latimer, A. M. (2011). Mechanical reproductive isolation facilitates parallel speciation in Western North American Scincid lizards. *The American Naturalist, 178*, 320-332.

Riley, J. L., Küchler, A., Damasio, T., Noble, D. W. A., Byrne, R. W., & Whiting, M. J. (2018). Learning ability is unaffected by isolation rearing in a family-living lizard. *Behavioral Ecology and Sociobiology, 72*, 20.

Riley, J. L., & Litzgus, J. (2014). Cues used by predators to detect freshwater turtle nests may persist late into incubation. *Canadian Field Naturalist, 128*, 179-188.

Riley, J. L., Noble, D. W. A., & Byrne, R. W. (2017). Early social environment influences the behavior of a family living lizard. *Royal Society Open Science, 4*, 161082.

Riley, J. L., Noble, D. W. A., Byrne, R. W., & Whiting, M. J. (2017). Does early social environment influence learning ability in a family-living lizard? *Animal Cognition, 20*, 449-458.

Rismiller, P. D., McKelvey, M. W., & Green, B. (2010). Breeding phenology and behavior of Rosenberg's goanna (*Varanus rosenbergi*) on Kangaroo Island, South Australia. *Journal of Herpetology, 44*, 399-408.

Rivas, J. A. (2020). *Anaconda: The secret life of the world's largest snake.* Oxford, UK: Oxford University Press.

Rivas, J. A., & Burghardt, G. M. (2001). Understanding sexual size dimorphism in snakes: wearing the snake's shoes. *Animal Behaviour, 62*, F1-F6.

Rivas, J. [A.], & Burghardt, G. M. (2002). Crotalomorphism: A metaphor for understanding anthropomorphism by omission. In Bekoff, M., Allen, C., & Burghardt, G. M. (Eds.), *The cognitive animal: Empirical and theoretical perspectives on animal cognition* (pp. 9-18). Cambridge, MA: MIT Press.

Rivas, J. A., & Burghardt, G. M. (2005). Snake mating systems, behavior, and evolution: The revisionary implications of recent findings. *Journal of Comparative Psychology*, 119, 447–454.

Rivas, J. A., & Levin, L. E. (2004). Sexually dimorphic antipredator behavior in juvenile green iguanas: Kin selection in the form of fraternal care? In A. C. Alberts, R. L. Carter, W. K. Hayes & E. P. Martins (Eds.), *Iguanas: Biology and conservation* (pp. 119–126). Berkeley: University of California Press.

Rivas, J. A., Molina, C. R., Corey, S. J., & Burghardt, G. M. (2016). Natural history of neonatal green anacondas (*Eunectes murinus*): A chip from the old block. *Copeia*, 2016, 402–410.

Rivas, J. A., Muñoz, M. de C., Burghardt, G. M., & Thorbjarnarson, J. B. (2007). Sexual size dimorphism and the mating system of the green anaconda (*Eunectes murinus*). In R. W. Henderson & G. W. Powell (Eds.), *Biology of the boas and pythons* (pp. 312–325). Eagle Mountain, UT: Eagle Mountain Publishing.

Robbins, T. R., Schrey, A., McGinley, S., & Jacobs, A. (2013). On the incidences of cannibalism in the lizard genus *Sceloporus*: Updates, hypotheses, and the first case of siblicide. *Herpetology Notes*, 6, 523–528.

Robinson C., & Bider, J. R. (1988). Nesting synchrony: A strategy to decrease predation of snapping turtles, *Chelydra serpentina*, nests. *Journal of Herpetology*, 22, 470–473.

Robinson, C. D., Kircher, B. K., & Johnson, M. A. (2014). Communal nesting in the Cuban Twig Anole (*Anolis angusticeps*) from south Bimini, Bahamas. *IRCF Reptiles & Amphibians*, 21, 71–72.

Rodda, G. H. (1984). The orientation and navigation of juvenile alligators: Evidence of magnetic sensitivity. *Journal of Comparative Physiology A*, 154, 649–658.

Rodda, G. H. (1992). The mating behavior of *Iguana iguana*. *Smithsonian Contributions to Zoology*, 534, 1–40.

Rodda, G. H. (2020). *Lizards of the world: Natural history and taxon accounts*. Baltimore: Johns Hopkins University Press.

Rodda, G. H., Bock, B. C., Burghardt, G. M., & Rand, A. S. (1988). New techniques for identifying lizard individuals at a distance reveal influences of handling. *Copeia*, 1988, 905–913.

Rodda, G. H., & Burghardt, G. M. (1985). *Iguana iguana* (green iguana) terrestriality. *Herpetological Review*, 16, 112.

Rodriguez, G., & Reagan, G. P. (1984). Bat predation by the Puerto Rican boa (*Epicrates inornatus*). *Copeia*, 1984, 219–220.

Rodriguez-Prieto, I., Ondo-Nguema, E., Sima T., Osa-Akara, L. B., & Abeso, E. (2010). Unusual nesting behaviour in reptiles: Egg-laying in solitary wasp mud-nests by the island endemic Annobon dwarf gecko (*Lygodactylus thomensis wermuthi* Pasteur 1962). *North-West. J. Zool.*, 6, 144–147.

Rogers, R. R., Krause, D. W., & Rogers, K. C. (2003). Cannibalism in the Madagascan dinosaur *Majungatholus atopus*. *Nature*, 422, 515–518.

Roggenbuck, M. E., & Jenssen, T. A. (1986). The ontogeny of display behaviour in *Sceloporus undulatus* (Sauria: Iguanidae). *Ethology*, 71, 153–165.

Rojas, S. J. S., Villacampa, J., & Whitworth, A. (2016). Notes on the reproduction of *Kentropyx altamazonica* (Squamata: Teiidae) and *Imantodes lentiferus* (Serpentes: Dipsadidae) from southeast Peru. *Phyllomedusa, 15,* 69–73.

Rootes, W. L., & Chabreck, R. H. (1993). Cannibalism in the American alligator. *Herpetologica, 67,* 99–107.

Rose, F. L. (1970). Tortoise chin gland fatty acid composition: Behavioral significance. *Comparative Biochemistry and Physiology, 32,* 577–580.

Roth, T. C., Krochmal, A. R., & LaDage. L. D. (2019). Reptilian cognition: A more complex picture via integration of neurological mechanisms, behavioral constraints, and evolutionary context. *BioEssays, 41,* 1900033. https://doi.org /10.1002/bies.201900033.

Rubenstein, D. I., & Wrangham, R. W. (Eds.). (1986). *Ecological aspects of social evolution: Birds and mammals.* Princeton, NJ: Princeton University Press.

Rubenstein, D. R., & Abbot, P. (2017). *Comparative social evolution.* Cambridge, UK: Cambridge University Press.

Ruby, D. E., & Niblick H. A. (1994). A behavioral inventory of the desert tortoise: Development of an ethogram. *Herpetological Monographs, 8,* 88–102.

Ruiz-Monachesi, M. R., Paz, A., & Quipildor, M. (2018). Hemipenes eversion behavior: A new form of communication in two *Liolaemus* lizards (Iguania: Liolaemidae). *Canadian Journal of Zoology, 97,* 187–194.

Rusli, M., Booth, D., & Joseph, J. (2016). Synchronous activity lowers the energetic cost of nest escape for sea turtle hatchlings. *Journal of Experimental Biology, 219,* 1505–1513.

Ruxton, G. D., Birchard, G. F., & Deeming, D. C. (2014). Incubation time as an important influence on egg production and distribution into clutches for sauropod dinosaurs. *Paleobiology, 40,* 323–330.

Ryan, M. J. (1982). Variation in iguanine social organization: Mating systems in chuckwallas (*Sauromalus*). In G. M. Burghardt & A. S. Rand (Eds.), *Iguanas of the world: Their behavior, ecology, and conservation* (pp. 380–390). Park Ridge, NJ: Noyes Publications.

Ryan, M. J. (1985). *The túngara frog: A study in sexual selection and communication.* Chicago: University of Chicago Press.

Sacchi, R., Galeotti, P., Fasola, M., & Ballasina, D. (2003). Vocalizations and courtship intensity correlate with mounting success in marginated tortoises *Testudo marginata. Behavioral Ecology and Sociobiology, 55,* 95–102.

Sacchi, R., Pupin, F., Gentilli, A., Rubolini, D., Scali, S., Fasola, M., & Galeotti, P. (2009). Male-male combats in a polymorphic lizard: Residency and size, but not color, affect fighting rules and contest outcome. *Aggressive Behavior, 35,* 274–283.

Sahney, S., & Benton, M. J. (2008). Recovery from the most profound mass extinction of all time. *Proceedings of the Royal Society B, Biological Sciences, 275,* 759–765.

Sahney, S., Benton, M. J., & Ferry, P. A. (2010). Links between global taxonomic diversity, ecological diversity and the expansion of vertebrates on land. *Biology Letters, 6,* 544–547.

Sakai, O. (2020). Do different food amounts gradually promote personality variation throughout the life stage in a clonal gecko species? *Animal Behaviour, 162,* 47–56.

Sakata, J. T., & Crews, D. (2003). Embryonic temperature shapes behavioural change following social experience in male leopard geckos, *Eublepharis macularius*. *Animal Behaviour*, *66*, 839–846.

Sale, A., & Luschi, P. (2009). Navigational challenges in the oceanic migrations of leatherback sea turtles. *Proceedings of the Royal Society B: Biological Sciences*, *276*, 3737–3745.

Sander, P. M., Peitz, C., Jackson, F. D., & Chiappe, L. M. (2008). Upper Cretaceous titanosaur nesting sites and their implications for sauropod dinosaur reproductive biology. *Palaeontographica Abteilung A*, *284*, 69–107.

Sanger, T. J., Gredler, M. L., & Cohn, M. J. (2015). Resurrecting embryos of the tuatara, *Sphenodon punctatus*, to resolve vertebrate phallus evolution. *Biology Letters*, *11*, 20150694.

Santacà, M., Miletto Petrazzini, M. E., Agrillo, C., & Wilkinson, A. (2019). Can reptiles perceive visual illusions? Delboeuf illusion in red-footed tortoise (Chelonoidis carbonaria) and bearded dragon (*Pogona vitticeps*). *Journal of Comparative Psychology*, *133*, 419–427.

Santos, X., Badiane, A., & Matos, C. (2016). Contrasts in short-and long-term responses of Mediterranean reptile species to fire and habitat structure. *Oecologia*, *180*, 205–216.

Sarvella, P. (1974). Testes structure in normal and parthenogenetic turkeys. *Journal of Heredity*, *65*, 287–290.

Sato, K. (2014). Body temperature stability achieved by the large body mass of sea turtles. *The Journal of Experimental Biology*, *217*, 3607–3614.

Savitzky, A. H., & Moon, B. R. (2008). Tail morphology in the western diamond-backed rattlesnake, *Crotalus atrox*. *Journal of Morphology*, *269*, 935–944.

Schmedes, A. (2002). Der giftige Koenig der Provence. *Abenteuer Wildnis*. Accessed from https://www.youtube.com/watch?v=2-JBNqnybB4.

Schmidt, W. 1993. Minisaurier aus dem Regenwald. *Aqua Geographia*, *1*, 62–69.

Schmitz, A., Mansfeld, P., Hekkala, E., Shine, T., Nickel, H., Amato, G., & Böhme, W. (2003). Molecular evidence for species level divergence in African Nile crocodiles *Crocodylus niloticus* (Laurenti, 1786). *Comptes Rendus Palevol*, *2*, 703–712.

Schmitz, L., & Motani, R. (2011). Nocturnality in dinosaurs inferred from scleral ring and orbit morphology. *Science*, *332*, 705–708.

Schoch, R. R., & Sues, H. D. (2015). A Middle Triassic stem-turtle and the evolution of the turtle body plan. *Nature*, *523*, 584–587.

Schoch, R. R., & Sues, H. D. (2018). Osteology of the Middle Triassic stem-turtle *Pappochelys rosinae* and the early evolution of the turtle skeleton. *Journal of Systematic Palaeontology*, *16*, 927–965.

Schott, R. K., Müller, J., Yang, C. G., Bhattacharyya, N., Chan, N., Xu, M., Morrow, J. M., Ghenu, A. H., Loew, E. R., Tropepe, V., & Chang, B. S. (2016). Evolutionary transformation of rod photoreceptors in the all-cone retina of a diurnal garter snake. *Proceedings of the National Academy of Sciences*, *113*, 356–361.

Schuett, G. W. (1992). Is long-term sperm storage an important component of the reproductive biology of temperate pitvipers? In J. A. Campbell & E. D. Brodie, Jr (Eds.), *Biology of the pitvipers* (pp. 169–184). Selva: Tyler.

Schuett, G. W. (1996). Fighting dynamics of male copperheads, *Agkistrodon contortrix* (Serpentes, Viperidae): Stress-induced inhibition of sexual behavior in losers. *Zoo Biology, 15*, 209–221.

Schuett, G. W. (1997). Body size and agonistic experience affect dominance and mating success in male copperheads, *Agkistrodon contortrix. Animal Behaviour, 54*, 213–224.

Schuett, G. W., Clark, R. W., Repp, R. A., Amarello, M., Smith, C. F., Douglas, M. R., Douglas, M. E., & Herrmann, H.-W. (2014). Communal winter denning in the western diamond-backed rattlesnake (*Crotalus atrox*): Molecular evidence for kin-related social structure. *Biology of the Pitvipers, 2*, Tulsa, Oklahoma: University of Tulsa.

Schuett, G. W., Clark, R. W., Repp, R. A., Amarello, M., Smith, C. F., & Greene, H. W. (2016). Social behavior of rattlesnakes: A shifting paradigm. In G. W. Schuett, M. J. Feldner, C. F. Smith, & R. S. Reiser (Eds.), *Rattlesnakes of Arizona. Vol. 2, Conservation, behavior, venom and evolution* (pp. 161–242). Rodeo, NM: ECO Publishing.

Schuett, G. W., & Duvall, D. (1996). Head lifting by female copperheads, *Agkistrodon contortrix*, during courtship: Potential mate choice. *Animal Behaviour, 51*, 367–373.

Schuett, G. W., Gergus, E., & Kraus, F. (2001). Phylogenetic correlation between male-male fighting and mode of prey subjugation in snakes. *Acta Ethologica, 4*, 31–49.

Schuett, G. W., & Gillingham, J. C. (1988). Courtship and mating of the copperhead, *Agkistrodon contortrix. Copeia, 1988*, 374–381.

Schuett, G. W., Höggren, M., Douglas, M. E., & Greene, H. W. (Eds.) (2002). *Biology of the vipers.* Eagle Mountain, UT: Eagle Mountain Publishing.

Schuett, G. W., Reiserer, R. S., & Earley, R. L. (2009). The evolution of bipedal postures in varanoid lizards. *Biological Journal of the Linnean Society, 97*, 652–663.

Schulte J. A., II, Valladares J. P., Larson A. (2003). Phylogenetic relationships within Iguanidae using molecular and morphological data and a phylogenetic taxonomy of iguanian lizards. *Herpetologica, 59*, 399–419.

Schutz, L., Stuart-Fox, D., & Whiting, M. (2007). Does the lizard *Platysaurus broadleyi* aggregate because of social factors? *Journal of Herpetology, 41*, 354–360.

Schwartz, J. M. (1985). Life History: *Thamnophis sirtalis. Herpetological Review, 16*, 112.

Schwartz, J. M., McCracken, G. F., & Burghardt, G. M. (1989). Multiple paternity in wild populations of the garter snake, *Thamnophis sirtalis. Behavioral Ecology and Sociobiology, 25*, 269–273.

Seale, D. B. (1982). Physical factors influencing oviposition by the woodfrog, *Rana sylvatica*, in Pennsylvania. *Copeia, 1982*, 627–635.

Seebacher, F., Elsworth, P. G., & Franklin, C. E. (2003). Ontogenetic changes of swimming kinematics in a semi-aquatic reptile (*Crocodylus porosus*). *Australian Journal of Zoology, 51*, 15–24.

Seigel, R. A., & Collins, J. T. (1993). *Snakes: Ecology and behavior.* New York: McGraw-Hill.

Seigel-Causey, D., & Kharitonov, S. P. (1990). The evolution of coloniality. In D. M. Power (Ed.), *Current ornithology* (Vol. 7, pp. 285-330). New York: Plenum Press.

Sellés, A. G., Bravo, A. M., Delclòs, X., Colombo, F., Martí, X., Ortega-Blanco, J., & Galobart, À. (2013). Dinosaur eggs in the Upper Cretaceous of the Coll de Nargó area, Lleida Province, south-central Pyrenees, Spain: Oodiversity, biostratigraphy and their implications. *Cretaceous Research, 40*, 10-20.

Senter, P. (2007). Necks for sex: Sexual selection as an explanation for sauropod dinosaur neck elongation. *Journal of Zoology, 271*, 45-53.

Senter, P. (2008). Homology between and antiquity of stereotyped communicatory behaviors in Crocodilians. *Journal of Herpetology, 42*, 354-360.

Senter, P., Harris, S. M., & Kent, D. L. (2014). Phylogeny of courtship and male-male combat behavior in snakes. *PLoS ONE, 9*, e107528.

Séon, N., Amiot, R., Martin, J. E., Young, M. T., Middleton, H., Fourel, F., Picot, L., Valentin, X., & Lécuyer, C. (2020). Thermophysiologies of Jurassic marine crocodylomorphs inferred from the oxygen isotope composition of their tooth apatite. *Philosophical Transactions of the Royal Society B: Biological Sciences, 375*, 20190139.

Sereno, P. C., Larsson, H. C., Sidor, C. A., & Gado, B. (2001). The giant crocodyliform *Sarcosuchus* from the Cretaceous of Africa. *Science, 294*, 1516-1519.

Sever, D. M., & Hamlett, W. C. (2002). Female sperm storage in reptiles. *Journal of Experimental Zoology, 292*, 187-199.

Sexton, O. J., Jacobson, P., & Bramble, E. (1992). Geographic variation in some activities associated with hibernation in Nearctic pitvipers. In J. D. Campbell and E. D. Brodie, Jr. (Eds.), *Biology of the pitvipers* (pp. 337-346). Tyler, Texas: Selva.

Seymour, R. S. (2013). Maximal aerobic and anaerobic power generation in large crocodiles versus mammals: Implications for dinosaur gigantothermy. *PLoS ONE, 8*, e69361.

Shaffer, S. A., Clatterbuck, C. A., Kelsey, E. C., Naiman, A. D., Young, L. C., VanderWerf, E. A. Bower, G. C. (2014). As the egg turns: Monitoring egg attendance behavior in wild birds using novel data logging technology. *PLoS ONE, 9*, e97898.

Shah, B., Shine, R., Hudson, S., & Kearney, M. (2003). Sociality in lizards: Why do thick-tailed geckos (*Nephrurus milii*) aggregate? *Behaviour, 140*, 1039-1052.

Sharp, S. P., McGowan, A., Wood, M. J., & Hatchwell, B. J. (2005). Learned kin recognition cues in a social bird. *Nature, 434*, 1127-1130.

Shea, G. M., & Sadlier, R. A. (2000). A mixed communal nest of skink eggs. *Herpetofauna, 30*, 46-47.

Shine, R. (1978). Sexual size dimorphism and male combat in snakes. *Oecologia, 33*, 269-277.

Shine, R. (1979). Sexual selection and sexual dimorphism in the Amphibia. *Copeia, 1979*, 297-306.

Shine, R. (1985). The evolution of reptilian viviparity: An ecological analysis. In C. G. F. Billett (Ed.), *Biology of the Reptilia* (Vol. 15, Developmental biology, pp. 605-694). New York: John Wiley.

Shine, R. (1988). Parental care in reptiles. In C. Gans and R. B. Huey (Eds.), *Biology of the Reptilia* (Vol. 16, pp. 275-330). New York: Alan R. Liss.

Shine, R. (1991). Strangers in a strange land: Ecology of the Australian colubrid snakes. *Copeia, 1991*, 120-131.

Shine, R. (1994). Sexual size dimorphism in snakes revisited. *Copeia, 1994*, 326-346.

Shine, R. (1995). *Australian snakes: A natural history*. Ithaca: Cornell University Press.

Shine, R. (2003). Reproductive strategies in snakes. *Proceedings of the Royal Society of London. Series B: Biological Sciences, 270*, 995-1004.

Shine, R. (2005). All at sea: Aquatic life modifies mate-recognition modalities in sea snakes (*Emydocephalus annulatus*, Hydrophiidae). *Behavioral Ecology and Sociobiology, 57*, 591-598.

Shine, R., & Bull, J. J. (1979). The evolution of live-bearing in lizards and snakes. *The American Naturalist, 113*, 905-923.

Shine, R., Cogger, H. G., Reed, R. R., Shetty, S., & Bonnet, X. (2003). Aquatic and terrestrial locomotor speeds of amphibious sea-snakes (Serpentes, Laticaudidae). *Journal of Zoology, 259*, 261-268.

Shine, R., & Fitzgerald, M. (1995). Variation in mating systems and sexual size dimorphism between populations of the Australian python *Morelia spilota* (Serpentes: Pythonidae). *Oecologia, 103*, 490-498.

Shine, R., & Harlow, P. O. (1993). Maternal thermoregulation influences offspring viability in a viviparous lizard. *Oecologia, 96*, 122-127.

Shine, R., Langkilde, T., & Mason, R. T. (2003a). Cryptic forcible insemination: Male snakes exploit female physiology, anatomy, and behavior to obtain coercive matings. *The American Naturalist, 162*, 653-667.

Shine, R., Langkilde, T., & Mason, R. T. (2003b). The opportunistic serpent: Male garter snakes adjust courtship tactics to mating opportunities. *Behaviour, 140*, 1509-1526.

Shine, R., & Lee, M. S. Y. (1999). A reanalysis of the evolution of viviparity and egg-guarding in squamate reptiles. *Herpetologica, 55*, 538-549.

Shine, R., LeMaster, M., Moore, I., Olsson, M., & Mason, R. (2001). Bumpus in the snake den: Effects of sex, size, and body condition on mortality of red-sided garter snakes. *Evolution, 55*, 598-604.

Shine, R., O'Connor, D., LeMaster, M., & Mason, R. (2001). Pick on someone your own size: Ontogenetic shifts in mate choice by male garter snakes result in size-assortative mating. *Animal Behaviour, 61*, 1133-1141.

Shine, R., O'Connor, D., & Mason, R. T. (2000). Sexual conflict in the snake den. *Behavioral Ecology and Sociobiology, 48*, 392-401.

Shine, R., Olsson, M. M., LeMaster, M. P., Moore, I. T., & Mason, R. T. (2000). Are snakes right-handed? Asymmetry in hemipenis size and usage in garter-snakes (*Thamnophis sirtalis*). *Behavioral Ecology, 11*, 411-415.

Shine, R., Phillips, B., Waye, H., LeMaster, M. P., & Mason, R. T. (2001). Advantage of female mimicry in snakes. *Nature, 414*, 267.

Shine, R., Phillips, B., Waye, H., LeMaster, M., & Mason, R. T. (2003). Chemosensory cues allow courting male garter snakes to assess body length and body condition of potential mates. *Behavioral Ecology and Sociobiology, 54*, 162-166.

Shine, R., Reed, R. N., Shetty, S., LeMaster, M. P., & Mason, R. T. (2002). Reproductive isolating mechanisms between two sympatric sibling species of sea snakes. *Evolution, 56*, 1655–1662.

Shuster, S., & Wade, M. (2003). *Mating systems and strategies*. Princeton, NJ: Princeton University Press.

Siegel-Causey, D., & Karitonov, S. P. (1990). The evolution of coloniality. In Power, D. M. (Ed.), *Current ornithology* (Vol. 7, pp. 285–330). New York: Plenum Press.

Sih, A., & Moore, R. D. (1993). Delayed hatching of salamander eggs in response to enhanced larval predation risk. *The American Naturalist, 142*, 947–960.

Sillman, A. J., Govardovskii, V. I., Röhlich, P., & Southard, J. A. (1997). The photoreceptors and visual pigments of the garter snake (*Thamnophis sirtalis*): A microspectrophotometric, scanning electron microscopic and immunocyto-chemical study. *Journal of Comparative Physiology A, 181*, 89–101.

Simões, B. F., Sampaio, F. L., Douglas, R. H., Kodandaramaiah, U., Casewell, N. R., Harrison, R. A., . . . Gower, D. J. (2016). Visual pigments, ocular filters and the evolution of snake vision. *Molecular Biology and Evolution, 33*, 2483–2495.

Sinervo, B., & Lively, C. (1996). The rock-paper-scissors game and the evolution of alternative male strategies. *Nature, 380*, 240.

Singh, L. A. K., & Rao, R. J. (1990). Territorial behavior of male gharial *Gavialis gangeticus* in the National Chambal Sanctuary, India. *Journal of Bombay Natural History Society, 87*, 149–151.

Sinn, D. L., While, G. M., & Wapstra, E. (2008). Maternal care in a social lizard: Links between female aggression and offspring fitness. *Animal Behaviour, 76*, 1249–1257.

Siviter, H., Deeming, D, C., van Giesen, M. F. T., & Wilkinson, A. (2017). Incubation temperature impacts the social cognition of adult lizards. *Royal Society Open Science, 4*, 170742. https://doi.org/10.1098/rsos.170742.

Siviter, H., Deeming, D. C., & Wilkinson, A. (2019). Egg incubation temperature influences the growth and foraging behaviour of juvenile lizards. *Behavioural Processes, 165*, 9–13.

Skinner, M., & Miller, N. (2020). Aggregation and social interaction in garter snakes (*Thamnophis sirtalis sirtalis*). *Behavioral Ecology & Sociobiology, 74*, 51. https://doi.org/10.1007/s00265-020-2827-0.

Smaldino, P. E., Palagi, E., Burghardt, G. M., & Pellis, S. M. (2019). The evolution of two types of play. *Behavioral Ecology, 30*, 1388–1397. https://doi.org/10.1093/beheco/arz090.

Smith, A. M. A. (1987). The sex and survivorship of embryos and hatchlings of the Australian freshwater crocodile, *Crocodylus johnstoni* [Unpublished PhD thesis]. Australian National University.

Smith, G., Trumbo, S. T., Sikes, D. S., Scott, M. P., & Smith, R. L. (2007). Host shift by the burying beetle, *Nicrophorus pustulatus*, a parasitoid of snake eggs. *Journal of Evolutionary Biology, 20*, 2389–2399.

Smith, K. T., Bhullar, B. A. S., Köhler, G., & Habersetzer, J. (2018). The only known jawed vertebrate with four eyes and the Bauplan of the pineal complex. *Current Biology, 28*, 1101–1107.

Smith, R. M., & Susan, E. E. (1995). An aggregation of juvenile *Youngina* from the Beaufort Group, Karoo Basin, South Africa. *Palaeontologia Africana, 32,* 86–92.

Smotherman, W. P., & Robinson, S. R. (1988). The uterus as environment: The ecology of fetal behavior. In E. M. Blass (Ed.), *Handbook of behavioral neurobiology* (Vol. 9, pp. 149–196). New York: Plenum.

Snow, D. W. (2004). Family Pipridae (Manakins). In J. del Hoyo, A. Elliott & D. A. Christie (Eds.), *Handbook of the birds of the world.* (Vol. 9. Cotingas to Pipits and Wagtails., pp. 110–169). Barcelona: Lynx Edicions.

Soares, D. (2002). Neurology: An ancient sensory organ in crocodilians. *Nature, 417,* 241.

Somaweera, R., Webb, J. K., & Shine, R. (2011). It's a dog-eat-croc world: Dingo predation on the nests of freshwater crocodiles in tropical Australia. *Ecological Research, 26,* 957–967.

Somma, L. (2003). *Parental behavior in lepidosaurian and testudinian reptiles. A literature survey.* Malabar: Krieger Publishing Company.

Spalding, D. A. (1872). On instinct. *Nature, 6,* 485–486.

Spalding, D. A. (1875). Instinct and acquisition. *Nature, 12,* 507–508.

Spencer, R. (2012). Embryonic heart rate and hatching behavior of a solitary nesting turtle. *Journal of Zoology, 287,* 169–174.

Spencer, R., & Janzen, F. (2011). Hatching behavior in turtles. *Integrative and Comparative Biology, 51,* 100–110.

Spencer, R., Thompson, M., & Banks, P. (2001). Hatch or wait? A dilemma in reptilian incubation. *Oikos, 93,* 401–406.

Spotila, J. R. (2004). *Sea turtles: A complete guide to their biology, behavior, and conservation.* Baltimore: Johns Hopkins University Press.

Sreekar, R., Srinivasulu, C., Seetharamaraju, M., & Srinivasulu, C. (2010). Selection of egg attachment sites by the Indian golden gecko *Calodactylodes aureus* (Beddome, 1870) (Reptilia: Gekkonidae) in Andhra Pradesh, India. *Journal of Threatened Taxa, 2*(11), 1268–1272.

Srivastava, R., Patnaik, R., Shukla, U. K., & Sahni, A. (2015). Crocodilian nest in a Late Cretaceous sauropod hatchery from the type Lameta Ghat locality, Jabalpur, India. *PLoS ONE, 10,* e0144369.

Stahlschmidt, Z. R., & DeNardo, D. F. (2011). Parental care in snakes. In R. D. Aldridge & D. M. Sever (Eds.), *Reproductive biology and phylogeny of snakes* (pp. 673–702). Enfield, NH: Science Publishers.

Stahlschmidt, Z. R., Shine, R., & DeNardo, D. F. (2012). The consequences of alternative parental care tactics in free-ranging pythons in tropical Australia. *Functional Ecology, 26,* 812–821.

Stamps, J. A. (1977). Social behaviour and spacing patterns in lizards. In C. Gans & D. W. Tinkle (Eds.), *Biology of the Reptilia, Vol. 7: Ecology and behaviour* (pp. 265–334). London: Academic Press.

Stamps, J. A. (1978). A field study of the ontogeny of social behavior in the lizard *Anolis aeneus. Behaviour, 66,* 1–31.

Stamps, J. A. (1983). Territoriality and the defence of predator-refuges in juvenile lizards. *Animal Behaviour, 31,* 857–870.

Stamps, J. A. (1984). Growth costs of territorial overlap: Experiments with juvenile lizards (*Anolis aeneus*). *Behavioral Ecology and Sociobiology*, 15, 115–119.

Stamps, J. A. (1988). Conspecific attraction and aggregation in territorial species. *The American Naturalist*, 131, 329–347.

Stamps, J. A. (2018). Polygynandrous anoles and the myth of the passive female. *Behavioral Ecology and Sociobiology*, 72, 107. https://doi.org/10.1007/s00265-018-2523-5.

Stamps, J. A., & Frankenhuis, W. E. (2016). Bayesian models of development. *Trends in Ecology and Evolution*, 31, 260–268.

Stanback, M. T., & Koenig, W. D. (1992). Cannibalism in birds. In M. A. Elgar & B. J. Crespi (Eds.), *Cannibalism: Ecology and evolution among diverse taxa* (pp. 277–298). New York: Oxford University Press

Stancyk, S. E. (1982). Non-human predators of sea turtles and their control. In K. A. Bjorndal (Ed.). *Biology and conservation of sea turtles* (pp. 139–152). Washington, DC: Smithsonian Institution.

Standing, K., Herman, T., Hurlburt, D., & Morrison, I. (1997). Post-emergence behaviour of neonates in a northern peripheral population of Blanding's turtle, *Emydoidea blandingii*, in Nova Scotia. *Canadian Journal of Zoology*, 75, 1387–1395.

Standora, E. A., Spotila, J. R., Keinath, J. A., & Shoop, C. R. (1984). Body temperatures, diving cycles, and movement of a subadult leatherback turtle, *Dermochelys coriacea*. *Herpetologica*, 40, 169–176.

Staniewicz, A. M. (2020). Acoustic communication of rare and threatened crocodilians and its use for population monitoring [Doctoral dissertation]. School of Biological Sciences, University of Bristol.

Stapley, J., & Whiting, M. (2006). Ultraviolet signals fighting ability in a lizard. *Biology Letters*, 2, 169–172.

Starck, J. M., & Ricklefs, R. E. (1998). *Avian growth and development: Evolution within the altricial-precocial spectrum*. Oxford: Oxford University Press on Demand.

Starin, E. D., & Burghardt, G. M. (1992). African rock pythons (*Python sebae*) in The Gambia: Observations on natural history and interactions with primates. *The Snake*, 24, 50–62.

Staton, M. A. (1978). "Distress calls" of crocodilians—whom do they benefit? *The American Naturalist*, 112, 327–332.

Staton, M. A., & Dixon, J. R. (1975). Studies on the dry season biology of *Caiman crocodilus crocodilus* from the Venezuelan Llanos. *Memoria de la Sociedad de Ciencias Naturales La Salle*, 35, 237–265.

Steer, D., & Doody, J. S. (2009). Dichotomies in perceived predation risk of drinking wallabies in response to predatory crocodiles. *Animal Behaviour*, 78, 1071–1078.

Steffen, J. E., Learn, K. M., Drumheller, J. S., Boback, S. M., & McGraw, K. J. (2015). Carotenoid composition of colorful body stripes and patches in the painted turtle (*Chrysemys picta*) and red-eared slider (*Trachemys scripta*). *Chelonian Conservation and Biology*, 14, 56–63.

Steiger, S. S., Fidler, A. E., Valcu, M., & Kempenaers, B. (2008). Avian olfactory receptor gene repertoires: Evidence for a well-developed sense of smell in birds? *Proceedings of the Royal Society of London B: Biological Sciences, 275,* 2309–2317.

Sterli, J. (2010). Phylogenetic relationships among extinct and extant turtles: The position of Pleurodira and the effects of the fossils on rooting crown-group turtles. *Contributions to Zoology, 79,* 93–106.

Stewart, B. S. (2014). Family Phocidae (true seals). In D. E. Wilson & Mittermeier, R. A. (Eds.), *Handbook of the mammals of the world* (Vol. 4: Sea Mammals, pp. 102–119). Barcelona: Lynx Edicions.

Stocker, M. R., Nesbitt, S. J., Criswell, K. E., Parker, W. G., Witmer, L. M., Rowe, T. B., . . . Brown, M. A. (2016). A dome-headed stem archosaur exemplifies convergence among dinosaurs and their distant relatives. *Current Biology, 26,* 2674–2680.

Stone, P. A., Dobie, J. L., & Henry, R. P. (1992). Cutaneous surface area and bimodal respiration in soft-shelled (*Trionyx spiniferus*), stinkpot (*Sternotherus odoratus*), and mud turtles (*Kinosternon subrubrum*). *Physiological Zoology, 65,* 311–330.

Storey, K. B., & Storey, J. M. (1992). Natural freeze tolerance in ectothermic vertebrates. *Annual Review of Physiology, 54,* 619–637.

Stow, A., & Sunnucks, P. (2004). High mate and site fidelity in Cunningham's skinks (*Egernia cunninghami*) in natural and fragmented habitat. *Molecular Ecology, 13,* 419–430.

Streicher, J. W., & Wiens, J. J. (2016). Phylogenomic analyses reveal novel relationships among snake families. *Molecular Phylogenetics and Evolution, 100,* 160–169.

Streicher, J. W., & Wiens, J. J. (2017). Phylogenomic analyses of more than 4000 nuclear loci resolve the origin of snakes among lizard families. *Biology Letters, 13,* 20170393.

Strickland, K., Gardiner, R., Schultz, A., & Frère, C. (2014). The social life of eastern water dragons: Sex differences, spatial overlap and genetic relatedness. *Animal Behaviour, 97,* 53–61.

Strine, C., Brown, A., Barnes, C., Major, T., Artchawakom, T., Hill, J., III, & Suwanwaree, P. (2018). Arboreal mating behaviors of the big-eyed green pit viper (*Trimeresurus macrops*) in Northeast Thailand (Reptilia: Viperidae). *Current Herpetology, 37,* 81–87.

Strüssmann, C. (1997). *Dracaena paraguayensis.* Courtship. *Herpetological Review, 28,* 151.

Stuart-Fox, D. M., Moussalli, A., Marshall, N. J., & Owens, I. P. (2003). Conspicuous males suffer higher predation risk: Visual modelling and experimental evidence from lizards. *Animal Behaviour, 66,* 541–550.

Suboski, M. (1992). Releaser-induced recognition in amphibians and reptiles. *Animal Learning and Behavior, 20,* 63–52.

Sues, H.-D. (2019). *The rise of reptiles: 320 million years of evolution.* Baltimore: Johns Hopkins University Press.

Sun, L., Shine, R., Debi, Z., & Zhengren, T. (2001). Biotic and abiotic influences on activity patterns of insular pit-vipers (*Gloydius shedaoensis*, Viperidae) from north-eastern China. *Biological Conservation, 97,* 387–398.

Surmik, D., & Pelc, A. (2012). Geochemical methods of inference the thermo-regulatory strategies in Middle Triassic marine reptiles–A pilot study. *Contemporary Trends in Geoscience*, *1*, 87–91.

Swain, T. A., & Smith, H. M. (1978). Communal nesting in *Coluber constrictor* in Colorado (Reptilia: Serpentes). *Herpetologica*, *34*, 175–177.

Sweeney, B. W., & Vannote, R. L. (1982). Population synchrony in mayflies: A predator satiation hypothesis. *Evolution*, *36*, 810–821.

Sweet, S. S., & Pianka, E. R. (2003). The lizard kings. *Natural History*, *112*, 40–45.

Szabo, B., Noble, D. W. A., & Whiting, M. J. (2020). Learning in non-avian reptiles 40 years on: Advances and promising new directions. *Biological Reviews*, doi: 10.1111/brv.12658.

Taborsky, M., & Brockmann, H. J. (2010). Alternative reproductive tactics and life history phenotypes. In P. M. Kappeler (Ed.), *Animal behaviour: Evolution and mechanisms* (pp. 537–586). Berlin: Heidelberg: Springer.

Takasu, K., & Hirose, Y. (1993). Host acceptance behavior by the host-feeding egg parasitoid, *Ooencyrtus nezarae* (Hymenoptera: Encyrtidae): Host age effects. *Annals of the Entomological Society of America*, *86*, 117–121.

Tallamy, D. W. (1985). "Egg dumping" in lace bugs (*Gargaphia solani*, Hemiptera: Tingidae). *Behavioral Ecology and Sociobiology*, *17*, 357–362.

Tallamy, D. W. (2005). Egg dumping in insects. *Annual Review of Entomology*, *50*, 347–370.

Tanaka, K., Kobayashi, Y., Zelenitsky, D. K., Therrien, F., Lee, Y. N., Barsbold, R., . . . Idersaikhan, D. (2019). Exceptional preservation of a Late Cretaceous dinosaur nesting site from Mongolia reveals colonial nesting behavior in a non-avian theropod. *Geology*, *47*, 843–847.

Tanaka, K., Zelenitsky, D. K., Therrien, F., & Kobayashi, Y. (2018). Nest substrate reflects incubation style in extant archosaurs with implications for dinosaur nesting habits. *Scientific Reports*, *8*, 3170.

Tarkhnishvili, D., Gavashelishvili, A., Avaliani, A., Murtskhvaladze, M., & Mumladze, L. (2010). Unisexual rock lizard might be outcompeting its bisexual progenitors in the Caucasus. *Biological Journal of the Linnean Society*, *101*, 447–460.

Tattersall, G. J., Leite, C. A., Sanders, C. E., Cadena, V., Andrade, D. V., Abe, A. S., & Milsom, W. K. (2016). Seasonal reproductive endothermy in tegu lizards. *Science Advances*, *2*, e1500951.

Taylor, M. P., Hone, D. W. E., Wedel, M. J., & Naish, D. (2011). The long necks of sauropods did not evolve primarily through sexual selection. *Journal of Zoology*, *285*, 150–161.

Terrick, T. D., Mumme, R. L., & Burghardt, G. M. (1995). Aposematic coloration enhances chemosensory recognition of noxious prey in the garter snake *Thamnophis radix*. *Animal Behaviour*, *49*, 857–866.

Test, F. H., & Heatwole, H. (1962). Nesting sites of the red-backed salamander, *Plethodon cinereus*, in Michigan. *Copeia*, *1962*, 206–207.

Thompson, M. B. (1988). Nest temperatures in the pleurodiran turtle, *Emydura macquarii*. *Copeia*, *1988*, 996–1000.

Thompson, M. B. (1989). Patterns of metabolism in embryonic reptiles. *Respiration Physiology*, *76*, 243–255.

Thompson, M. B., & Daugherty, C. H. (1998). Metabolism of tuatara, *Sphenodon punctatus*. *Comparative Biochemistry and Physiology Part A: Molecular & Integrative Physiology*, *119*, 519–522.

Thompson, M. B., Packard, G. C., Packard, M. J., & Rose, B. (1996). Analysis of the nest environment of tuatara *Sphenodon punctatus*. *Journal of Zoology*, *238*, 239–251.

Thompson, M. B., Speake, B. K., & Deeming, D. C. (2004). Egg morphology and composition. In D. C. Deeming (Ed.). *Reptilian incubation: Environment, evolution and behaviour* (pp. 45–74). Nottingham: Nottingham University Press.

Thomson, R. C., & Shaffer, H. B. (2010). Sparse supermatrices for phylogenetic inference: Taxonomy, alignment, rogue taxa, and the phylogeny of living turtles. *Systematic Biology*, *59*(1), 42–58.

Thorbjarnarson, J. B., & Hernández, G. (1993a). Reproductive ecology of the Orinoco crocodile (*Crocodylus intermedius*) in Venezuela. I. Nesting ecology and egg and clutch relationships. *Journal of Herpetology*, *27*, 363–370.

Thorbjarnarson, J. B., & Hernández, G. (1993b). Reproductive ecology of the Orinoco crocodile (*Crocodylus intermedius*) in Venezuela. II. Reproductive and social behavior. *Journal of Herpetology*, *27*, 371–379.

Tiatragul, S., Hall, J. M., & Warner, D. A. (2020). Nestled in the city heat: Urban nesting behavior enhances embryo development of an invasive lizard. *Journal of Urban Ecology*, *6*, juaa001.

Timberlake, W. (2007). Anthropomorphism revisited. *Comparative Cognition and Behavior Reviews*, *2*, 139–144.

Tinbergen, N. (1951). *The study of instinct*. Oxford: Oxford University Press.

Tinbergen, N. (1953a). *The herring gull's world: A study of the social behaviour of birds*. Oxford: Frederick A. Praeger.

Tinbergen, N. (1953b). *Social behaviour in animals with especial reference to vertebrates*. London: Methuen.

Tinbergen, N. (1963). On aims and methods of ethology. *Zeitschrift für Tierpsychologie*, *20*, 410–433.

Tinbergen, N., Impekoven, M., & Frank, D. (1967). An experiment on spacing out as a defense against predation. *Behaviour*, *28*, 307–321.

Tinkle, D. W., & Gibbons, J. W. (1977). *The distribution and evolution of viviparity in reptiles*. Ann Arbor: Museum of Zoology, University of Michigan.

Todd, N. P. M. (2007). Estimated source intensity and active space of the American alligator (*Alligator mississippiensis*) vocal display. *The Journal of the Acoustical Society of America*, *122*, 2906–2915.

Tokarz, R. R. (1988). Copulatory behaviour of the lizard *Anolis sagrei*: Alternation of hemipenis use. *Animal Behaviour*, *36*, 1518–1524.

Tokarz, R. R. (2002). An experimental test of the importance of the dewlap in male mating success in the lizard *Anolis sagrei*. *Herpetologica*, *58*, 87–94.

Tokarz, R. R., & Slowinski, J. B. (1990). Alternation of hemipenis use as a behavioural means of increasing sperm transfer in the lizard *Anolis sagrei*. *Animal Behaviour*, *40*, 374–379.

Tolley, K. A., Chauke, L. F., Jackson, J. C., & Feldheim, K. A. (2014). Multiple paternity and sperm storage in the Cape dwarf chameleon (*Bradypodion pumilum*). *African Journal of Herpetology, 63*, 47–56.

Tomkins, J. L., Le Bas, N. R., Witton, M.P., Martill, D.M., & Humphries, S. (2010). Positive allometry and the prehistory of sexual selection. *The American Naturalist, 176*, 141–148.

Tompkins, S. (2016). Copperhead snakes engage in nightly summertime feeding congregation. *Houston Chronicle*, July 20, 2016, https://www.chron.com /sports/outdoors/article/Copperhead-engage-in-nightly-summertime -feeding-8399696.php.

Touchon, J., McCoy, M., Vonesh, J., & Warkentin, K. (2013). Effects of plastic hatching timing carry over through metamorphosis in red-eyed treefrogs. *Ecology, 94*, 850–860.

Townsend, T. M., Larson, A., Louis, E., & Macey, J. R. (2004). Molecular phylogenetics of Squamata: The position of snakes, amphisbaenians, and dibamids, and the root of the squamate tree. *Systematic Biology, 53*, 735–757.

Toxopeus, A. G, Kruijit, J. P., & Hillenius, D. (1988). Pair-bonding in chameleons. *Naturwissenschaften, 75*, 268–269.

Trivers, R. L. (1974). Parent-offspring conflict. *Integrative and Comparative Biology, 14*, 249–264.

Trumbo, T. (2012). Patterns of parental care. In N. J. Royle, P. T. Smiseth & M. Kölliker (Eds.), *The evolution of parental care*. Oxford: Oxford University Press.

Trut, L. N. (1999). Early canid domestication: The fox-farm experiment. *American Scientist, 87*, 160–169.

Tsellarius, A. Yu. (1994). Поведение и образ жизни серого варана в песчаной пустыне. [Behavior and life history of desert monitor in sandy deserts] (in Russian). *Priroda, 5*, 26–35.

Tsellarius, A. Yu., & Tsellarius, E. Yu. (1996). Courtship and mating in *Varanus griseus* of Western Kyzylkum (in Russian). *Russian Journal of Herpetology, 3*, 16–18.

Tu, A. (1976). Investigation of the sea snake, *Pelamis platurus* (Reptilia, Serpentes, Hydrophiidae), on the Pacific coast of Costa Rica, Central America. *Journal of Herpetology, 10*, 13–18.

Tuberville, T., Norton, T., Waffa, B., Hagen, C., & Glenn, T. (2011). Mating system in a gopher tortoise population established through multiple translocations: Apparent advantage of prior residence. *Biological Conservation, 144*, 175–183.

Tucker, J. K. (1997). Natural history notes on nesting, nests, and hatchling emergence in the red-eared slider turtle, *Trachemys scripta elegans*, in west-central Illinois, *Illinois Natural History Survey Biological Notes, 140*, 1–13.

Tucker, J. K., Filoramo, N. I., Paukstis, G. L., & Janzen, F. J. (1998). Residual yolk in captive and wild-caught hatchlings of the red-eared slider turtle (*Trachemys scripta elegans*). *Copeia, 1998*, 488–492.

Tucker, J. K., Paukstis, G. L., & Janzen, F. J. (2008). Does predator swamping promote synchronous emergence of turtle hatchlings among nests? *Behavioral Ecology, 19*, 35–40.

Tullberg, B. S., Ah-King, M., & Temrin, H. (2002). Phylogenetic reconstruction of parental-care systems in the ancestors of birds. *Philosophical Transactions of the Royal Society of London. Series B: Biological Sciences, 357,* 251-257.

Turner, G., & James, B. (2010). Field observations of male combat in brown tree snakes *Boiga irregularis* (Colubridae) from north Queensland. *Herpetofauna, 39,* 109-114.

Turner, G. F., & Pitcher, T. J. (1986). Attack abatement: A model for group protection by combined avoidance and dilution. *The American Naturalist, 128,* 228-240.

Turtle Taxonomy Working Group. (2017). Turtles of the world: Annotated checklist and atlas of taxonomy, synonymy, distribution, and conservation status. *Chelonian Research Monographs, 7,* 1-292.

Uetz, P. (2010). The original descriptions of reptiles. *Zootaxa, 2334,* 59-68.

Uetz, P. (2015). Worldwide diversity of reptiles (2000-2015). Retrieved October 15, 2015, from http://www.reptile-database.org/db-info/diversity.html.

Uetz, P., Freed, P., & Hošek, J. (Eds.). (2020). *The Reptile Database.* http://www.reptile-database.org, accessed March 10, 2020.

Uller, T., & Olsson, M. (2008). Multiple paternity in reptiles: Patterns and processes. *Molecular Ecology, 17,* 2566-2580.

Uller, T., Schwartz, T., Koglin, T., & Olsson, M. (2013). Sperm storage and sperm competition across ovarian cycles in the dragon lizard, *Ctenophorus fordi. Journal of Experimental Zoology Part A: Ecological Genetics and Physiology, 319,* 404-408.

Umbers, K., Osborne, L., & Keogh, J. (2012). The effects of residency and body size on contest initiation and outcome in the territorial dragon, *Ctenophorus decresii. PLoS ONE, 7,* e47143.

Unwin, D. M., & Deeming, D. C. (2019). Prenatal development in pterosaurs and its implications for their postnatal locomotory ability. *Proceedings of the Royal Society B, 286,* 20190409.

Urra, H., Marín, J. F., Páez-Silva, M., Taki, M., Coulibaly, S., Gordillo, L., & García-Ñustes, M. A. (2019). Localized Faraday patterns under heterogeneous parametric excitation. *Physical Review E, 99,* 033115.

Valenzuela, N., & Lance, V. (2004). *Temperature-dependent sex determination in vertebrates.* Washington, DC: Smithsonian Books.

Valladas, H., Clottes, J., Geneste, J.-M., Garcia, M. A., Arnold, M., Cachier, H., & Tisnérat-Laborde, N. (2001). Evolution of prehistoric cave art. *Nature, 413,* 479.

Vallarino, O., & Weldon, J. P. (1996). Reproduction in the yellow-bellied sea snake (*Pelamis platurus*) from Panama: Field and laboratory observations. *Zoo Biology, 15,* 309-314.

Valverde, R. A., Orrego, C. M., Tordoir, M. T., Gómez, F. M., Solís, D. S., Hernández, R. A., & Spotila, J. R. (2012). Olive ridley mass nesting ecology and egg harvest at Ostional Beach, Costa Rica. *Chelonian Conservation and Biology, 11,* 1-11.

Valverde, R. A., Wingard, S., Gómez, F., Tordoir, M. T., & Orrego, C. M. (2010). Field lethal incubation temperature of olive ridley sea turtle *Lepidochelys olivacea* embryos at a mass nesting rookery. *Endangered Species Research, 12,* 77-86.

van der Kooi, C. J., & Schwander, T. (2015). Parthenogenesis: Birth of a new lineage or reproductive accident? *Current Biology, 25,* 659-651.

Vandewege, M. W., Mangum, S. F., Gabaldón, T., Castoe, T. A., Ray, D. A., & Hoffmann, F. G. (2016). Contrasting patterns of evolutionary diversification in the olfactory repertoires of reptile and bird genomes. *Genome Biology and Evolution, 8*, 470–480.

van Dijk, P. P., Iverson, J. B., Rhodin, A. G. J., Shaffer, H. B., & Bour, R. (2014). Turtles of the world, 7th edition: Annotated checklist of taxonomy, synonymy, distribution with maps, and conservation status. In A. G. J. Rhodin, P. C. H. Pritchard, P. P. van Dijk, R. A. Saumure, K. A. Buhlmann, J. B. Iverson & R. A. Mittermeier (Eds.), Conservation biology of freshwater turtles and tortoises: A compilation project of the IUCN/SSC Tortoise and Freshwater Turtle Specialist Group. *Chelonian Research Monographs, 5*, 329–479.

Van Doren, M. (1928). *Travels of William Bartram.* New York: Dover Publishing House.

van Veelen, M., García, J., & Avilés, L. (2010). It takes grouping and cooperation to get sociality. *Journal of Theoretical Biology, 264*, 1240–1253.

Varela, S. A. M., Danchin, E., & Wagner, R. H. (2007). Does predation select for or against avian coloniality? A comparative analysis. *Journal of Evolutionary Biology, 20*, 1490–1503.

Varricchio, D. J., Jackson, F., Borkowski, J. J., & Horner, J. R. (1997). Nest and egg clutches of the dinosaur *Troodon formosus* and the evolution of avian reproductive traits. *Nature, 385*, 247–250.

Varricchio, D. J., Martin, A. J., & Katsura, Y. (2007). First trace and body fossil evidence of a burrowing, denning dinosaur. *Proceedings of the Royal Society B: Biological Sciences, 274*, 1361–1368.

Varricchio, D. J., Moore, J. R., Erickson, G. M., Norell, M. A., Jackson, F. D., & Borkowski, J. J. (2008). Avian paternal care had dinosaur origin. *Science, 322*, 1826–1828.

Vaz-Ferreira, R., De Zolessi, L. C., & Achaval, F. (1973). Oviposicion y desarrollo de ofidios y lacertilios en hormigueros de *Acromyrmex* II. *Trabajos de V Congreso Latinoamericano Zoologico, 1*, 232–244.

Vehrencamp, S. L. (1978). The adaptive significance of communal nesting in groove-billed anis (*Crotophaga sulcirostris*). *Behavioral Ecology and Sociobiology, 4*, 1–33.

Vergne, A. L., Avril, A., Martin, S., & Mathevon, N. (2007). Parent-offspring communication in the Nile crocodile *Crocodylus niloticus*: Do newborns' calls show an individual signature? *Naturwissenschaften, 94*, 49–54.

Vergne, A. L., & Mathevon, N. (2008). Crocodile egg sounds signal hatching time. *Current Biology, 18*, R513-R514.

Vergne, A. L., Pritz, M. B., & Mathevon, N. (2009). Acoustic communication in crocodilians: From behaviour to brain. *Biological Reviews, 84*, 391–411.

Vicente, N. S. (2018). Headbob displays signal sex, social context and species identity in a *Liolaemus* lizard. *Amphibia-Reptilia, 39*, 203–218. https://doi.org /10.1163/15685381-17000163.

Vicente, N. S., & Halloy, M. (2017). Interaction between visual and chemical cues in a *Liolaemus* lizard: A multimodal approach. *Zoology, 125*, 24–28.

Vijaya, J. (1983a). Auditory cues as possible stimuli for hatching eggs of the flap-shell turtle *Lissemys punctata granosa*. *Hamadryad*, *8*, 23.

Vijaya, J. (1983b). Freshwater turtle survey in India: 1982-83. *Hamadryad*, *8*, 21-22.

Vince, M. A., & Chinn, S. (1971). Effect of accelerated hatching on the initiation of standing and walking in the Japanese quail. *Animal Behaviour*, *19*, 62-66.

Virlich, (1984). Памятники природы Херсонской области. [*Natural monuments of Kherson Oblast*] (in Russian). Simferopol, Ukraine: Tavria.

Visagie, L., Mouton, P., & Bauwens, D. (2005). Experimental analysis of grouping behaviour in cordylid lizards. *Herpetological Journal*, *15*, 91-96.

Vitt, L. J. (1991). Ecology and life history of the scansorial arboreal lizard *Plica plica* (Iguanidae) in Amazonian Brazil. *Canadian Journal of Zoology*, *69*, 504-511.

Vitt, L. J., Caldwell, J. P., Zani, P. A., & Titus, T. A. (1997). The role of habitat shift in the evolution of lizard morphology: Evidence from tropical *Tropidurus*. *Proceedings of the National Academy of Sciences*, *94*, 3828-3832.

Vitt, L. J., Pianka, E. R., Cooper, Jr, W. E., & Schwenk, K. (2003). History and the global ecology of squamate reptiles. *The American Naturalist*, *162*, 44-60.

Vleck, C., Hoyt, D., & Vleck, D. (1979). Metabolism of avian embryos: Patterns in altricial and precocial birds. *Physiological Zoology*, *52*, 363-377.

Vliet, K. A. (1989). Social displays of the American alligator. *American Zoologist*, *29*, 1019-1031.

Vliet, K. A. (2001). Courtship behaviour of American alligators *Alligator mississippiensis*. In G. C. Grigg, F. Seebacher & C. E. Franklin (Eds.). *Crocodilian biology and evolution* (pp. 383-408). Sydney: Surrey Beatty & Sons.

Vogt, R. C. (1979). Cleaning/feeding symbiosis between grackles (Quiscalus: Icteridae) and map turtles (*Graptemys*: Emydidae). *Auk*, *96*, 608-609.

Vogt, R. C. (2008). *Turtles of the Amazon*. Lima: Wust, Ediciones.

von Uexküll, J., & Kriszat, G. (1934/1957). A stroll through the worlds of animals and men. In C. H. Schiller (Ed.), *Instinctive behavior* (pp. 5-80, translated from the original German). New York: International Universities Press.

Wade, J. (2002). Sexual dimorphisms in avian and reptilian courtship: Two systems that don't play by mammalian rules. *Brain, Behavior and Evolution*, *54*, 15-27.

Waldman, B. (1982). Adaptive significance of communal oviposition in wood frogs (*Rana sylvatica*). *Behavioral Ecology and Sociobiology*, *10*, 169-174.

Waldman, B., & Ryan, M. J. (1983). Thermal advantages of communal egg mass deposition in wood frogs (*Rana sylvatica*). *Journal of Herpetology*, *17*, 70-72.

Wallace, D. R. (2011). *Chuckwalla land: The riddle of California's desert*. Berkeley: University of California Press.

Walls, G. L. (1940) Ophthalmological implications for the early history of snakes. *Copeia*, *1940*, 1-8.

Wang, H., Yan, P., Zhang, S., Sun, L., Ren, M., Xue, H., & Wu, X. (2017). Multiple paternity: A compensation mechanism of the Chinese alligator for inbreeding. *Animal Reproduction Science*, *187*, 124-132.

Wang, M., O'Connor, J. K., Xu, X., & Zhou, Z. (2019). A new Jurassic scansoriopterygid and the loss of membranous wings in theropod dinosaurs. *Nature*, *569*, 256.

Wang, X., Kellner, A. W. A., Jiang, S., Cheng, X., Wang, Q., Ma, Y., . . . Li, N. (2017). Egg accumulation with 3D embryos provides insight into the life history of a pterosaur. *Science, 358*, 1197–1201.

Wang, X., Kellner, A. W. A., Jiang, S., Paidoula, Y., Meng, X., Wang, Q., & Zhou, Z. (2014). Sexually dimorphic tridimensionally preserved pterosaurs and their eggs from China. *Current Biology, 24*, 1–8.

Wang, X., Miao, D., & Zhang, Y. (2005). Cannibalism in a semi-aquatic reptile from the Early Cretaceous of China. *Chinese Science Bulletin, 50*, 282–284.

Wang, Z., Pascual-Anaya, J., Zadissa, A., Li, W., Niimura, Y., Huang, Z., Li, C., & Irie, N. (2013). The draft genomes of soft-shell turtle and green sea turtle yield insights into the development and evolution of the turtle-specific body plan. *Nature Genetics, 45*, 701–706.

Wapstra, E., & Olsson, M. (2014). The evolution of polyandry and patterns of multiple paternity in lizards. In J. L. Rheubert, D. S. Siegel & S. E. Trauth (Eds.), *Reproductive biology and phylogeny of lizards and tuatara* (pp. 564–589). Boca Raton, FL: CRC Press.

Ward, S. A., & Kukuk, P. F. (1998). Context-dependent behavior and the benefits of communal nesting. *The American Naturalist, 152*, 249–263.

Warkentin, K. (1995). Adaptive plasticity in hatching age: A response to predation risk trade-offs. *Proceedings of the National Academy of Sciences, 92*, 3507–3510.

Warkentin, K. (2000). Wasp predation and wasp-induced hatching of red-eyed treefrog eggs. *Animal Behaviour, 60*, 503–510.

Warkentin, K. (2011a). Environmentally cued hatching across taxa: Embryos respond to risk and opportunity. *Integrative and Comparative Biology, 51*, 14–25.

Warkentin, K. (2011b). Plasticity of hatching in amphibians: Evolution, trade-offs, cues and mechanisms. *Integrative and Comparative Biology, 51*, 111–127.

Warkentin, K., & Caldwell, M. (2009). Assessing risk: Embryos, Information, and escape hatching. In R. Dukas & J. M. Ratcliffe (Eds.), *Cognitive ecology II* (pp. 177–200). Chicago: University of Chicago Press.

Waters, R. M., Bowers, B. B., & Burghardt, G. M. (2017). Personality and individuality in reptile behavior. In J. Vonk, A. Weiss, & S. A. Kuczaj (Eds.), *Personality in non-human animals* (pp. 153–154). New York: Springer Nature.

Watson, D. M. S. (1957). On *Millerosaurus* and the early history of the sauropsid reptiles. *Philosophical Transactions of the Royal Society of London, Series B, Biological Sciences, 240*, 325–400.

Weaver, L. N., Varricchio, D. J., Sargis, E. J., Chen, M., Freimuth, W. J., & Mantilla, G. P. W. (2020). Early mammalian social behaviour revealed by multituberculates from a dinosaur nesting site. *Nature Ecology & Evolution,* https://doi.org/10.1038/s41559-020-01325-8.

Weaver, W. G. (1970). Courtship and combat behavior in *Gopherus berlandieri*. *Bulletin of Florida State Museum of Biological Sciences, 15*, 1–43.

Webb, G. J. W., Choquenot, D., & Whitehead, P. J. (1986). Nests, eggs, and embryonic development of *Carettochelys insculpta* (Chelonia: Carettochelyidae) from Northern Australia. *Journal of Zoology, 1*, 521–550.

Webb, G. J. W., & Manolis, S. C. (1989). *Crocodiles of Australia*. Sydney: Reed Books.

Webb, G. J. W., Manolis, S. C., & Buckworth, R. (1983). *Crocodylus johnstoni* in the McKinlay River Area N. T, VI. Nesting Biology. *Wildlife Research, 10,* 607–637.

Webb, G. J. W., & Smith, A. M. A. (1984). Sex ratio and survivorship in the Australian freshwater crocodile *Crocodylus johnstoni. Symposium of the Zoological Society of London, 52,* 319–355.

Webb, J. K., Pike, D. A., & Shine, R. (2008). Population ecology of the velvet gecko, *Oedura lesueurii* in south eastern Australia: Implications for the persistence of an endangered snake. *Austral Ecology, 33,* 839–847.

Webber, P. (1979). Burrow density, position and relationship of burrows to vegetation cover shown Rosen's desert skink *Egernia inornata* (Lacertilia: Scincidae). *Herpetofauna, 10,* 16–20.

Webster, T., & Bums, J. M. (1973). Dewlap color variation and electrophoretically detected sibling species in a Haitian lizard, *Anolis brevirostris. Evolution, 27,* 368–377.

Weekes, H. C. (1935). A review of placentation among reptiles with, particular regard to the function and evolution of the placenta. *Journal of Zoology, London, 105,* 625–645.

Weldon, P. J., & Burghardt, G. M. (1984). Deception divergence and sexual selection. *Zeitschrift für Tierpsychologie, 65,* 89–102.

Weldon, P. J., & Burghardt, G. M. (2015). Evolving détente: The origin of warning signals via concurrent reciprocal selection. *Biological Journal of the Linnean Society, 116,* 239–246.

Weldon, P. J., Flachsbarth, B., & Schulz, S. (2008). Natural products from the integument of nonavian reptiles. *Natural Products Reports, 25,* 738–756.

Weldon, P. J., & Wheeler, J. W. (2001). The chemistry of crocodilian skin glands. In G. C. Grigg, F. Seebacher & C. E. Franklin (Eds.). *Crocodilian Biology and evolution* (pp. 286–296). Sydney: Surrey Beatty & Sons.

Wermuth, H., & Fuchs, K. (1978). *Bestimmen von Krokodilen und ihrer Häute: e. Anleitung zum Identifizieren d. Art-u. Rassen-Zugehörigkeit d. Krokodile für Behörden (im Zusammenhang mit d. Artenschutz-Übereinkommen von Washington), für Reptilleder-Industrie u. Reptilien-Handel, für Reptilien-Liebhaber sowie für zoolog.* Berlin: Fischer.

Werner, D. I. (1983). Reproduction in the iguana *Conolophus subcristatus* on Fernandina Island, Galapagos: Clutch size and migration costs. *The American Naturalist, 121,* 757–775.

Werner, Y. L. (1980). Apparent homosexual behaviour in an all-female population of a lizard, *Lepidodactylus lugubris* and its probable interpretation. *Zeitschrift für Tierpsychologie, 54,* 144–150.

West-Eberhard, M. J. (2003). *Developmental plasticity and evolution.* New York: Oxford University Press.

Wever, E. G. (1974). The evolution of vertebrate hearing. In W. D. Keidel & W. D. Neff (Eds.), *Auditory system. Handbook of sensory physiology* (Vol, 5, pp. 423–454). Springer Berlin Heidelberg.

Weygoldt, P. (1980). Complex brood care and reproductive behaviour in captive poison-arrow frogs, *Dendrobates pumilio* O. Schmidt. *Behavioral Ecology and Sociobiology, 7,* 329–332.

While, G. M., Gardner, M. G., Chapple, D. G., & Whiting, M. J. (2019). Stable social grouping in lizards. In V. L. Bels & A. P. Russell (Eds.), *Behavior of lizards: Evolutionary and mechanistic perspectives* (pp. 321–339). Boca Raton: CRC Press.

While, G. M., Sinn, D. L., & Wapstra, E. (2009). Female aggression predicts mode of paternity acquisition in a social lizard. *Proceedings of the Royal Society B: Biological Sciences, 276,* 2021–2029.

While, G. M., Uller, T., & Wapstra, E. (2009a). Family conflict and the evolution of sociality in reptiles. *Behavioral Ecology, 20,* 245–250.

While, G. M., Uller, T., & Wapstra, E. (2009b). Within-population variation in social strategies characterize the social and mating system of an Australian lizard, *Egernia whitii. Austral Ecology, 34,* 938–949.

Whishaw, I. Q., Pellis, S. M., & Gorny, B. P. (1992). Skilled reaching in rats and humans: Evidence for parallel development or homology. *Behavioural Brain Research, 47,* 59–70.

Whitaker, N. (2007). Extended parental care in the Siamese crocodile (*Crocodylus siamensis*). *Russian Journal of Herpetology, 14,* 203–206.

Whitaker, P. B., & Shine, R. (2002). Thermal biology and activity patterns of the eastern brownsnake (*Pseudonaja textilis*): A radiotelemetric study. *Herpetologica, 58,* 436–452.

Whitaker, P. B., & Shine, R. (2003). A radiotelemetric study of movements and shelter-site selection by free-ranging brownsnakes (*Pseudonaja textilis,* Elapidae). *Herpetological Monographs, 17,* 130–144.

Whitaker, R., & Basu, D. (1982). The gharial (*Gavialis gangeticus*): A review. *Journal of Bombay Natural History Society, 79,* 531–548.

Whitear, A. K., Wang, X., Catling, P., McLennan, D. A., & Davy, C. M. (2016). The scent of a hatchling: Intra-species variation in the use of chemosensory cues by neonate freshwater turtles. *Biological Journal of the Linnean Society, 120,* 179–188.

Whitehead, P. J. (1987). Respiration of *Crocodylus johnstoni* embryos. In G. J. W. Webb, S. C. Manolis & P. J. Whitehead (Eds.), *Wildlife management: Crocodiles and alligators* (pp. 473–497). Sydney: Surrey Beatty.

Whitehead, P. J., & Seymour, R. (1990). Patterns of metabolic rate in embryonic crocodilians *Crocodylus johnstoni* and *Crocodylus porosus. Physiological Zoology, 63,* 334–352.

Whitford, M. D., Freymiller, G. A., Higham, T. E., & Clark, R. W. (2020). The effects of temperature on the defensive strikes of rattlesnakes. *Journal of Experimental Biology, 223,* jeb223859.

Whiting, M. J., & Greeff, J. M. (1999). Use of heterospecific cues by the lizard *Platysaurus broadleyi* for food location. *Behavioral Ecology and Sociobiology, 45,* 420–423.

Whiting, M. J., Stuart-Fox, D., O'Connor, D., Firth, D., Bennett, N., & Blomberg, S. (2006). Ultraviolet signals ultra-aggression in a lizard. *Animal Behaviour, 72,* 353–363.

Whiting, M. J., Webb, J., & Keogh, J. (2009). Flat lizard female mimics use sexual deception in visual but not chemical signals. *Proceedings of the Royal Society B: Biological Sciences, 276,* 1585–1591.

Whiting, M. J., & While, G. M. (2017). Sociality in lizards. In D. R. Rubenstein & P. Abbott (Eds.), *Comparative social evolution* (pp. 390–426). Cambridge, UK: Cambridge University Press.

Wickramasinghe, M., & Somaweera, R. (2002). Endemic geckos of Sri Lanka. *Gekko, 3*, 1–3.

Wiens, J. A. (1966). On group selection and Wynne-Edwards' hypothesis. *American Scientist, 54*, 273–287.

Wiens, J. J., Hutter, C. R., Mulcahy, D. G., Noonan, B. P., Townsend, T. M., Sites, J. W., & Reeder, T. W. (2012). Resolving the phylogeny of lizards and snakes (Squamata) with extensive sampling of genes and species. *Biology Letters, 8*, 1043–1046.

Wiewandt, T. A. (1982). Evolution of nesting patterns in iguanine lizards. In G. M. Burghardt & A. S. Rand (Eds.), *Iguanas of the world: Their behavior, ecology and conservation* (pp. 119–141). Park Ridge, NJ: Noyes Publications.

Wikelski, M., & Bäurle, S. (1996). Pre-copulatory ejaculation solves time constraints during copulations in marine iguanas. *Proceedings of the Royal Society of London. Series B: Biological Sciences, 263*, 439–444.

Wilbur, H. M., & Collins, J. P. (1973). Ecological aspects of amphibian metamorphosis. *Science, 182*, 1305–1314.

Wilgers, D., & Horne, E. (2009). Discrimination of chemical stimuli in conspecific fecal pellets by a visually adept iguanid lizard, *Crotaphytus collaris*. *Journal of Ethology, 27*, 157–163.

Wilkinson, A., & Huber, L. (2012). Cold-blooded cognition: Reptilian cognitive abilities. In J. Vonk & T. K. Shackelford (Eds.), *The Oxford handbook of comparative evolutionary psychology* (pp. 129–143). New York: Oxford University Press.

Wilkinson, A., Kuenstner, K., Mueller, J., & Huber, L. (2010). Social learning in a non-social reptile (*Geochelone carbonaria*). *Biology Letters, 6*, 614–616.

Wilkinson, A., Mueller-Paul, J., & Huber, L. (2013). Picture–object recognition in the tortoise *Chelonoidis carbonaria*. *Animal Cognition, 16*, 99–107.

Williams, G. C. (1966). Natural selection, the costs of reproduction, and a refinement of Lack's principle. *The American Naturalist, 100*, 687–690.

Williams, M. F. (1997). The adaptive significance of endothermy and salt excretion amongst the earliest archosaurs. *Speculations in Science and Technology, 20*, 237–247.

Williamson, E. A., Maisels, F. G., & Groves, C. P. (2013). Family Hominidae (great apes). In R. A. Mittermeier, A. B. Rylands & D. E. Wilson (Eds.), *Handbook of the mammals of the World* (Vol. 3: Primates, pp. 792–856). Barcelona: Lynx Edicions.

Willis, K. L. (2016). Underwater hearing in turtles. In A. N. Popper & A. Hawkins (Eds.), *The effects of noise on aquatic life II* (pp. 1229–1235). New York: Springer.

Wilmes, A. J., Siegel, D. S., & Aldridge, R. D. (2011). Premature sperm ejaculation in captive African brown house snake *Lamprophis fuliginosus*. *African Journal of Herpetology, 60*, 177–180.

Wilmshurst, J. M., Anderson, A. J., Higham, T. F., & Worthy, T. H. (2008). Dating the late prehistoric dispersal of Polynesians to New Zealand using the commensal Pacific rat. *Proceedings of the National Academy of Sciences, 105*, 7676–7680.

Wilson, A. C., Bush, G. L., Case, S. M., & King, M. C. (1975). Social structuring of mammalian populations and rate of chromosomal evolution. *Proceedings of the National Academy of Sciences, 72*, 5061-5065.

Wilson, D. E., & Mittermeier, R. A. E. (Eds.) (2009-2019). *Handbook of mammals of the world*. Barcelona: Lynx Edicions.

Wilson, D. S., & Wilson, E. O. (2007). Rethinking the theoretical foundation of sociobiology. *Quarterly Review of Biology, 82*, 327-348.

Wilson, E. O. (1975). *Sociobiology: The new synthesis*. Cambridge, MA: Harvard University Press.

Wilson, V. (1968). The leopard tortoise, *Testudo pardalis babcocki*, in eastern Zambia. *Arnoldia, 3*, 1-11.

Wiltschko, W., & Wiltschko, R. (2005). Magnetic orientation and magnetoreception in birds and other animals. *Journal of Comparative Physiology A, 191*, 675-693.

Wiman, C. (1931). *Parasaurolophus tubicen*, n. sp. aus der Kreide in New Mexico. *Nova Acta Regia Societas Scientarum Upsaliensis* series 4, *7*, 1-11.

Winck, G. R., & Cechin, S. Z. (2008). Hibernation and emergence pattern of *Tupinambis merianae* (Squamata: Teiidae) in the Taim Ecological Station, southern Brazil. *Journal of Natural History, 42*, 239-247.

Winkler, J. D., & Sánchez-Villagra, M. R. (2006). A nesting site and egg morphology of a Miocene turtle from Urumaco, Venezuela: Evidence of marine adaptations in Pelomedusoides. *Palaeontology, 49*, 641-646.

Wipatayotin, P. (2014, January 12). Pattaya thieves. *Bangkok Post*.

Witmer, L. M., Chatterjee, S., Franzosa, J., & Rowe, T. (2003). Neuroanatomy of flying reptiles and implications for flight, posture and behaviour. *Nature, 425*, 950-953.

Witmer, L. M., & Ridgely, R. C. (2009). New insights into the brain, braincase, and ear region of tyrannosaurs (Dinosauria, Theropoda), with implications for sensory organization and behavior. *The Anatomical Record: Advances in Integrative Anatomy and Evolutionary Biology, 292*, 1266-1296.

Wittenberger, J. F. (1981). *Animal social behavior*. Boston: Duxbury Press.

Woodland, D. J., Jaafar, Z., & Knight, M. L. (1980). The "pursuit deterrent" function of alarm signals. *The American Naturalist, 115*, 748-753.

Wörner, L. L. B. (2009). Aggression and competition for space and food in captive juvenile tuatara (*Sphenodon punctatus*) [Master's thesis]. Victoria University of Wellington.

Wright, K. M. (1993). Captive husbandry of the Solomon Island prehensile-tailed skink, *Corucia zebrata*. *Bulletin of the Association of Reptilian and Amphibian Veterinarians, 3*, 18-21.

Wright, L., Fuller, W., Godley, B., McGowan, A., Tregenza, T., & Broderick, A. (2012). Reconstruction of paternal genotypes over multiple breeding seasons reveals male green turtles do not breed annually. *Molecular Ecology, 21*, 3625-3635.

Wrona, F. J., & Dixon, R. J. (1991). Group size and predation risk: A field analysis of encounter and dilution effects. *The American Naturalist, 137*, 186-201.

Wu, X., & Hu, Y. (2010). Multiple paternity in Chinese alligator (*Alligator sinensis*) clutches during a reproductive season at Xuanzhou Nature Reserve. *Amphibia-Reptilia, 31*, 419-424.

Wusterbarth, T., King, R., Duvall, M., Grayburn, W., & Burghardt, G. (2010). Phylogenetically widespread multiple paternity in new world natricine snakes. *Herpetological Conservation and Biology, 5*, 86-93.

Wymann, M. N., & Whiting, M. J. (2003). Male mate preference for large size overrides species recognition in allopatric flat lizards (*Platysaurus broadleyi*). *Acta Ethologica, 6*, 19-22.

Wyneken, J. (2007). Reptilian neurology: Anatomy and function. *Veterinary Clinics of North America: Exotic Animal Practice, 10*, 837-853.

Wynn, A. H., Cole, C. J., & Gardner, A. L. (1987). Apparent triploidy in the unisexual Brahminy blind snake, *Ramphotyphlops braminus. American Museum Novitates, 2868*, 1-7.

Wynne-Edwards, V. C. (1962). Animal dispersion: In relation to social behaviour. Edinburgh, UK: Oliver and Boyd.

Xu, X., Zheng, Z., & You, H. (2010). Exceptional dinosaur fossils show ontogenetic development of early feathers. *Nature, 464*, 1338-1341.

Yeager, C. P., & Burghardt, G. M. (1991). Effect of food competition on aggregation: Evidence for social recognition in the plains garter snake (*Thamnophis radix*). *Journal of Comparative Psychology, 105*, 380-386.

York, D. S., & Burghardt, G. M. (1988). Brooding in the Malayan pit viper, *Calloselasma rhodostoma*: Temperature, relative humidity, and defensive behaviour. *Herpetological Journal, 1*, 210-213.

Yosef, R., & Yosef, N. (2010). Cooperative hunting in brown-necked raven (*Corvus rufficollis*) on Egyptian mastigure (*Uromastyx aegyptius*). *Ethology, 28*, 385-388.

Young, B. A. (1997). A review of sound production and hearing in snakes, with a discussion of intraspecific acoustic communication in snakes. *Journal of the Pennsylvania Academy of Science, 71*, 39-46.

Young, B. A., Mathevon, N., & Tang, Y. (2014). Reptile auditory neuroethology: What do reptiles do with their hearing? In C. Köppl, G. A. Manley, A. N. Popper & R. R. Fay (Eds.), *Insights from comparative hearing research* (pp. 323-346). New York: Springer.

Zacariotti, R. L., & del Rio do Valle, R. (2010). Observation of mating in the calico snake *Oxyrhopus petola* Linnaeus, 1758. *Herpetology Notes, 3*, 139-140.

Žagar, A., & Carretero, M. A. (2012). A record of cannibalism in *Podarcis muralis* (Laurenti, 1768) (Reptilia, Lacertidae) from Slovenia. *Herpetology Notes, 5*, 211-213.

Zaher, H., Murphy, R. W., Arredondo, J. C., Graboski, R., Machado-Filho, P. R., Mahlow, K., . . . Zhang, Y. P. (2019). Large-scale molecular phylogeny, morphology, divergence-time estimation, and the fossil record of advanced caenophidian snakes (Squamata: Serpentes). *PLoS ONE, 14*, e0216148.

Zamudio, K., & Sinervo, B. (2000). Polygyny, mate-guarding, and posthumous fertilization as alternative male mating strategies. *Proceedings of the National Academy of Sciences, 97*, 14427-14432.

Zardoya, R., & Meyer, A. (2004). Molecular evidence on the origin and phylogenetic relationships among the major groups of vertebrates. In A. Moya & E. Font (Eds.), *Evolution from molecules to ecosystems* (pp. 209-217). Oxford: Oxford University Press.

Zhang, F., Zhou, Z., Xu, X., Wang, X., & Sullivan, C. (2008). A bizarre Jurassic maniraptoran from China with elongate ribbon-like feathers. *Nature, 455,* 1105-1108.

Zhang, Y.-P., Li, S.-R., Ping, J., Li, S.-W., Zhou, H.-B., Sun, B.-J., & Du, W.-G. (2016). The effects of light exposure during incubation on embryonic development and hatchling traits in lizards. *Scientific Reports, 6,* 38527. https://doi.org/10.1038/srep38527.

Zhang, Z., Zhang, Y., Zhang, Q., Cheng, T., & Wu, X. (2015). Bionic research of pit vipers on infrared imaging. *Optics Express, 23,* 19299-19317.

Zhao, E., & Adler, K. (1993). *Herpetology of China.* Oxford: Society for the Study of Amphibians and Reptiles.

Zhao, Q., Benton, M. J., Xu, X., & Sander, P. M. (2013). Juvenile-only clusters and behaviour of the Early Cretaceous dinosaur *Psittacosaurus. Acta Palaeontologica Polonica, 59,* 827-833.

Zheng, X., Bi, S., Wang, X., & Meng, J. (2013). A new arboreal haramiyid shows the diversity of crown mammals in the Jurassic period. *Nature, 500,* 199-202.

Zimmerman, K., & Heatwole, H. (1990). Cutaneous photoreception: A new sensory mechanism for reptiles. *Copeia, 1990,* 860-862.

Zimmerman, L. C., & Tracy, C. R. (1989). Interactions between the environment and ectothermy and herbivory in reptiles. *Physiological Zoology, 62,* 374-409.

Zink, A. G. (2000). The evolution of intraspecific brood parasitism in birds and insects. *The American Naturalist, 155,* 395-405.

Zink, A. G. (2001). The optimal degree of parental care asymmetry among communal breeders. *Animal Behaviour, 61,* 439-446.

Zink, A. G. (2005). The dynamics of brood desertion among communally breeding females in the treehopper, *Publilia concava. Behavioral Ecology and Sociobiology, 58,* 466-473.

Zitterbart, D. P., Wienecke, B., Butler, J. P., & Fabry, B. (2011). Coordinated movements prevent jamming in an emperor penguin huddle. *PLoS ONE, 6,* e20260.

Zug, G. R. (1966). The penial morphology and the relationships of cryptodiran turtles. *Occasional Papers of the Museum of Zoology, University of Michigan, 647,* 1-24.

Zug, G. R., Vitt, L. J., & Caldwell, J. P. (2001). *Herpetology: An introductory biology of amphibians and reptiles.* London: Academic Press.

Zykova, L., & Panov, E. (1993). Notes on social organization and behavior of Khorosan agama, *Stellio erythrogaster,* in Badkhyz. *Zoologichesky Zhurnal, 72,* 148-151.

Page numbers in *italics* refer to illustrations.

Sarcosuchus, 85
Sceloporus (fence lizards), 29, 259
Sceloporus jarrovii (Yarrow's spiny lizard), 180, 226
Sceloporus mucronatus (cleft lizard), 56
Sceloporus occidentalis (western fence lizard), 35, 80
Sceloporus undulatus (eastern fence lizard), 226, 240
Shinisaurus crocodilurus (crocodile lizard), 30
Sistrurus (pygmy rattlesnakes), 100, 129
Sitana ponticeriana (fan-throated lizard), 92, plate 4.2
Sotalia guianensis (Guiana dolphin), 46
Sphaerodactylus nicholsi (Puerto Rican dwarf gecko), 97
Sphenodon punctatus (tuatara), 20, 41, 44, 48–49, 55, 57, 92, 119, 120, 136, 142, 156–58, 191, 237–38, 258
Spinolestes, 19
Stegonotus cucullatus (slaty-grey snake), 172
Sternotherus minor (loggerhead musk turtle), 205
Storeria (North American brown snakes), 180
Storeria dekayi (brown snake), 245, 264
Sylviornis, 184

Tachyglossus aculeatus (short-beaked echidna), 20
Tadarida brasiliensis (Mexican free-tailed bat), 144, 235–36
Taeniopygia guttata (zebra finch), 108
Tamias sibiricus (Siberian chipmunk), 80
Taxidea taxus (American badger), 251
Terrapene (box turtles), 26, 36
Terrapene carolina (common box turtle), 36, 119
Terrapene ornata (eastern box turtle), 90
Testudo, 92
Testudo graeca (Mediterranean spur-thighed tortoise), 119
Testudo hermanni (Hermann's tortoise), 92
Thamnophis (gartersnakes), 56, 180
Thamnophis butleri (Butler's gartersnake), 246

Thamnophis melanogaster (Mexican black-bellied gartersnake), 285
Thamnophis radix (plains gartersnake), 246, 265
Thamnophis sauritus (eastern ribbon snake), 245
Thamnophis sirtalis (common gartersnake), 48, 65, 104–5, 108, 120, 132, 245–46, 255, 262, 264, 268, 278
Thecadactylus rapicauda, 172
Thermophis baileyi (hot springs keelback snake), 21
Tiliqua, 73
Tiliqua rugosa (sleepy lizard), 57, 68, 69, 76, 248, plate 3.1
Tomistoma schlegelii (Sunda gharial), 24, 61, 187
Trachemys, 222
Trachemys scripta (red-eared slider), 89, 108, 166, 209, 287–88
Trachemys stejnegeri malonei (Inagua slider), 189
Trimeresurus macrops (big-eyed pitviper), 132
Trioceros hoehnelii (helmeted chameleon), 94
Trionyx triunguis (Nile soft-shelled turtle), 280
Trogonophis wiegmanni (checkerboard worm lizard), 194
Troodon, 182
Troodon formosus, 52
Tropidonophis mairii (common keelback snake), 160, 166
Tropidurus montanus (upland calango), 261
Tupinambis merianae (Argentine giant tegus), 99–100
Tursiops truncatus (bottlenose dolphin), 251
Tyrannosaurus, 44, 250

Urosaurus ornatus (tree lizard), 264
Uta stansburiana (side-blotched lizard), 62, 95

Varanus, 192
Varanus (monitor lizards), 30, 60

Page numbers in *italics* refer to illustrations.

bipedal motion, 39, 220, 229
birds: aggregations of, 68-69; anatomy of, 38, 47, 49; cannibalism in, 257; communal egg-laying by, 138, 143, 163; courtship displays of, 81-82, 110; feathers of, 45, 81, 82, 109; hatching of, 199, 218; lineage of, 19, 21, 24; parental care by, 51-52, 183-84; senses of, 41, 43, 45; social behavior of, 2-3, 229-30, 233, 250; thermoregulation by, 2, 33, 34, 230
blind lizards, 27, 29, 121
blind snakes, 31, 44, 61
boas, 31, 46, 128-29, 130, 197, 255
body temperature, 20, 189, 245; maintaining, 33-37. *See also* thermoregulation
brain structure and size, 46-48, 275, 284
breathing, 37-38
burrowing, 70-71, 74-75, 269

Caenophidia, 129, 130, 131
caimans, 24, 83, 88, 110-11, *plate 4.1. See also* crocodylians
cannibalism, 256-60
Carboniferous period, 17, 178
catch-up hypothesis, 217-18
chain fishing, 251-53
chameleons, 29, 40, 53, 121, 124, 144, 241, *plate 4.3*; communication by, 94-95, 99
Cheloniidae, 26, 140
chemical communication, 80-81, 90; by lizards, 80, 96-98; by snakes, 80, 100, 103-5
chemical cues, 44, 73, 80, 264; for forming aggregations, 105, 161, 264-65; in mating, 98, 104
chemical senses, 43-45
choristoderes, 18, 19, 190
classification frameworks, 4-6
climate change, 158, 239, 267
cobras, 100-101, 127, 194, 195-96
cognition, 47, 275-76
colonial roosting, 1
coloration, 62, 94-95, 97, 162
color vision, 102
colubrids, 32, 37, 128, 130, 140, 142
communal egg-laying, 13, 134-77; benefits to eggs of, 164-67; benefits to

hatchlings of, 171; by birds, 138, 143, 163; costs of, 161-64; by crocodylians, 143, 156; cues for, 147, 161, 163, 166, 174; defined, 135-36; future research on, 174-77; and green iguanas, 149-51; hypotheses on, 158-61, 165-67; interspecific, 172-73; and intraspecific competition, 164; by lizards, 139, 140, 144-45, 151-52; macroevolution of, 146-47; maternal benefits of, 168-70; reasons for, 136-38, 158-71; by snakes, 71-72, 139, 141-42, 154-56; and social evolution, 64, 286; and Sri Lankan Golden Gecko, 153-54; taxonomic distribution of, 138-46; traditional-site, 171-72; by tuataras, 142, 156-58, 162; by turtles, ii, 142-43, 147-49, 162, 173, *plate 6.3*
communication, 78-105; acoustic, 79-80, 81, 83, 86-87, 90-91, 98-100, 101; by archosaurs, 81-86; chemical, 80-81, 90, 96-98, 100, 103-5; by crocodylians, 43, 82-89, 110-11, 113; definitions of, 78-79; embryo-to-embryo, 201, 205, 218-19, 221; between hatchlings, 221-22; by lizards and tuataras, 92-100, *plate 6.6*; by snakes, 80, 100-105, 131; tactile, 79-80, 82, 89, 91, 103, 116, 131; by turtles and tortoises, 89-92, 116, 218-19, 235-36; types of, 79-81; visual, 79, 90, 92-96, 100-103, 269
conspecific attraction, 64, 67-68, 74, 134, 139, 166, 173-76; vs. habitat saturation hypothesis, 158-61
cooperative hunting. *See* social hunting
copulation: by crocodylians, 112; by lizards, 49, 59, 121, 125, *126*, 127; by snakes, 130-31, 132-33; by turtles and tortoises, 115, 117-18
Corytophanidae, 144
courtship and mating, 106-33; by crocodylians, 85-86, 110-14; fossilized evidence on, 108-9; by lizards, *70*, 119-27; mating systems, 55-58; research studies on, 107-8, 226, 268-70; by snakes, 48, 50, 66, 105, 127-33; by tuataras, 48-49, 92, 119, *120*,

hatching and emergence, 198–222; asynchronous, 199, 206, 207–8, 210, 212, 213, 222, 241; delayed, 202–6; distinguishing hatching and emergence, 208; early, 201–2; environmentally cued, 199–200, 201–3, 204, 205–6; and flooding/hypoxia, 202–3, 204, 205–6; research needed on, 206, 221–22; synchronous, 204–5, 206–20; vibration-induced, 91, 219. *See also* synchronized hatching and emergence

head oblique tail arched (HOTA) posture, 84, *84*, 85, 86, 88, 110, 112

hearing, sense of, 43, 47–48. *See also* acoustic communication

heart anatomy, 38

heart rates, 104, 219

heat-sensing organs, 46, 197

Heinroth, Oskar, 7

Helodermatidae, 144

herbivory, 36–37

Herrera, Andres, 147–48

hibernation, 36, 197, 258, 264, 265

homosexual mating, 122

Hoplocercidae, 144

hormones, 284–85

humans, 42, 81, 101, 251, 287; reptile bonds with, 279

Huxley, Julian, 7

hypoxia, 202–3, 204, 205–6

ichthyosaurs, 18, 35, 39, 42, 52–53

iguanas, 29, 66, 191, 220, 256, 279; communal egg-laying by, 144, 149–51, 161–62, 169; copulation by, 59, *126*; displays by, 93, 96, 97, 240; social behavior by, 59–60, 243–44, 256

incubation temperature, 199, 239–40, 273; and sex determination, 52, 239

infanticide, 257

infrasound, 43, 83, 86, 88–89, 110

insects, 3, 5, 224

integumentary sense organs (ISOs), 45

intraspecific brood parasitism hypothesis, 170

intraspecific competition, 61, 164, 258–59

invasive species, 49, 287–88

Jacobson's organ, 44

James, William, 231

Jurassic period, 19, 35, 143, 237

kin aggregations, 69

kin recognition, 70, 72, 246, 272

kin selection hypothesis, 170–71

kleptoparasitism, 256, 259, 276

Lanthanotidae, 144

lepidosauromorphs, 17, 19, 32–33, 190

limbs: dexterity and precision of, 41; position of, 38–39

Linnaeus, Carl, xv

lizards: aggregations of, 69–70, 262–63, 264, 265; chemical communication by, 80, 96–98; communal egg-laying by, 139, 140, 144–45, 151–52; courtship and mating by, *70*, 119–27; dominance hierarchies and displays of, 62–63, 95–96; embryos and hatchlings, 200; family groups of, 28–29, 69; headbobbing by, 93, 240, 276, 281; limbs of, 41; multiple paternity in, 123, 269; parental care by, 137, 190–91, 192, 193–94, 241, 271–72; reproduction by, 48, 56–57, 123; senses of, 42, 44; social behavior by, 54, 238–44; social learning by, 242–43; territoriality in, 58–59, 60; visual displays by, 92–96, 240, 281; vocal communication by, 98–100

longevity, 74, 76

long-term memory, 47

Lorenz, Konrad, 7, 8

lung anatomy, 37–38

Madtsoiidae, 31–32

magnetoreception, 46

males mimicking females, 59, 62, 97, 105, 243–44

male-to-male combat, 60, 61, 66, 101–2, 105, *125*, 127, 128–30

mammals: anatomy of, 33, 34, 41, 44, 53; cannibalism in, 256–57; and communal breeding, 143–44; egg-laying, 20, 189; emergence of, 18, 20–21; and parental care, 179; senses of, 41, 42, 43; social behavior by, 224, 230–31, 250–51; and viviparity, 52–53, 180

marsupials, 20, 53, 180, 251

mass extinctions, 17–18, 19, 20, 24, 34

mate choice copying, 176

mate guarding, 56, 62, 124–25, 127, 131–32, 268

maternal: assistance, 204; behavior, 279; benefits, 161, 168; care, 156, 179; defense, 72; diet, 271; effects, 247, 274, 285; experience, 179; investment, 247; lineage, 170; milk, 230; provisioning, 271; return to nest sites, 235; temperature effects, 245

mating. *See* courtship and mating

mating balls, 66, 105, 130, 132

megapodes, 24, 51, 184, 185

metabolic heating, 167, 200, 222

monitor lizards, 29, 30, 47, 50, 100; communal egg-laying by, 151–52; courtship and mating by, 124, 125, 127

monogamy, 5, 55, 75, 114, 127, 268, 270; and social organization, 54, 56–57

mosasaurs, 19, 35

motor functions, 38–41

mud and musk turtles, 27

multiple defenders hypothesis, 170

multiple mating, 48, 55–56, 57, 104, 121, 169

multiple paternity, 50–51, 55–56, 114; in lizards, 123, 269; in snakes, 55, 132, 268

Neoaves, 24, 49

nest guarding, 157, 185, 187–88, 191, 194, 233

nesting burrows, 150, 151–52, 154

nest site choice, 137–38

nest warming, 166–67

neuroscience, 284–85

null hypothesis, 232

object play, 124, 227, 284

olfactory signals and displays, 147, 163, 172; by crocodylians, 84–85; by lizards, 96–98; and synchronized hatching, 209, 214

Opluridae, 144

oviparity, 180, 272

pair bonding, 2, 50, 57, 69, 74–75, 123, 137–38. *See also* monogamy

Palaeognathae, 24

parasites, 76, 180, 270

parental care, 178–97, 225; and birds, 51–52, 183–84; by crocodylians, 156, 179, 184–88, 233–34, 272, *plate 7.1*; and delayed dispersal, 74–75, 76; fossil evidence of, 178; frequency of in reptiles, 56, 75, 284; in lizards, 137, 190–91, 192, 193–94, 241, 271–72; research needs on, 227, 270–72; by snakes, 72, 137, 190–91, 194–97, 227, 247, 270–71; and turtles, 137, 188, 189–90, 227, 271–72; and viviparity, 134–35, 193

parent-offspring conflict, 175

parthenogenesis, 49–50, 121

paternal: care, *cover photo*, 75, 179, 182; defense, 270

penis structure, 49

Permian period, 17–18, 34, 178, 180

pets, reptiles as, 28, 274, 281

Phasianidae, 82

pheromones: and aggregations, 172–73; for communication, 80, 92, 104–5

pitvipers, 32, 37, 40, 46, 135; displays by, 101, 103; mating by, 131, 132; and parental care, 72, 194, 196, 197

plasticity: environmentally induced, 273–75; social, 272–73

plated lizards, 28

Platysternidae, 26

play, 124, 227, 279–84

plesiosaurs, 18, 35, 52–53

Pleurodira, 25

polyandry (polygynandry), 55, 56–57, 75, 269

polygamy, 5, 268, 270

predation avoidance: and aggregations, 75, 260–61; and communal egg-laying, 137, 148, 163, 164–65, 169, 171, 173

predator dilution, 1, 148, 164–65, 171, 173, 206–7

predator preclusion hypothesis, 213–14

predator swamping hypotheses, 206-7, 208-9, 212, 214, 220, 222

preparatory hypothesis, 232

pterosaurs, 18-19, 40, 42, 46, 47, 49, 51, 81, 109; nesting by, 143, 180-81, 233

pursuit-deterrent signals (PDS), 96

pythons, 31, 46, 128-29, 189, 249; and parental care, 194, 195-96, 197

Rathke's glands, 90

rattlesnakes, 40, 50, 102, 104, 127, 259; courtship and mating by, 129, 268; and parental care, 194-95, 196-97; photos, *103, 125, plates 7.2, 7.3*; social behavior of, 47, 67-68, 69, 72-73; tail rattling by, 100, 101, 260

reproduction, reptile, 48-53. *See also* courtship and mating

reproductive-success-based hypothesis, 165-66

Reptile Database, 21, 22

reptile-human bonds, 279

reptile term, 16-17

rhynchocephalians, 19, 20, 22, 92, 237. *See also* tuataras

sauropods, 182, 183

Scolecophidia, 31, 269

sea snakes, 22, 37, 39, 105, 255; aggregations of, 66-67. *See also* snakes

sea turtles, 26, 35, 46, 190, 271-72; communal egg-laying arribadas by, 65-66, 91, 142, 147-48, 165, *plate 6.1*; courtship and mating by, 56, 114-16; green, 36-37, 56, 115, 166, 199-200, 212, 219; hatching and emergence of, 171, 173, 209, 210, 212-15, 219-20; Kemp's, 65, 147, 218, 258; leatherback, 22, 35, 91, 114, 151, 162, 218; loggerhead, 22, 114, 212; migrating, 39, 65; olive ridley, 65, 147-48, 165, 209, *plates 6.1, 6.2*; species of, 22. *See also* turtles

sensitization, 232-33

sensory organs, 41-46

sex determination, temperature-dependent, 52, 185, 231-32, 239

sexual dimorphism, 66, 108, 109, 227, 270

sexual selection, 6, 268, 269; and courtship displays, 81-82, 109; and phenotypic diversity, 61-62

siblicide, 259-60

signaling. *See* communication

skinks: communal egg-laying by, 141, 144, 159-60, 166; lineage of, 28-29; mating by, 57, 62, 122; parental care by, 190, 191, 192, 241; placentae of, 53, 180; social behavior of, 70, 73, 75, 190, 241-42

smell, sense of, 43-45, 104-5. *See also* olfactory signals

snake-lizards, 98-99, 138, 140

snakes: aggregations of, 66, 67-68, 105, 132, 244-46, 262, 264, 265, *plate 5.4*; cannibalism in, 259-60; chemical communication by, 80, 100, 103-5; communal egg-laying by, 71-72, 139, 141-42, 154-56; courtship and mating by, 48, 50, 66, 105, 127-33; emergence of, 19; hunting by, 36, 67, 254, 255; motion by, 39-40; multiple paternity in, 55, 132, 268; parental care in, 72, 137, 190-91, 194-97, 227, 247, 270-71; phylogeny of, 31-32; senses of, 40, 41-42, 43, 44, 45, 46, 47-48; social behavior in, 67, 244-47, 254, 255; social groups of, 71-73; and social recognition, 275-76, 277-78; sperm storage by, 133, 268; tactile communication by, 103, 131; and territoriality, 60-61; visual communication by, 100-103

snapping turtles, 26, 140

social behavior definitions, x, 1-3

social evolution, 285-86

social experience, 242, 273, 277

social facilitation hypothesis, 214-15

social feeding, 249-56, *plates 10.1, 10.2*

social hunting, 67, 249-50, 251-55

social learning, 47, 76, 176, 242-43, 276

social network analysis, 273

social-nonsocial dichotomy, 3-4, 9-10, 137, 158

social recognition, 276-78
sociobiology, 7-8
softshell turtles, 27, 37, 116, 140
sounds: by crocodylians, 43, 83, *84,*
86-87, 88-89, 110-11, 113; infrasound,
43, 83, 86, 88-89, 110; by lizards,
98-100; by tortoises, 91-92. *See also*
acoustic communication
sperm storage, 50, 55, 114, 123, 133, 268
spinytail lizards, 28, 35-36
squamate evolution, 19
squamate phylogeny, 27-32
stable aggregations, 69-77
steroids, 98
stimuli, 64, 191, 226-27, 232, 243, 276-77
surplus resource theory, 281
synapsids, 34, 39, 80, 108, 178-79;
emergence and extinction, 17-18, 19
synchronized hatching and emergence,
206-20, *plates 8.1, 8.2, 8.3;* adaptive
hypotheses on, 208-16, 222; among-
clutch, 220-21; defining, 207-8;
delaying, 204-5; evidence on, 208-16;
nonadaptive explanation for, 216-17;
and predator dilution, 206-7, 208-9;
social and physiological mechanisms
of, 217-20

tactile communication, 79-80, 82; by
snakes, 103, 131; by turtles, 89, 91,
116
taste, sense of, 43, 97
teiids, 29, 37, 49, 140
temperature-dependent sex determination, 52, 185, 231-32, 239
temporal buffering, 203
territoriality, 97, 269-70, 284, 287; and
social organization, 58-60, 61, 62, 272
thermal gradients, 206-7, 212, 215,
216-17, 222
thermogenesis, 33
thermoregulation, 2, 74, 75, 86, 179, 229,
230; and basking, 35-36, 189, 263; of
birds, 2, 33, 34, 230; and body size,
33-34; and habitat choice, 263-64; as
parental duty, 189, 192, 193
threadsnakes, 31, 140, 194
Tinbergen, Nikolaas, 7

titillation, 89, 116-17, 236, 263, 273, 281,
282
tortoises, 26, 114, 188, 256, 266; aggregations of, 67, 264, 265; cognitive
abilities of, 47, 276; communication by,
91-92; courtship and mating by, 115,
118-19; families and lineage of, 25-26
touch, sense of, 45-46. *See also* tactile
communication
trailing behavior, 246
Triassic period, 18, 34, 40, 181, 237
tuataras, 38, 41, 61, 66, 191, 267; colorful
displays by, 92; communal egg-laying
by, 142, 156-58, 162; courtship and
mating by, 48-49, 92, 119, *120,* 121-22;
monogamy among, 55, 57; senses of,
42, 44; social behavior in, 237-38
turtles: aggregations of, 65-66, 91,
147-48, 165, 263; anatomy of, 37, 38,
49; communal egg-laying by, ii,
142-43, 147-49, 162, 173, *plate 6.3;*
communication by, 89-92, 116, 218-19,
235-36; courtship and mating by, 56,
91-92, 114-18, 236; evolutionary
emergence of, 18-19; lineages of,
24-26; and parental care, 137, 188,
189-90, 227, 271-72; senses of, 43, 44;
social organization of, 236, 256; tactile
communication by, 89, 91, 116; and
territoriality, 61; titillation by, 89,
116-17, 236, 263, 273, 281, *282. See also*
sea turtles; tortoises

Uexküll, Jakob von, 78
ultraviolet (UV) light, 41, 95
underwater vocalization, 148-49
urbanization, 287
urine, 38, 80

varanids, 29, 37, 96, 144
vipers, 37, 40, 46, 129, 131, 132; displays
by, 101, 103, 129; lineage of, 32;
parental care by, 72, 189, 194, 196-97,
270; social behavior by, 72, 253, 254, 255
vision, sense of, 41-43
visual communication, 79, 90, 269; by
lizards and tuataras, 92-96; by snakes,
100-103

viviparity, 52–53, 73–74, 134–35, 179–81, 189, 286; and parental care, 134–35, 193

vocalization: by crocodylians, 83, 86–87, 234; by embryos, 186–87, 199–200, 201–2, 218–19; by hatchlings, 82, 149, 218; by turtles, 90–91, 235–36; underwater, 148–49. *See also* sounds

vomerolfaction, 44, 161, 190, 246
vomeronasal organ (VNO), 44

wall lizards, 29, 50, 98, 140
Whitman, Charles Otis, 7
worm lizards, 22, 29, 43, 107, 139, 194

Xenosauridae, 144